Fortschritte der Chemie organischer Naturstoffe

# Progress in the Chemistry of Organic Natural Products

# 39

Founded by L. Zechmeister

Edited by W. Herz, H. Grisebach, G. W. Kirby

Authors:
R. C. Anderson, B. Fraser-Reid,
H. Jones, T. Kasai, P. O. Larsen,
S. Liaaen-Jensen, G. H. Rasmusson

Springer-Verlag
Wien New York 1980

Dr. W. Herz, Professor of Chemistry, Department of Chemistry,
The Florida State University, Tallahassee, Florida, U. S. A.

Prof. Dr. H. Grisebach, Biologisches Institut II, Lehrstuhl für Biochemie der Pflanzen,
Albert-Ludwigs-Universität, Freiburg i. Br., Federal Republic of Germany

G. W. Kirby, Sc. D., Regius Professor of Chemistry, Chemistry Department,
The University, Glasgow, Scotland

With 5 Figures

© 1980 by Springer-Verlag/Wien

Softcover reprint of the hardcover 1st edition 1980

Library of Congress Catalog Card Number AC 39-1015

ISSN 0071-7886

ISBN-13: 978-3-7091-8553-7      e-ISBN-13: 978-3-7091-8551-3
DOI: 10.1007/978-3-7091-8551-3

# Contents

Contents

# List of Contributors

ANDERSON, R. C., Ph. D., B. Sc., Faculty of Science, Department of Chemistry, University of Waterloo, Waterloo, Ontario, Canada N2L 3G1.

FRASER-REID, Prof. B., Faculty of Science, Department of Chemistry, University of Waterloo, Waterloo, Ontario, Canada N2L 3G1.

JONES, Dr. H., Director of Medicinal Chemistry, USV Pharmaceutical Corporation, Tuckahoe, NY 10707, U.S.A.

KASAI, Dr. T., Department of Agricultural Chemistry, Faculty of Agriculture, Hokkaido University, Kita 9, Nishi 9, Kita-Ku, Sapporo 060, Japan.

LARSEN, Prof. P. O., Chemistry Department, Royal Veterinary and Agricultural University, 40, Thorvaldsensvej, DK-1871 Copenhagen V, Denmark.

LIAAEN-JENSEN, Prof. S., Organic Chemistry Laboratories, Norwegian Institute of Technology, University of Trondheim, N-7034 Trondheim, Norway.

RASMUSSON, Dr. G. H., Senior Research Fellow, Synthetic Chemical Research, Merck Sharp & Dohme Research Laboratories, Rahway, NJ 07065, U.S.A.

# Carbohydrate Derivatives in the Asymmetric Synthesis of Natural Products

By B. Fraser-Reid and R. C. Anderson,
Guelph-Waterloo Centre for Graduate Work in Chemistry,
University of Waterloo, Ontario, Canada

**Contents**

**Acknowledgements**

We are greatly indebted to Professors S. Hanessian of the University of Montreal, A. S. Perlin of McGill University, and G. Stork of Columbia University for making available to us unpublished information from their laboratories, and to Professor R. B. Woodward of Harvard University for the loan of his personal copies of the Theses of Drs. R. D. Sitrin and J. Upeslacis.

# I. Introduction

Traditionally, carbohydrate derivatives have been utilized in the mainstream of organic chemistry primarily for studies relating to stereo-chemical and conformational problems. This is not surprising since these aspects have always been a prime area of concern to sugar chemists because of the role they play in the reactions of sugars. Indeed it was in this connection that the term "conformation" was coined by Haworth in 1929 (34). The work of Barton and Hassell in the late 1940's on the development of conformational analysis of six membered rings is, of course, well known. However, even prior to this, Reeves had begun his systematic work on the shapes of sugar molecules in solution (78, 79), and it was further concern with the latter that led Lemieux to undertake an examination of sugar acetates by $^1$H n.m.r. (pmr) spectroscopy in 1957 (53). Thus n.m.r. spectroscopy as practiced in organic chemistry today, owes its origins and much of its early development (54) to the availability of a wealth of well characterized sugar derivatives. $^{13}$C n.m.r. (cmr) spectroscopy is similarly indebted (74).

The foregoing indicates that the stereochemical properties of sugar derivatives have for long been well known or are amenable to elucidation. Nevertheless synthetic organic chemists have been slow to take advantage of their virtues, in spite of the fact that assignment of relative stereo-chemistry in most of these molecules is tantamount to assignment of absolute configuration. Indeed sugars are the most abundant natural source of chirality, and D-glucose, the most economical per chiral centre.

The traditional reluctance to dabble in sugar chemistry is undoubtedly associated with the accusation, unfortunately sometimes deserved, that sugars are difficult to crystallize. Since they are also non-volatile, purification by distillation is not a plausible alternative. Furthermore, the relatively delicate nature of many sugar derivatives precludes the use of many classical reagents.

However in the light of the present state of the art, the prospect of working with sugars is no longer as daunting as it once was. The advent of sophisticated chromatographic techniques now allows for ready purification without having to resort to crystallization or distillation and the emergence of high resolution mass spectrometry permits the determination of molecular composition. The current panoply of gentle, specific reagents available to organic chemists is also of timely significance.

Thus within the past few years there has been an increasing interest in the use of sugars as chiral synthons*. In this article we describe some recent achievements but since the emphasis here is in the area of natural product synthesis, some elegant studies such as those of INCH on chiral phosponates (29, 42) and STODDART on chiral crown ethers as enzyme analogues (13) will have to be omitted.

However, it should be noted that the concept of carbohydrates as chiral synthons has always fascinated sugar chemists as the excellent 1972 review by INCH will attest (42). Thus in order to preserve historical perspective three antecedents for the use of sugars in asymmetric syntheses are noted in Scheme 1. First is WOLFROM's proof of the structure of (+)-alanine by synthesis from D-glucosamine (102). Second is the synthesis of the optically pure mandelic acid derivatives from 2,3-4,5-di-O-isopropylidene-aldehydo-D-arabinose (1) by BONNER (9), and third is LEMIEUX's synthesis of ethanol-1-d from "diacetone glucose" (6a, Scheme 2) via the deuterated xylose (2) (51).

*Sources of Sugar Derivatives*

The sugar derivatives which are appropriate for use in synthesis must either be available commercially or be prepared readily from cheap precursors. Apart from the usual chemical suppliers, PFANSTIEHL (Waukegan, Illinois, U.S.A.) and RAYLO (Edmonton, Alberta, Canada) carry excellent stocks of sugars and their derivatives. Many of the latter can frequently be prepared rapidly and economically, and for this purpose the series Methods in Carbohydrate Chemistry (Academic Press), particularly volumes 1, 2 and 6, is unsurpassed.

---

* The intensity of developments in this area can be gauged by the fact that the Specialist Periodical Reports (Carbohydrates) of the Chemical Society, beginning with Volume 9, now include a section on "The Synthesis of Optically Active Non-Carbohydrate Compounds".

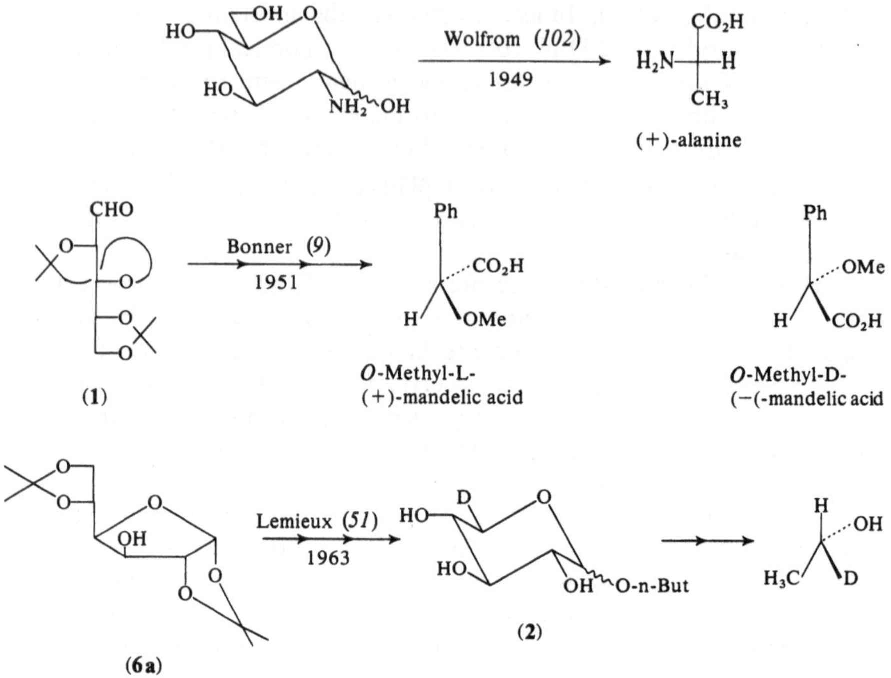

$$\text{(+)-alanine}$$

O-Methyl-L-
(+)-mandelic acid

O-Methyl-D-
(−(-mandelic acid

*Scheme 1*

Of course, the more recent literature should always be checked for improved procedures. In this connection The Specialist Periodical Reports of the Chemical Society are a convenient starting point.

## Nomenclature

As in other areas of organic chemistry, carbohydrate nomenclature is under constant review. However, the compendium of rules published in 1963 (*96*) is a good place to begin.

## Ready Sources of Information

In the following section we will be discussing some elementary aspects of sugar chemistry. The treatment will of necessity be general and cursory, and the reader is directed to other sources for more definitive information. "The Monosaccharides" by Staněk, Černý, Kocourek and Pacák (Academic Press, 1963) contains, in the authors' view, the most complete documentation of the traditional reactions of simple sugars, and the bibliography up to 1963 is excellent. On the other hand, "Monosaccharide Chemistry" by Ferrier and Collins (Penguin, 1972), which unfortunately lacks a bibliography, gives a modern mechanistic interpretation of many of the reactions which are catalogued in the text by Staněk *et al.* "Stereochemistry of Carbohydrates" by Stoddard (Wiley-Interscience, 1971) gives an expert handling of its subject matter and has a very complete bibliography. "The Carbohydrates" edited by Pigman and Horton (Academic Press, 1970 *et seq.*) is a series made up of several chapters contributed by eminent workers in the field. They may be consulted for in-depth discussions.

The annual series "Advances in Carbohydrate Chemistry" (latterly "Advances in Carbohydrate Chemistry and Biochemistry") contains scholarly and critical reviews of key topics. One would do well to consult these volumes for information dealing with compounds or reactions of interest. Fortunately there are cummulative indices to facilitate search.

## II. Some Background Information

Recurring areas of concern in contemplating syntheses with sugars are the form of the molecule (*i. e.* linear or cyclic) and if cyclic, the size and conformation of the ring. Accordingly these areas will be examined briefly.

A typical sugar e. g. glucose (Scheme 2) exists in solution as an equilibrium mixture primarily of the acyclic [*aldehydo* (4)] and cyclic [furanose (3), and pyranose (5)] forms. Both cyclic forms are sometimes discernible by pmr (*50*) and cmr (*99*) spectroscopy; however the *aldehydo*-form (4)

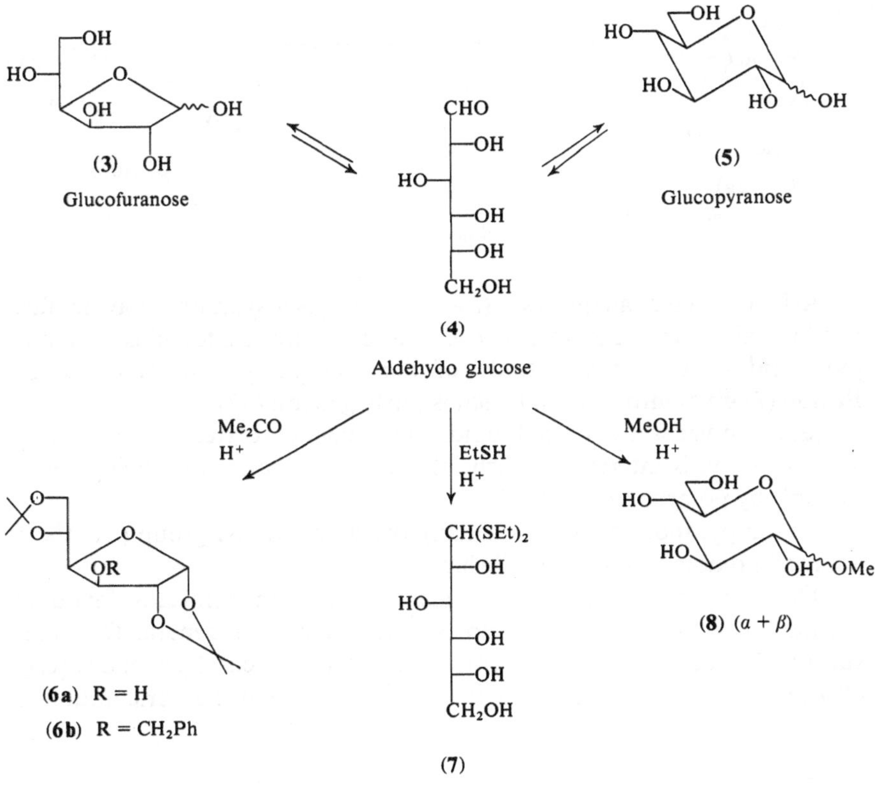

(3)   Glucofuranose

(4)
Aldehydo glucose

(5)
Glucopyranose

(6a)  R = H
(6b)  R = CH₂Ph

(7)

(8) (α + β)

*Scheme 2*

(0.0026% in the case of glucose) (59) is usually not detected spectroscopically. Nevertheless it is often possible to trap each form. Thus in the case
of glucose, the furanose form (3) is selectively trapped by acetonation
to "diacetone glucose" * (6a). The pyranose form (5) is trapped as the
glucopyranosides (8) (α + β) by prolonged treatment with an alcohol
and acid, and the *aldehydo* form (4) is trapped as the dithioacetal (7) by
treatment with mercaptans (16).

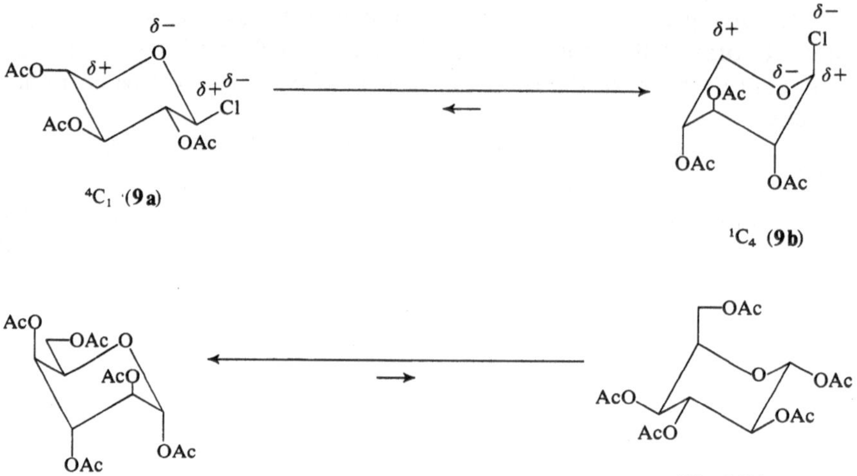

*Scheme 3*

Acid catalysed alcoholysis (*i.e.* Fischer glycosylation) may in fact
produce all four cyclic forms, *i.e.* α and β furanosides plus α and β
pyranosides. A thorough study of the methanolysis of **D**-xylose by
Bishop (7) gives information which is fairly general (57):

(a) furanosides are formed rapidly and may be isolated predominantly
if the reaction is interrupted early, the most judicious point being deter-
minable by polarimetry (57);

(b) the pyranoside with the axially oriented alkoxyl group (*i.e.* the α
anomer) predominates at equilibrium.

The phenomenon in (b), *viz.* that the thermodynamically favoured
orientation of the alkoxyl group is axial, is in fact general for polar
substituents at the anomeric centre. Thus with glycosyl bromides [e. g.
(77) Scheme 19] the anomer with bromine in equatorial orientation is

---

* 1,2-5,6-di-*O*-isopropylidene-α-*D*-glucofuranose.

too unstable to be isolated. The phenomenon which has come to be known as "the anomeric effect" occurs in a variety of organic molecules (*17, 52*). It may be interpreted in terms of a favourable dipolar interaction (Scheme 3) between the partial charges on the aglycone and on C-5. This interpretation has been supported by *ab initio* calculations (*101*); however there are alternative explanations (*15, 18, 43, 77*).

In addition to its role on the equilibrium concentrations of two possible carbon-1 isomers (*i. e.* anomers), the anomeric effect also influences the conformational populations of a given isomer. Thus tri-O-acetyl-β-D-xylopyranosyl chloride exists in the $^1C_4$ conformation* (**9b**) having the benefit of the anomeric effect, Scheme 3, in spite of severe 1,3-syn diaxial interactions (*40*). Similarly α-D-idopyranose pentacetate exists in the $^4C_1$ conformation (**10a**) rather than the $^1C_4$ conformation (**10b**) (*6*).

The foregoing indicates that the anomeric effect can be relied upon to confer a high degree of stereochemical and conformational homogeneity upon pyranose structures. Furanose derivatives also exhibit preferences (*87*), although to a lesser degree.

## III. Some Attributes of Carbohydrate Synthons

In view of the foregoing considerations, there are four attributes possessed by carbohydrate derivatives which make them ideal chiral synthons. These are discussed in general terms here, specific examples being encountered in the syntheses reviewed below.

### 1. Enantiomeric Purity

Sugars usually occur in nature as polysaccharides containing one or other enantiomeric form, *D* or *L*. There are however, exceptions, fortunately few, such as the galactan (*i. e.* poly-galactose) from snails which produces both *D*- and *L*-galactose upon hydrolysis (*93*).

### 2. Conformational Bias-High Stereoselectivity

Reactions of carbohydrate derivatives are usually attended by a high degree of stereoselectivity and many instances of this will be encountered. This is frequently attributable to the anomeric effect which, because it keeps the aglycon rigidly axial, can be relied upon for very efficient stereocontrol.

---

* This description indicates the atoms which lie above and below the plane containing the other four. See Scheme 3 and reference (*83*) for further details.

## 3. Ready Proof of Structure

The outcome of these stereoselective reactions can frequently be deduced with confidence by rational prognosis, as in predicting the formation of (84) from (83b) (Scheme 21). However, for independent proof, elementary pmr analysis is frequently all that is required. Normal hexose derivatives have only seven skeletal protons and although 60 MHz spectra usually provide adequate information, 220 MHz spectra are frequently first order (6, 40). In these analyses the location and characteristics of the anomeric proton (H-1) are of singular importance. In the first place H-1 normally occurs at lower field (> 5 ppm) than the other ring protons, and generally an equatorial H-1 resonates at lower field than does the corresponding axial counterpart (52, 54). From this vantage point, it is possible to make assignments by moving around the ring from one proton to the next, with the aid of decoupling experiments. For an interesting example see reference (40).

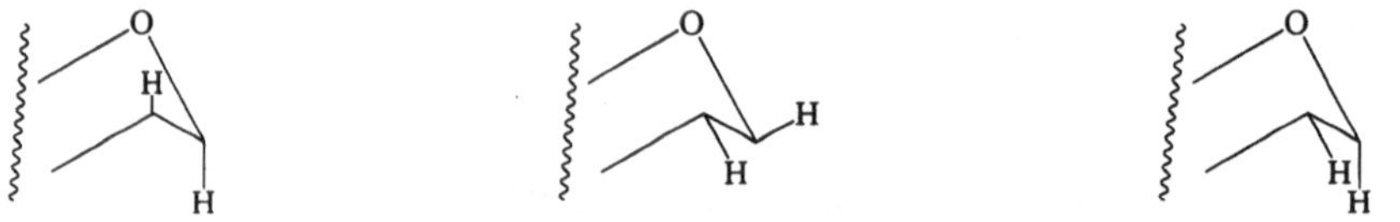

$J_{12} \sim 8$ Hz            $J_{12} \leq 1$ Hz            $J_{12} = 1.0$—$1.5$ Hz

*Scheme 4*

The coupling constant for H-1 also provides definitive evidence as indicated in Scheme 4. The values, although broadly in keeping with the Karplus relationship (45) are usually subject to variations arising from the electronegativities of substituents (55). Nevertheless, as in the example in Scheme 21, it is possible to tell by inspection of the 60 MHz spectrum that (84) is the product since there is only one signal for an anomeric proton, and it appears as a singlet.

For the information on cmr spectroscopy in sugars, a recent review by Perlin is recommended (74).

## 4. Latent Functionalities

To the uninitiated, simple sugars seem to consist of an aldehydo (or equivalent) group and an array of hydroxyl groups. However carbohydrate chemists are adept at exploiting subtle differences between the latter, thereby opening the way for the creation of a wide variety of functional groups.

Specific examples of the creation of such functional groups appear below along with the appropriate syntheses.

Attributes (1) and (2) of course enable the synthetic organic chemist to incorporate the absolute stereochemistry of his target molecule into the initial planning. Resolution of racemates is therefore avoided. This feature would of course be shared by any other optically pure starting material; however we would contend that only carbohydrates combine all four attributes.


## IV. Syntheses

In the examples which follow we have chosen to divide the syntheses into three groups:

### 1. Acyclic Transfer:

In these examples a fragment of known chirality obtained from a sugar or sugar derivative is degraded to the required product, or is transferred into an intermediate thereof. Historically this has been the most common employment of sugars as chiral synthons (42) as is exemplified in the syntheses summarized in Scheme 1.

Closely allied with this is the preparation of chiral components for the purpose of making stereochemical correlations. A notable example (although not of an acyclic entity) is found in the work of BELLEAU (Scheme 5) related to the potent cholinergic methiodide (11). In the pursuit of determining the relative and absolute configurations of members of this series, 1,6-anhydro-D-galactose (12)* was degraded to the chiral dioxolane (13) (94).

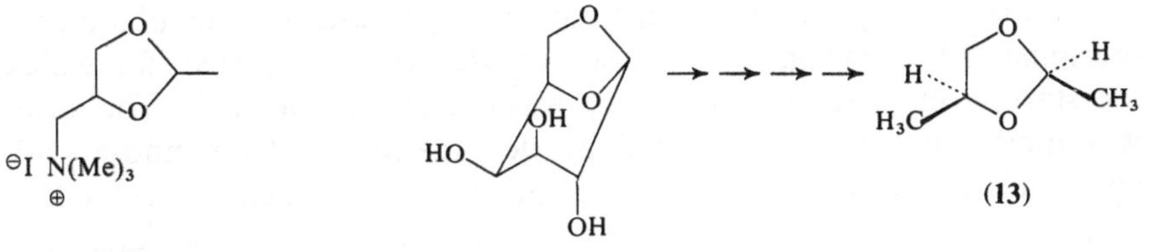

Scheme 5

---

* For further discussion of these anhydro systems see comments relating to Scheme 8 below.

The chiral fragments are normally excised by the use of periodate salts (*8*) and are thus well functionalized for incorporation.

### 2. Cyclic Transfer:

There are several natural products which contain tetrahydropyranoid or tetrahydrofuranoid rings and these are logical synthetic targets. Thus we will discuss several examples where the sugar has been in effect built into the target molecule.

### 3. Transcription:

The rings of carbohydrate derivatives are excellent templates upon which a wealth of stereochemical features may be assembled, as was implied above in discussing the attributes of carbohydrate synthons. Having served its stereochemical purpose the sugar moiety may be destroyed so that it does not exist as a discrete entity in the product, although sections of it may survive.

Although the template effect is encountered most commonly with ring systems of sugars, there are examples where acyclic sugar derivatives have been employed. In either case stereochemical information is *transcribed* from the sugar to the product.

The lines of demarcation between these classifications are, of course, not rigid since many of the syntheses to be discussed could fit into more than one of these categories.

### 1. Acyclic Transfer

#### a) L-erythro-Biopterin (22)

In order to provide an unequivocal synthesis of the ubiquitous pteridine (**22**), Taylor and Jacobi required the osone (**18**), for which the starting material is 5-deoxy-L-arabinose (**17**) (Scheme 6). The latter was most conveniently obtained by degradation of L-rhamnose (**14**), an abundant naturally-occurring sugar, by an improvement of the method originated by Hough and Taylor (*41*). This method involves formation of the dithioacetal (**15**) and oxidation of this to the disulfone (**16**). Upon gentle treatment with base, this undergoes a retro-aldol reaction giving the lower homologue.

"Osones", 2-keto-aldehydo sugars such as (**18**), were first prepared by Fischer by hydrolysis of the corresponding osazones (*5*). More recent methods involve careful oxidation of the aldose, and this was the route pursued by TAYLOR and JACOBI. The aldoxime (19) was obtained chemospecifically by judicious exchange with acetoneoxime, thereby making the 2-keto group available for reaction with the aminomalono-nitrile (20). The product arising therefrom (21) was then transformed into the title substance (22) by an established sequence of reactions (90).

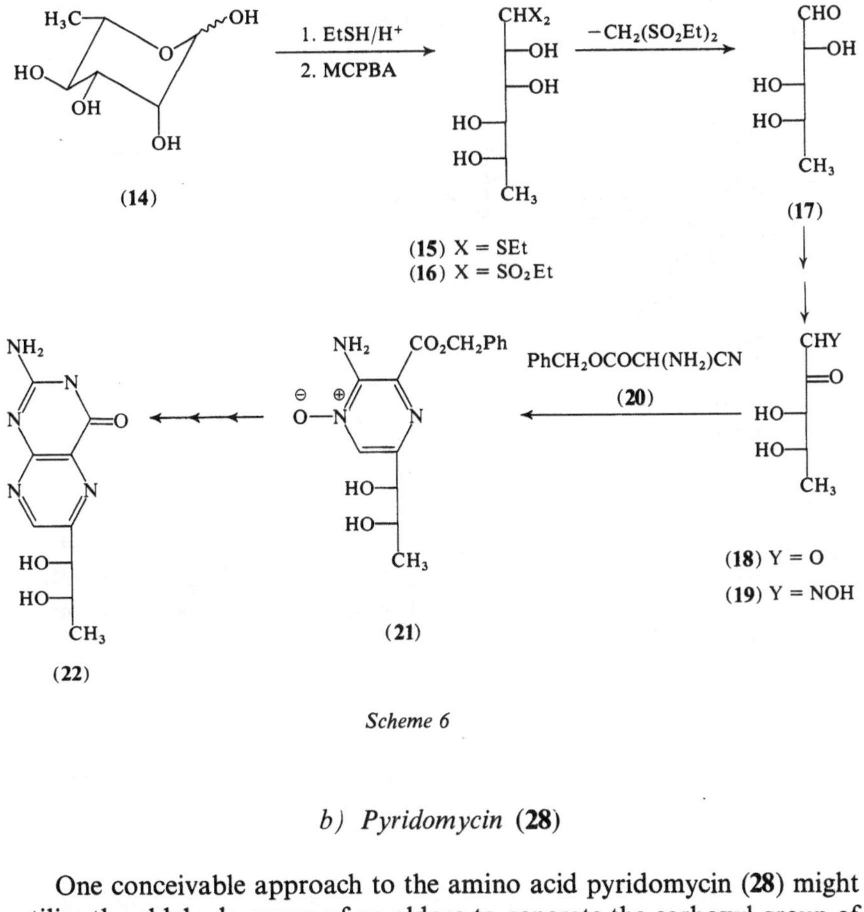

Scheme 6

### b) Pyridomycin (28)

One conceivable approach to the amino acid pyridomycin (28) might utilise the aldehydo group of an aldose to generate the carboxyl group of the product. However KINOSHITA and MARIYAMA wisely chose to utilize the glucose skeleton (6a) in a different way (Scheme 7).

In order to obtain the tertiary methyl group of pyridomycin, KINO-SHITA and MARIYAMA took advantage of the fact that additions at the trigonal centre in derivatives such as (24) occur with complete stereo-selectivity (82, 91) from the exo-face. The latter (24) was readily obtained from "diacetone glucose" (6a) (Scheme 2) via the ketone (23).

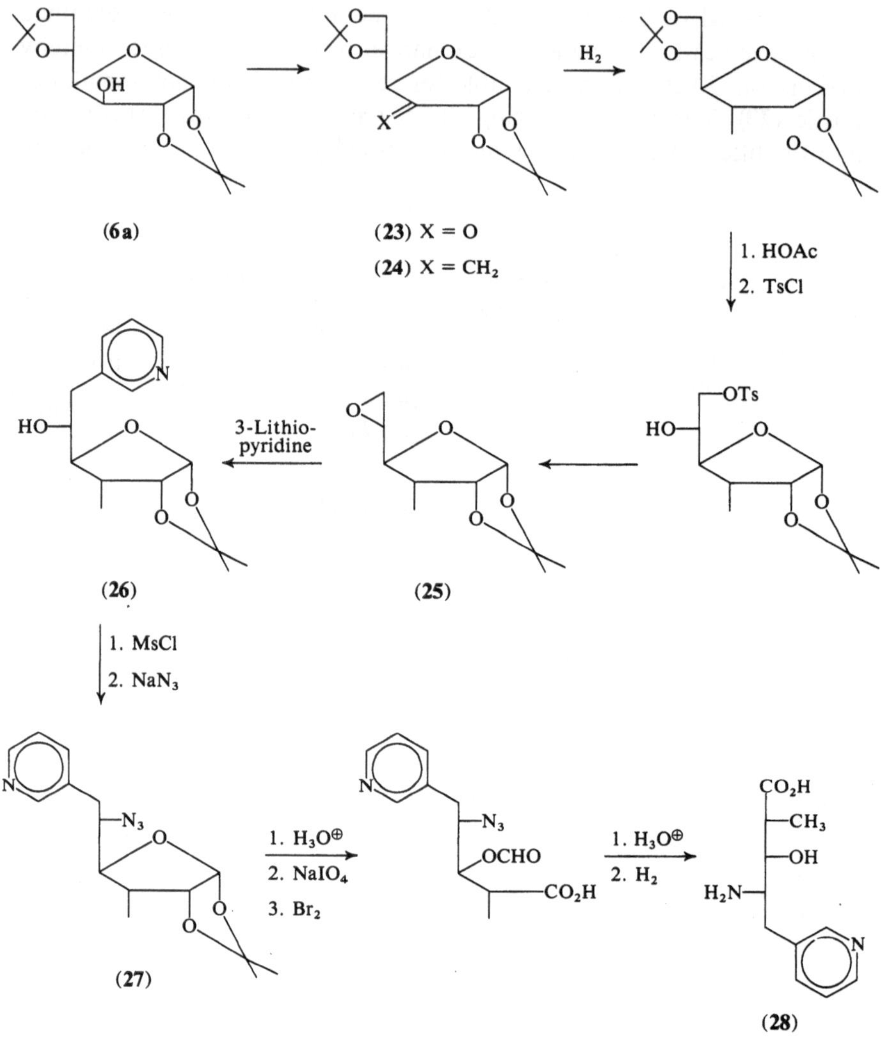
OH
X
H₂
(6a)
(23) X = O
(24) X = CH₂
1. HOAc
2. TsCl
N
HO
3-Lithio-
pyridine
OTs
HO
(26)
(25)
1. MsCl
2. NaN₃
N₃
1. H₃O⊕
2. NaIO₄
3. Br₂
OCHO
CO₂H
1. H₃O⊕
2. H₂
CO₂H
CH₃
OH
H₂N
(27)
(28)

*Scheme 7*

The 5,6-O-isopropylidene ring in systems such as these can be cleanly and specifically removed by gentle acid hydrolysis. Selective sulfonylation of the primary hydroxyl group precedes formation of the oxirane (**25**). These reactions preserve the chirality at C-5 of (**25**) as does the succeeding regio-specific cleavage leading to (**26**). The subsequent azidolysis giving (**27**) therefore sets the stage for an aminomethylene group of verified chirality in pyridomycin (**28**).

*References, pp. 56—61*

## c) (+)- and (−)-Frontalin (29)

Frontalin (29) (47), brevicomin (30) (48), and multistriatin (31) (72) pheromonal constituents of a number of beetles, all possess the dioxabicyclo (3.2.1) octyl skeleton (Scheme 8). These ring systems are frequently encountered in carbohydrate chemistry where they are known, most commonly, as 1,6-anhydro sugars (97) since they are formally obtained by cyclodehydration of hydroxyl groups at positions 1 and 6. Actually 1,6-anhydro-D-glucose or levoglucosan (32) is best prepared by pyrolysis of starch (97). However with sedoheptulose (33) dehydration occurs spontaneously giving sedoheptulosan (34).

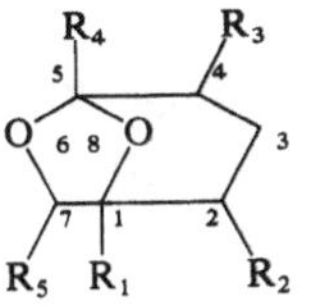

| | R₁ | R₂ | R₃ | R₄ | R₅ |
|---|---|---|---|---|---|
| (29) | CH₃ | H | H | CH₃ | H |
| (30) | H | H | H | CH₃ | C₂H₅ |
| (31) | H | CH₃ | CH₃ | C₂H₅ | H |

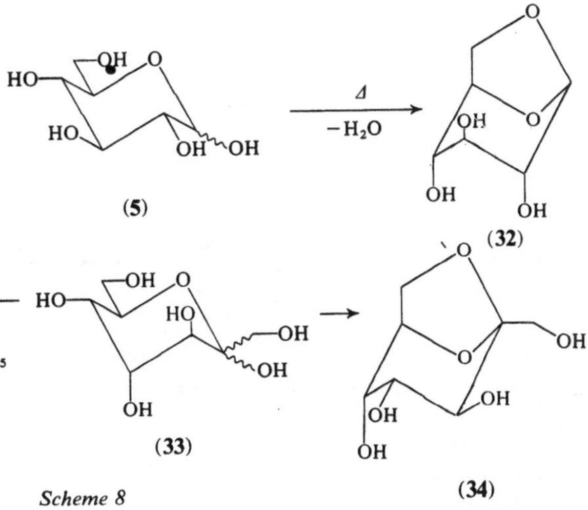

Scheme 8

A route to frontalin developed by HICKS and FRASER-REID (36) can be adapted to provide both enantiomers in pure (+)- and (−)-forms or as a mixture of both (Scheme 9). Although frontalin (29) has two chiral centres, only that at C-1 needs to be considered in a synthetic route since that at C-5 is controlled by the mode of cyclisation. Thus, as is evident from the acyclic precursor (42), a fragment such as (41), bearing the tertiary hydroxyl group in known chiral form would be a suitable objective. This was achieved by creating this centre stereoselectively on the pyranose ring and subsequently excising the chiral fragment.

Benzylidination of methyl α-D-glucopyranoside (8 α) gives the diol (35) which can be converted to the oxirane (37) in excellent yield (37)*. This process is accomplished in one step, but the 2-O-tosyl ester (36)

---

* For the conversion of (35) into the diastereomeric oxirane (54) see Scheme 13.

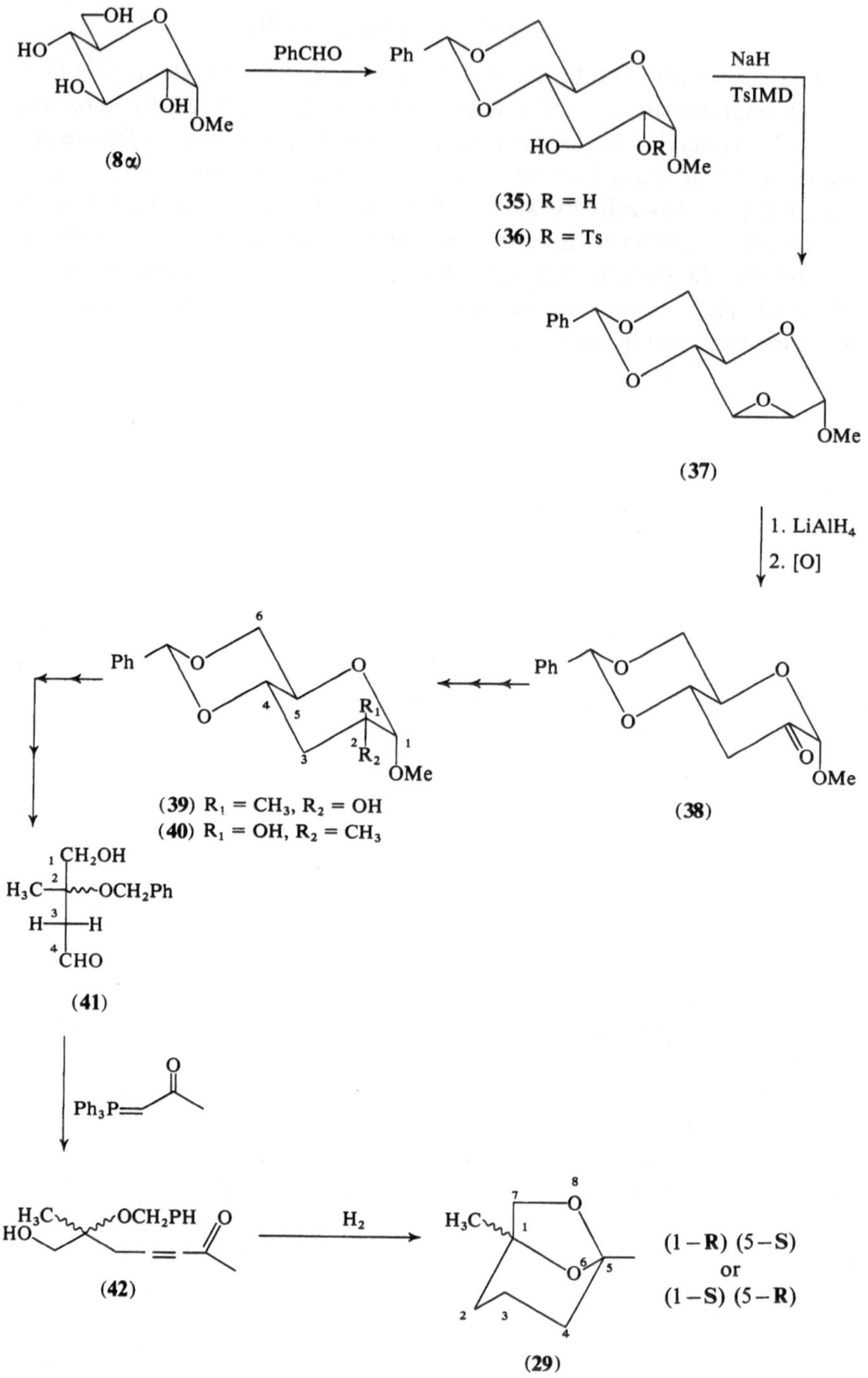

(8α)

(35) R = H
(36) R = Ts

(37)

1. LiAlH₄
2. [O]

(39) R₁ = CH₃, R₂ = OH
(40) R₁ = OH, R₂ = CH₃

(38)

(41)

Ph₃P=

(42)

H₂

(29)

(1−R) (5−S)
or
(1−S) (5−R)

*Scheme 9*

is known to be an intermediate. (The preferential esterification at C-2 is discussed in connection with Scheme 18.) Lithium aluminum hydride reduction occurs with complete stereo- and regio-selectivity giving an alcohol, from which the desired ketone (**38**) is obtained. The latter provides pure (**39**), pure (**40**) or a 3 : 2 mixture of both, and with this achieved, the fragment (**41**) is excised and elaborated to frontalin as indicated in Scheme 9.

A much more lengthy route from glucose giving only the (−)-enantiomer of frontalin has been reported by OHRUI and EMOTO (66). This synthesis, as well as that reviewed above share the common defect that some of the original carbons of the sugar are lost during periodate oxidations. If it is considered that the formation of C-C bonds is nine-tenths of synthetic organic chemistry, then the need for reticence in the use of sodium meta-periodate is warranted. This has been of prime consideration in the approaches to exo-brevicomin which follow.

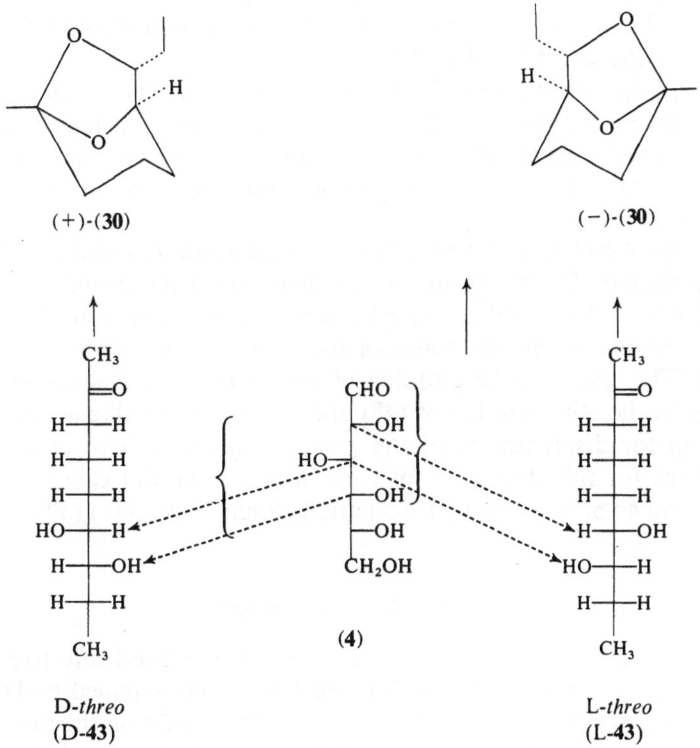

*Scheme 10*

### d) (+)- and (−)-exo-Brevicomin (30)

The acyclic hydrolysed forms (43) of (+)- and (−)-exo-brevicomin shown in Scheme 10 reveal that the chiral components are D-*threo* and L-*threo* vicinal glycols respectively. Thus the four-carbon fragments embodying the required glycol residues could be derived from D-glucose as indicated, and used to form chiral components of (+) and (−), (30).

For (+) (30), the benzyl ether (6b) is an obvious synthon since the 3-OH is protected, the 4-OH is in the ring, and the acetonide groups can be hydrolysed and cleaved selectively (see discussion above on pyrido-mycin, Scheme 7). The olefinic group in (44), which is readily generated from the di-O-methanesulfonyl precursor (44), is a convenient synthon for the ethyl residue. A major obstacle in contemplating the use of all six carbons was the required deoxygenation at C-2 of (45). With glycosides, nucleophilic displacements at C-2 are usually impossible and reasons for this have been adduced (80). Thus sulfonylation of (45), followed by hydride reduction did not give (46), but regenerated the alcohol (45) indicating that hydride attack had occurred at the sulfur. However the free-radical deoxygenation procedure of Barton and McCombie (4) accomplished the task with great facility.

With this obstacle overcome, the remaining steps were conveniently pursued as indicated in Scheme 11.

From the correlation shown in Scheme 10, the L.*threo* moiety encompassing carbons 2 and 3 of glucose is to be transferred into (−) exo-brevicomin. At the time of submission of this article, the synthesis of Fraser-Reid and Sherk was incomplete but the route pursued is sum-marized in Scheme 12.

Methyl 4,6-O-benzylidine-α-D-glucopyranoside (35 (Scheme 9) is an ideal precursor for the proposed synthesis since the 2 and 3 hydroxyl groups can be preserved as benzyl ethers (47) in which form they will be carried through to the final stages of synthesis — as was done in Schemes 9 and 11. The plan calls for addition of one carbon to the anomeric centre leading to the ethyl residue of (48) and thence to the alkene (49), which is set up for displacement of the primary hydroxyl. Use of the acetyl equivalent for this displacement then provides the full carbon skeleton in (50) and as previously, hydrogenation should lead to (–) (30).

### e) Erythronolide A (51)

A novel and ingenious approach to the stereochemically frightening macrolide erythronolide A (64, Scheme 15) is being pursued by Hannes-sian, Rancourt and Guindon. The approach is based on the recognition of "hidden symmetry" in the molecule (33) a conception which enables

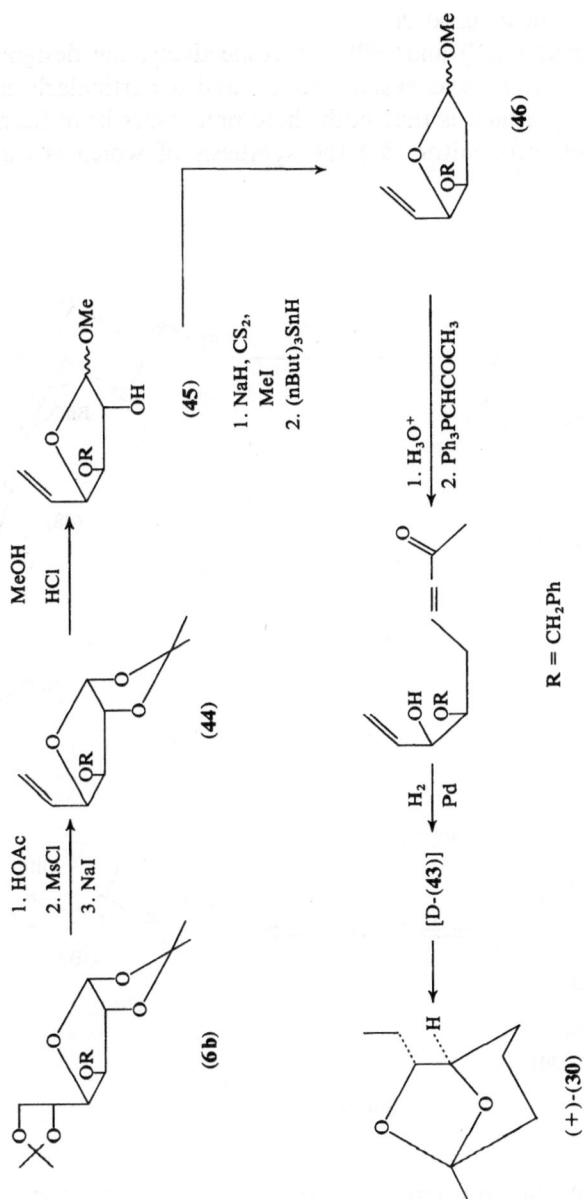

*Scheme 11*

its dissection into western and eastern zones (*32*) numbered here for convenience as carbons 1-6 and 1'-6' respectively. These zones embody all but one of the chiral centres and all but two of the fourteen carbons which constitute the macrocycle.

The pyranosides (**58**) and (**59**) are, respectively, the designated precursors of the western and eastern zones, and a particularly appealing feature of this approach is that both these precursors have been derived from a common progenitor (**57**) the synthesis of which is outlined in Scheme 13.

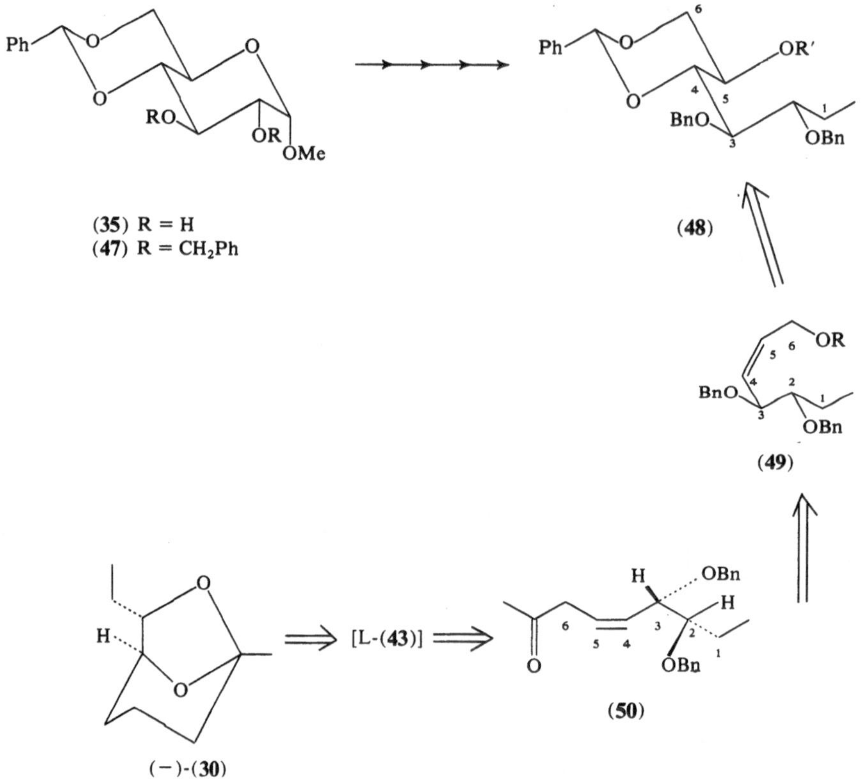

(**35**) R = H
(**47**) R = CH₂Ph

(**48**)

(**49**)

(**50**)

(−)-(**30**)    [L-(**43**)]

*Scheme 12*

In Scheme 9, the conversion of (**35**) into the *manno* ("up") oxirane (**37**) *via* the mono-O-sulfonate ester was discussed. The same substrate (**35**) leads conveniently to the *allo* ("down") oxirane (**51**) if the di-O-sulfonate is treated with base (*98*).

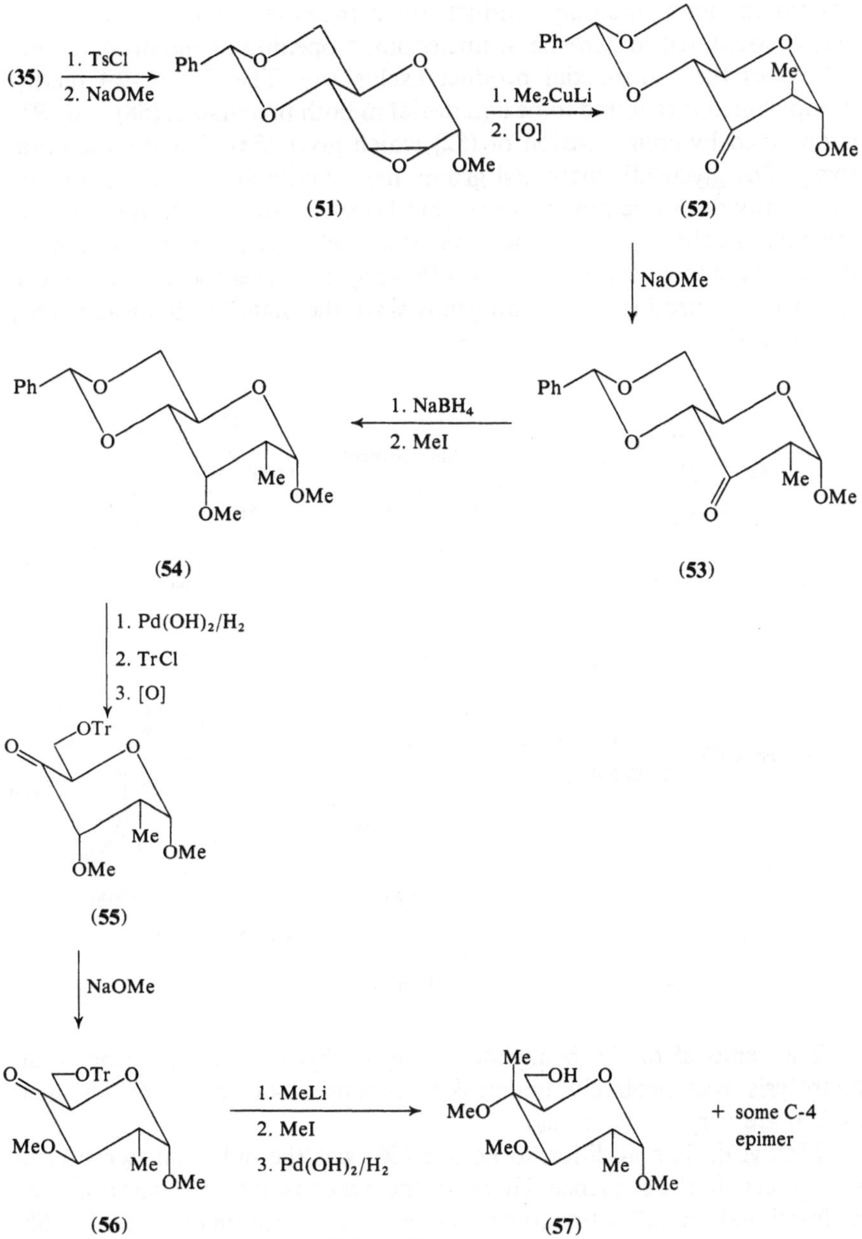

(35) $\xrightarrow[\text{2. NaOMe}]{\text{1. TsCl}}$ (51) $\xrightarrow[\text{2. [O]}]{\text{1. Me}_2\text{CuLi}}$ (52)

NaOMe

(54) $\xleftarrow[\text{2. MeI}]{\text{1. NaBH}_4}$ (53)

1. Pd(OH)$_2$/H$_2$
2. TrCl
3. [O]

(55)

NaOMe

(56) $\xrightarrow[\text{3. Pd(OH)}_2/\text{H}_2]{\substack{\text{1. MeLi} \\ \text{2. MeI}}}$ (57) + some C-4 epimer

*Scheme 13*

2*

Beginning with the epoxide **(51)**, each stereochemical feature is introduced sequentially, using procedures which give excellent yields of the desired epimer as the only product. As in the case of the *manno*-epoxide **(37)** encountered in Scheme 9, nucleophilic opening of the diastereomer **(51)** gives the transdiaxial product exclusively. The C-2 (C-2′) methyl group, which is required to be equatorial in both precursors, **(58)** and **(59)**, is obtained by epimerisation of **(52)** which gives **(53)** virtually quantitatively. The glycosidic methoxyl group, held rigidly axial by the anomeric effect provides steric hindrance so that borohydride reduction gives only the axial alcohol. Actually the latter stereoselectivity is not crucial since this centre is to be epimerized, *i. e.* **(55→56)**. However the stereochemical purity of the product is advantageous since the material **(55)** was readily crystallized.

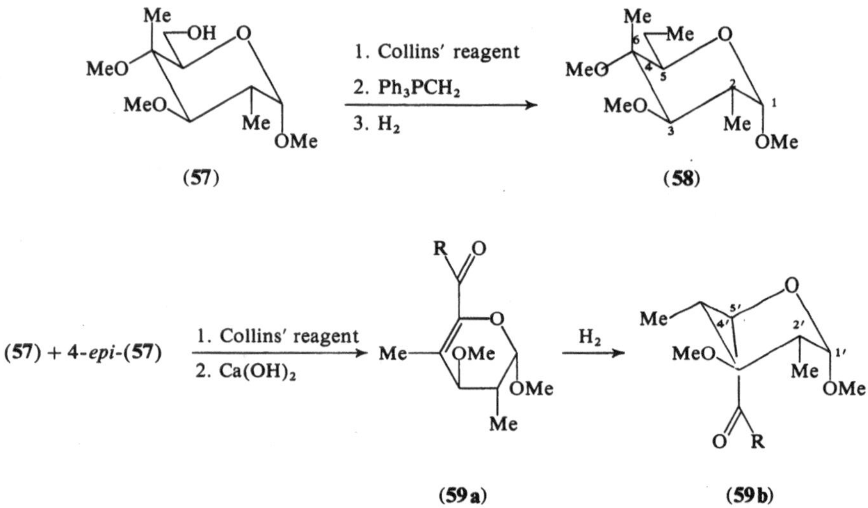

*Scheme 14*

The removal of the benzylidene ring by hydrogenolysis rather than hydrolysis was probably undertaken because 2-deoxy-glycosides such as **(54)** are very labile to acid.

Methyl lithium addition to ketone **(56)** was the only non-stereoselective process in the sequence. However the major isomer **(57)** was obtained by fractional crystallisation and was used for preparation of precursor **(58)** as indicated in Scheme 14. Residual **(57)** and its 4-epimer, contained in the mother liquors, were both used for preparation of precursor **(59)**, since one step involves β-elimination of methanol.

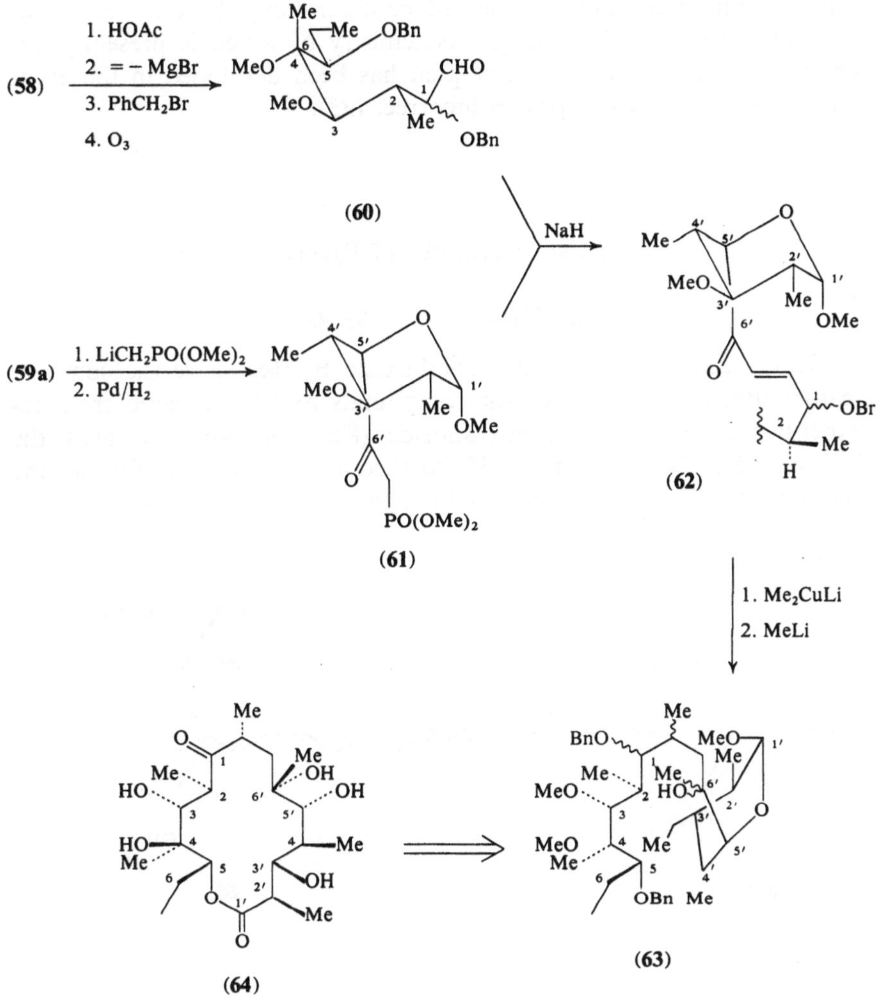

*Scheme 15*

Assembly of the carbon framework of (64) from (58) and (59) requires the addition of two carbons to bridge C-1 and C'-6, and the introduction of two methyl groups. The ingenious way in which this was achieved (*33*) is shown in Scheme 15. The two bridging carbons were attached, one to each of the precursors, in such a way that electrophilic (60) and nucleophilic (61) partners are produced, the condensation of which yields a substance (62) capable of receiving both the required methyl groups. Since the C-benzyloxy group of the assembled unit (63) is destined to become a carbonyl group by virtue of which the adjacent methyl group can be epimerized into the desired configuration, the only centre of concern is that at C'-6, the stereochemistry of which is presently unknown. However, a contingency plan has been developed in the event that the stereochemistry proves incorrect (33).

## 2. Cyclic Transfer of Pyranoids

### a) Thromboxane B₂ (66)

Three recent syntheses of thromboxane B₂ (66) share the common feature that they all begin with compounds readily prepared from inexpensive methyl α-D-glucopyranoside (8α) and work towards the lactone (65). The route from (65) to thromboxane B₂ (66) follows the standard methodology developed by Corey (12).

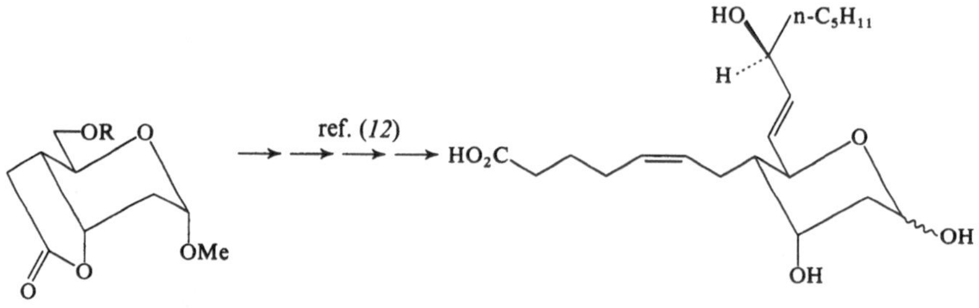

Scheme 16

The route of Hanessian and Lavallee (31, Scheme 17) utilizes the oxirane (51) encountered in Scheme 13. Reduction with lithium aluminum hydride gives (67) exclusively. After cleavage of the benzylidene ring, the diol is treated with tert-butyldiphenylsilyl chloride, a reagent developed in Hanessian's laboratory for selective silylation of a primary hydroxyl group. This enables the sequence leading to the unsaturated ester (68). The high degree of stereocontrol in the hydrogenation of (68) arises, undoubtedly, from the fact that the $^4C_1$ conformation is strictly maintained in spite of the syn-diaxial disposition of the methoxyl and benzoate groups, the latter effectively hindering approach from "below". With this crucial stereochemical problem solved, treatment of (69) with base leads to the lactone (65) which is converted into (66) by similar, although not identical methods to those used previously (*12, 63*).

The two other routes to thromboxanes from the laboratories of COREY (*11*) and HERNANDEZ (*35*) are similar in strategy in that they utilize a Claisen rearrangement of a hex-3-enopyranoside to attach the two-carbon residue. In this context it is appropriate to note that the Claisen rearrangement as a method of making branched chain sugars was first investigated by FERRIER and VETHAVIYASAR in 1973 (*20*). The vinyl ether derived from the allylic alcohol 82b (Scheme 20) and its 4-epi analogue, were successfully rearranged to the unsaturated 2-formyl-methyl isomers in yields of ~ 70%.

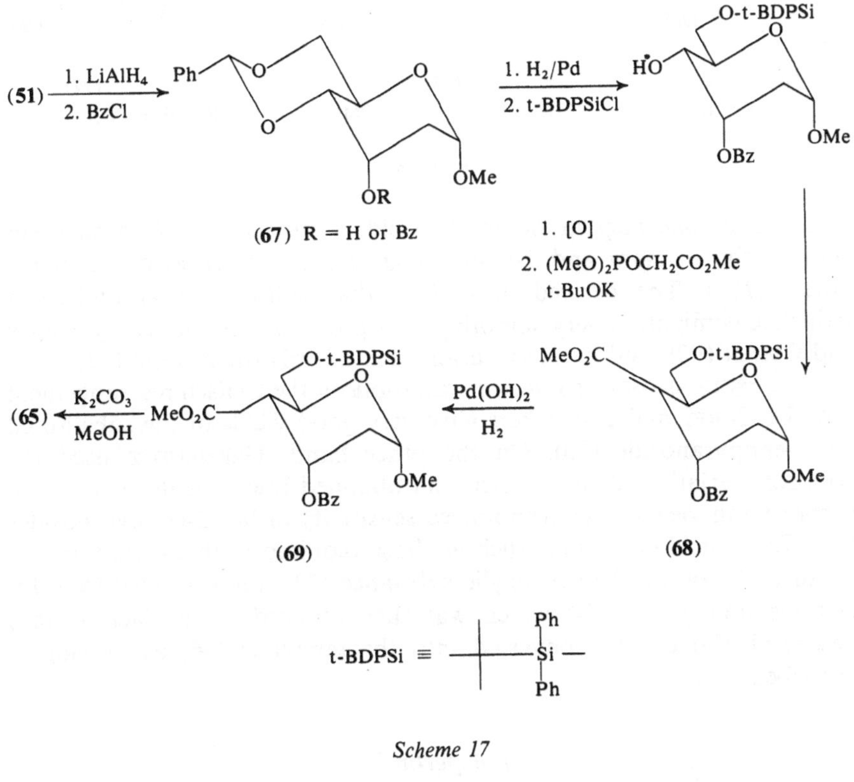

*Scheme 17*

The routes of COREY, SHIBASAKI and KNOLLE (*11*) and of HERNANDEZ (*35*) begin (Scheme 18) with the unsaturated sugar (72) whose synthesis from (8α) had been developed by HOLDER and FRASER-REID (*39*). The key substance in the preparation of (72) is the dibenzoate (70), which may be

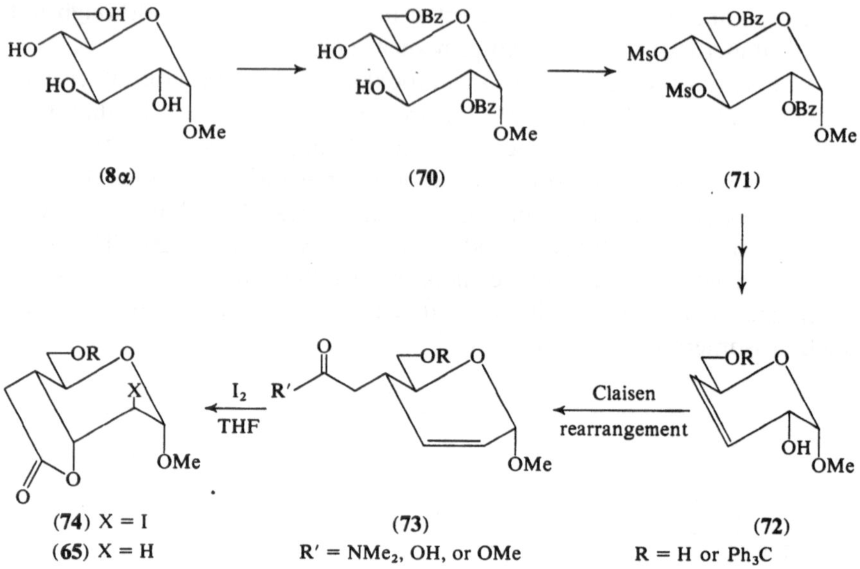

Scheme 18

prepared in one step in moderate yield, a procedure which succeeds because the primary and 2-hydroxy groups are more acidic than the others (100). The derived 3,4-di-O-methanesulfonate (71) undergoes reductive elimination very smoothly by a procedure developed by Tipson and Cohen (92), and deesterification then affords the alcohol (72).

The Eschenmoser or Ireland variations of the Claisen rearrangement used by Corey and co-workers gave very good yields of the substituted hex-2-enopyranoside (73). On the other hand, Hernandez used the Johnson variation of the Claisen and obtained lower yields. This is not surprising in view of the pronounced sensitivity of hex-2-enopyranosides (e. g. 73) to acid conditions such as those required in this variation.

In both routes, the carboxylic substance (73) was subjected to iodo-lactonisation giving (74) which was then reduced to the lactone (65) described above, and converted into thromboxane (66) as previously described (63).

### b) Asperlin (80)

Asperlin (80), an antibiotic from *Aspergillus nidulans,* was recently synthesized by Perlin and Lesage (56) by a route (Scheme 19) designed to establish the stereochemistries of the oxirane and acetoxy groups. Their spectroscopic studies indicated axial orientation for the latter and hence galactose, the most abundant sugar with the 4-hydroxy group axial, was chosen as starting material. Fortunately the diacetonide* thereof (75) has the primary hydroxyl free so that the alkene (76) can be prepared for eventual epoxidation.

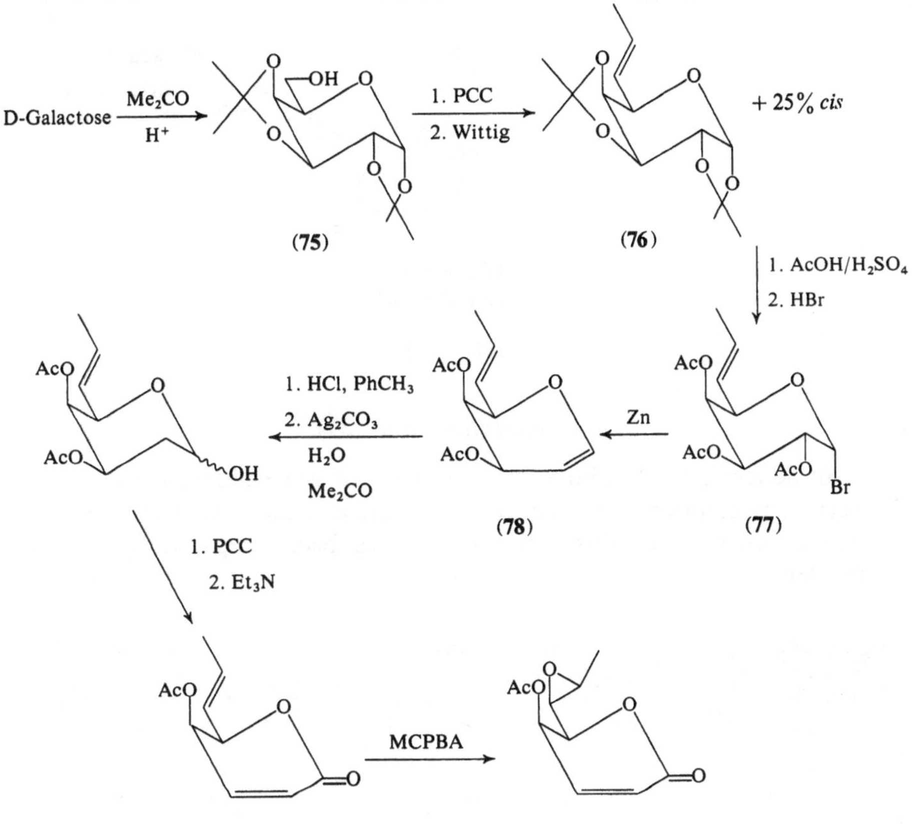

*Scheme 19*

Acetolysis of (76) gives a tetraacetate which reacts with hydrobromic acid at the activated centre giving the acetobrom sugar (77). Conversion of the derived glycal (78) into lactone (79) proved more problematic than had been anticipated and the indicated steps were therefore required. The epoxidation was non-stereospecific but the presence of asperlin was deduced spectroscopically. The chirality of the epoxide was not determined at the time of writing.

---

* The course of diacetonation of glucose, galactose and mannose (Schemes 2, 19 and 30 respectively) form an interesting contrast. MILLS has presented a rationalization for these annomalies (62).

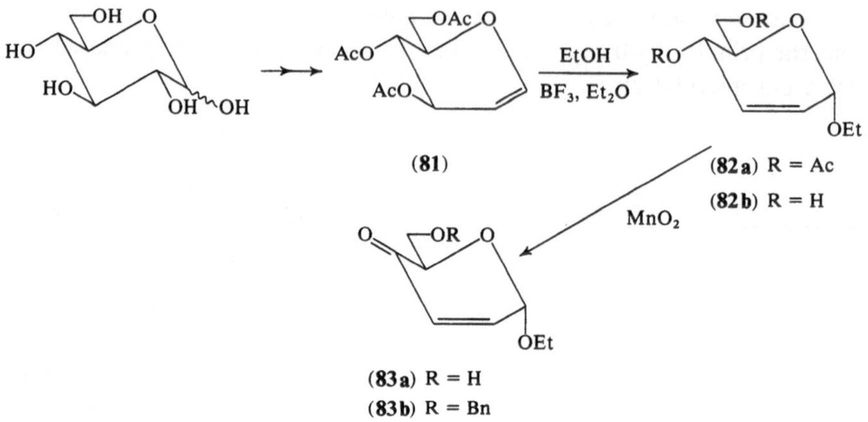

(81)

(82a) R = Ac
(82b) R = H

(83a) R = H
(83b) R = Bn

*Scheme 20*

## c) α-Multistriatin (31)

In Schemes 8—12 above, the syntheses of some dioxabicyclo (3.2.1) octyl pheromones by "acyclic transfer" were discussed. We now consider another member of this series which has been prepared by "cyclic transfer".

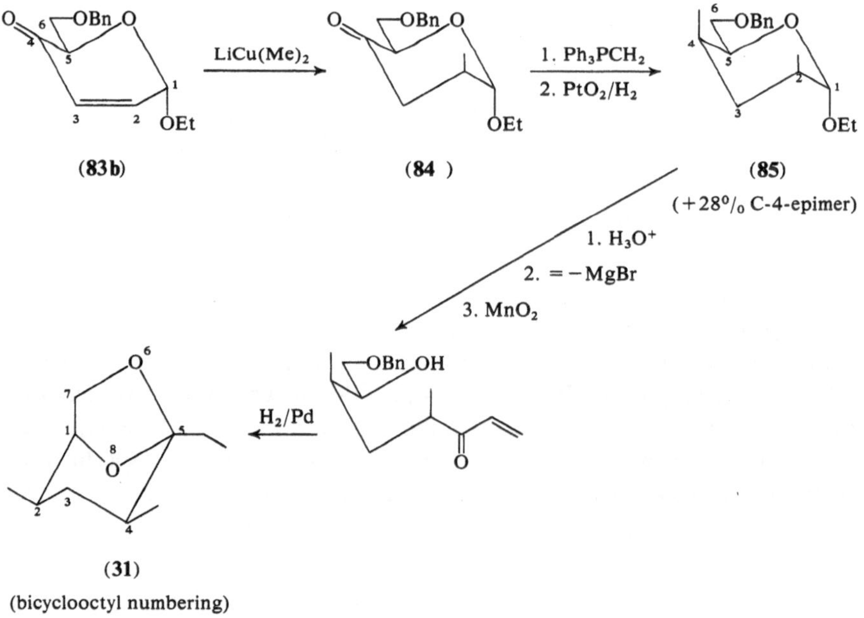

(83b)

(84)

(85)

(+28% C-4-epimer)

1. $H_3O^+$
2. $= -MgBr$
3. $MnO_2$

$H_2/Pd$

(31)
(bicyclooctyl numbering)

*Scheme 21*

For a recent synthesis of α-multistriatin (**31**) by PLAUMANN and FRASER-REID (*75*) the starting material was the enone (**83**) which is readily prepared from glucose as outlined in Scheme 20 (*26*). However triacetyl glucal (**81**) is commercially available. The Ferrier reaction with ethanol gives (**82a**) as a crystalline substance in excellent yield, and the derived diol (**82b**) may be oxidized with manganese dioxide to the hydroxy enone (**83a**). Alternatively the primary hydroxyl of the diol may be protected to allow for oxidation with Collins' reagent.'

A feature of the synthesis of (**31**) is that it avoids the use of acids for the final intramolecular cyclisation, which in previous syntheses had caused isomerization at C-4 (*28*). For the approach in Scheme 21, this chiral centre was established early at C-2 of (**84**) and was preserved by using compatible transformations.

Conjugate addition to (**83b**) gives (**84**) exclusively (*106*). However hydrogenation of the resulting methylene adduct proved to be less stereoselective than desired, giving a mixture of the C-4 epimers of (**85**). Because of the absence of an electron withdrawing group at C-2, hydrolysis of (**85**) proceeds under gentle conditions without causing epimerization of the fractious C-2 substituent. The succeeding steps proceeded smoothly. In the final step, saturation of the double bond and cleavage of the benzyl ether occur in tandem, with cyclization.

### 3. Cyclic Transfer of Furanoids

The fungal metabolites avenaciolide (**86**), isoavenaciolide (**87a**) (*cf.* ethisolide, **87b**), and canadensolide (**88**) (Scheme 22) form an interesting trio of structurally related di-γ-lactones. Avenaciolide was the forerunner

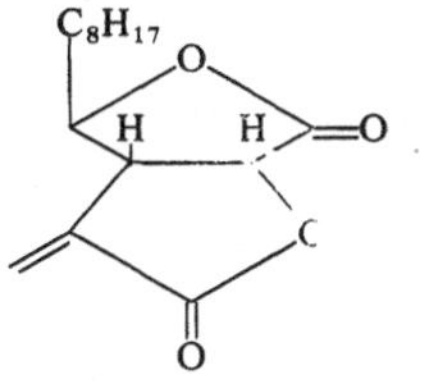

(−)-(**86**)
Avenaciolide

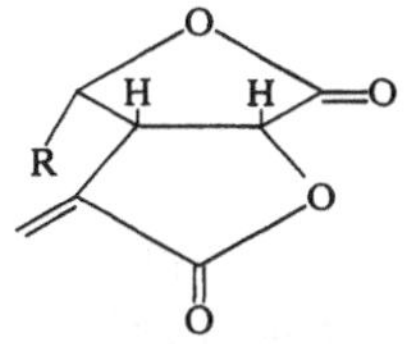

(−)-(**87a**)
R = $C_8H_{17}$
Isoavenaciolide

(**87b**)
R = $C_2H_5$
Ethisolide

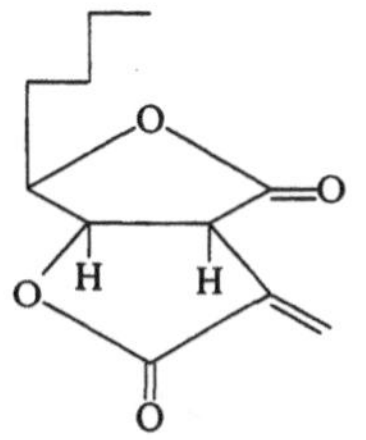

(−)-(**88**)
Canadensolide

*Scheme 22*

and its absolute stereochemistry was initially deduced by degradation and correlation with known chiral species (*10*). It transpires (see below) that this initial assignment was incorrect. It is an indication of the versatility of D-glucose that all three metabolites are preparable from it, and additionally, *both* the (+) and (−) enantiomers of avenaciolide.

### a) (−)-Avenaciolide (−**86**)

Juxtaposition of the skeleta of 1,2 : 5,6-di-O-isopropylidene ("diacetone") glucofuranose (**6a**) and avenaciolide reveals a number of intriguing relationships: (a) the protected anomeric centre which is a precursor for one of the lactones; (b) the free 3-hydroxy group which provides for the creation of the other lactone; (c) the presence of two of the three asymmetric centres at the proper oxidation level; and (d) the 5,6-O-isopropylidene group from which the n-octyl chain may be elaborated. With regard to objective (b) Rosenthal and Nguyen had already

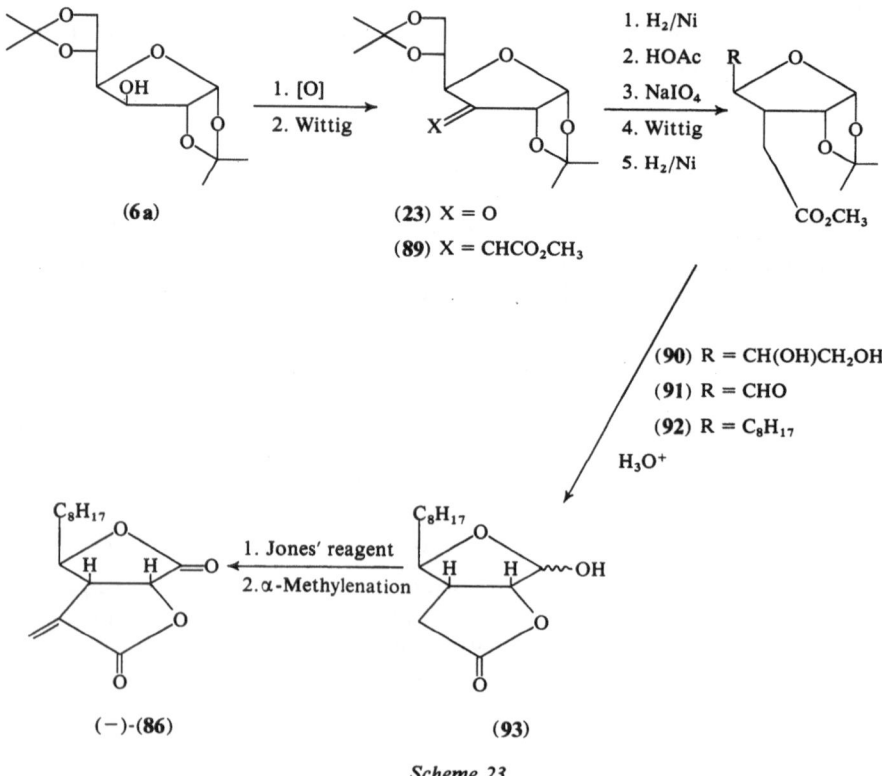

(**6a**)

1. [O]
2. Wittig

(**23**) X = O
(**89**) X = CHCO$_2$CH$_3$

1. H$_2$/Ni
2. HOAc
3. NaIO$_4$
4. Wittig
5. H$_2$/Ni

(**90**) R = CH(OH)CH$_2$OH
(**91**) R = CHO
(**92**) R = C$_8$H$_{17}$

H$_3$O$^+$

CO$_2$CH$_3$

1. Jones' reagent
2. α-Methylenation

(−)-(**86**)          (**93**)

*Scheme 23*

shown (*82*) that hydrogenation of the ester (**89**) was completely stereo-selective. Selective hydrolysis of the 5,6-acetonide gives (**90**), and cleavage then affords the aldehyde (**91**).

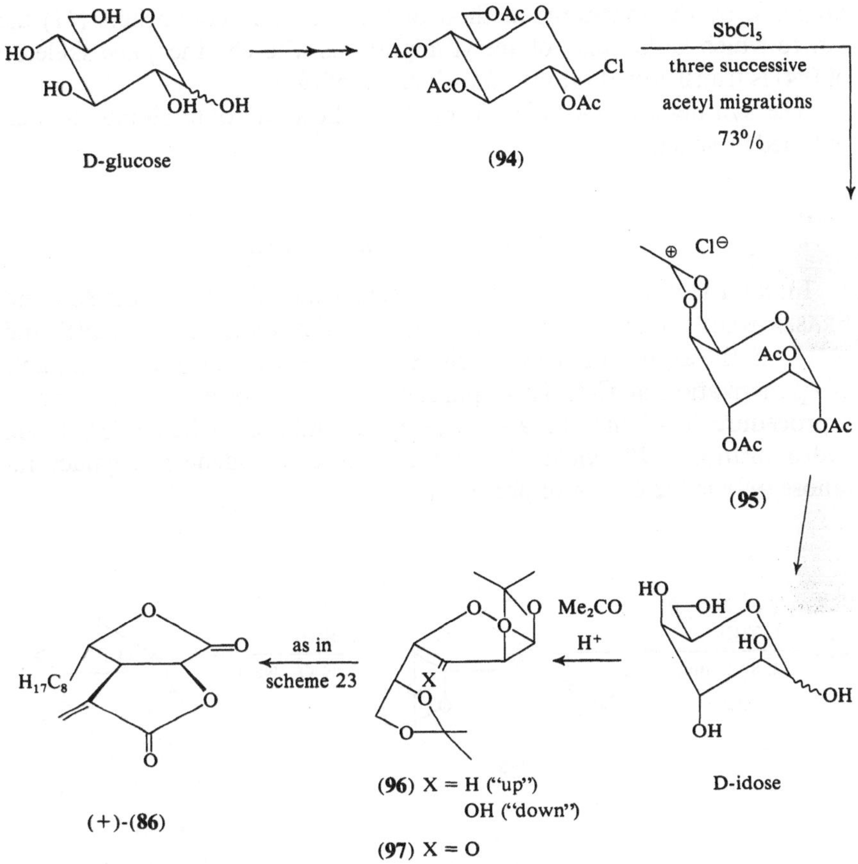

D-glucose

(**94**)

SbCl$_5$

three successive
acetyl migrations

73%

(**95**)

as in
scheme 23

Me$_2$CO
H$^+$

D-idose

H$_{17}$C$_8$

(+)-(**86**)

(**96**) X = H ('up')
        OH ('down')

(**97**) X = O

*Scheme 24*

Independent syntheses of avenaciolide by ANDERSON and FRASER-REID (*1*), and by OHRUI and EMOTO (*67*) capitalised upon the aldehyde (**91**). The route in Scheme 23 is that of the former authors (*1*) and had been designed to produce the unnatural enantiomer, based upon the early assignments (*10*). However the synthesis produced the natural material, indicating that the absolute stereochemistry deduced by degradation had been incorrect.

## b) (+)-Avenaciolide (+86)

In contrast to the foregoing, Ohrui had set out to prepare the presumed natural enantiomer *i.e.* (+)-(86). His starting material (Scheme 24) was the ketone (97) (1,2 : 5,6-di-O-isopropylidene-β-D-*lyxo*-hexofuran-3-ulose), for which diacetone lyxose (96) is required. The latter is readily prepared by the ingenious method of Paulsen and co-workers (*71*) the key to which is the relay of acetyl migrations whereby the *gluco* skeleton of (94) is transformed into the *ido*-skeleton of (96).

The synthesis of (+)-(86) from (97) follows a route similar to that outlined in Scheme 23.

## c) (−)-Isoavenaciolide (−87a)

Initial attempts at the synthesis of isoavenaciolide by Anderson and Fraser-Reid (*2*) sought to employ intermediates such as (23), (91) and (93) from the avenaciolide synthesis (Scheme 23) which should be capable of epimerization at C-4. This approach, however, proved futile, and so a procedure involving the alkene (98) was utilized (Scheme 25). Upon hydroboration, (98) yields 1,2 : 5,6-di-O-isopropylidene-α-D-galactofuranose (99) as the exclusive product (*70*).

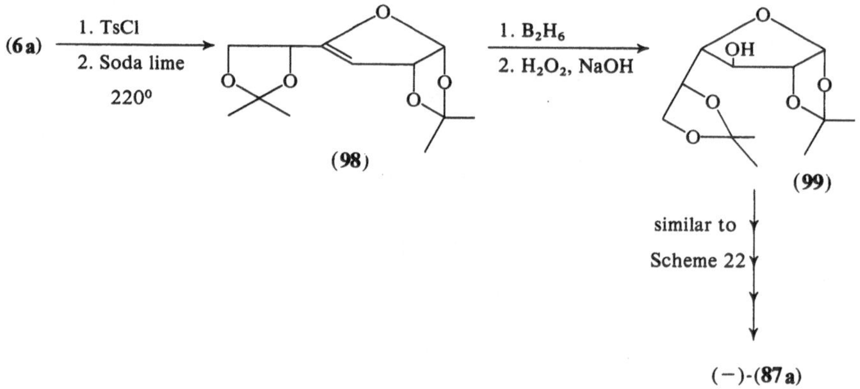

$$(6a) \xrightarrow[\substack{2.\ \text{Soda lime} \\ 220°}]{1.\ \text{TsCl}} (98) \xrightarrow[\substack{2.\ H_2O_2,\ NaOH}]{1.\ B_2H_6} (99)$$

similar to
Scheme 22

(−)-(87a)

*Scheme 25*

The route from (99) to isoavenaciolide (−)-(87a) follows a procedure similar to that outlined in Scheme 23. It was thereby confirmed that the previous assignment (*10*) of absolute stereochemistry was also incorrect.

## d) (−)-Canadensolide (−88)

The synthesis of canadensolide from "diacetone glucose" by ANDER-SON and FRASER-REID (3) differs from the cases described above in that the α-acrylate residue must be attached at C-2 and not C-3 and by corrollary that the C-3 oxygen must be preserved. The latter objective is readily achieved *via* the benzyl ether (6b), which is converted into (100) by the now familiar route (Scheme 23) and thence to the ketone, (101a) (see Scheme 26).

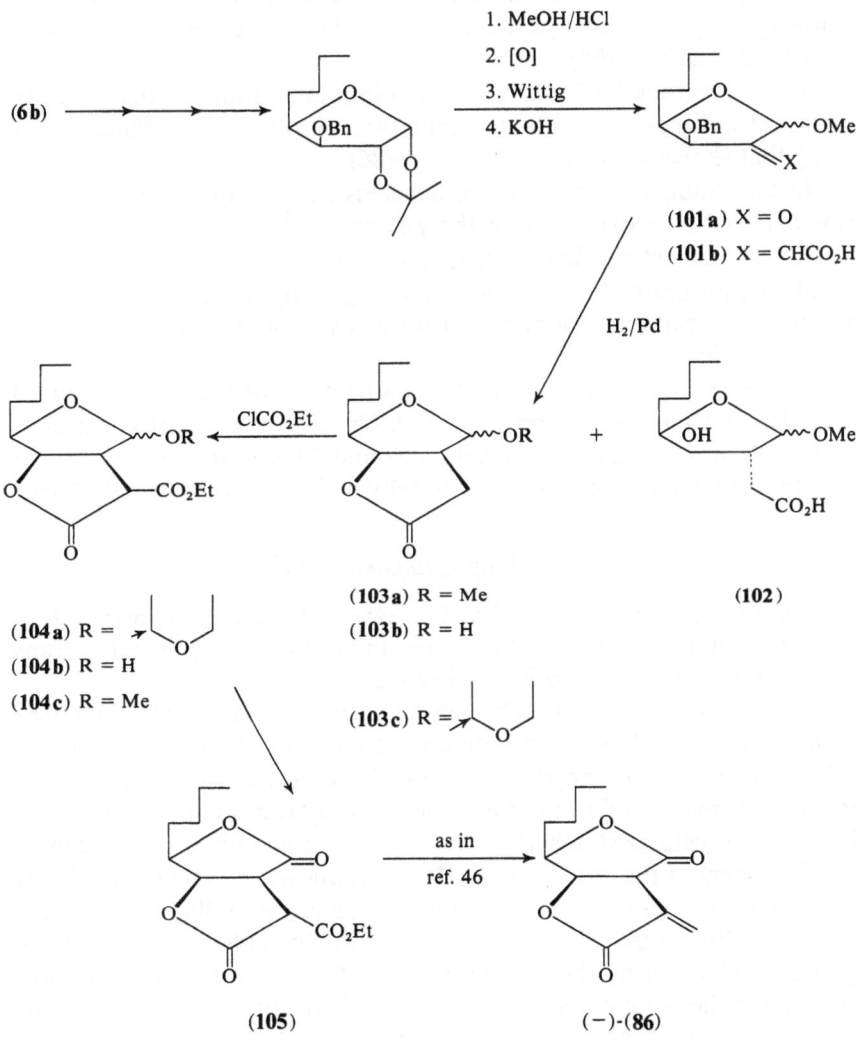

*Scheme 26*

Apart from its protective role, the bulky benzyl ether provides some hindrance for the "upper" face of the derived unsaturated acid (101 b) so that hydrogenation from "below" proceeds more readily, and upon *in situ* hydrogenolysis of the benzyl ether, γ-lactonization occurs spontaneously. This enables isolation of the lactone (103 a) from the *trans* hydroxy acid (102).

The success of the remaining steps hinged upon the simple expedient of replacing the methyl of (103 a) with ethoxyethyl in (103 c) prior to reaction with ethyl chloroformate. Thus attempts to selectively hydrolyse the acetal of (104 c) resulted in concurrent decarboxylation; however, gentle hydrolysis of (104 a) proceeded smoothly giving (104 b) which could then be oxidized to (105).

The bis-lactone (105) was converted into levorotatory canadensolide, which establishes the hitherto unknown absolute stereochemistry as being that shown in Scheme 26 for (−)-(86).

In the examples considered in Schemes 23—26, the anomeric acetal was oxidized to provide one of the γ-lactones. The reduced form of the acetal, *i. e.* the tetrahydrofuran, is also readily prepared *via* the corresponding unsaturated furan derivatives (64) [*i. e.* glycals *cf.* (81)]. Reductive reactions of suitable anomeric substituents (e. g. Br or SR) are also possible as in Scheme 28.

However a more direct route takes advantage of the ease with which cyclodehydration occurs between suitably disposed hydroxyl groups, giving an anhydro sugar. In Schemes 27 and 29 we consider case studies in which the latter approach to the tetrahydrofuran ring has been utilized.

### e) 11-Oxaprostaglandin (110)

The first synthesis of a chiral 11-oxaprostaglandin was achieved by Hanessian and coworkers (30), the tetrahydrofuranoid ring being transferred from an anhydroglucitol (Scheme 27).

Sorbitol (or D-glucitol) (106), an inexpensive, commercially available compound, is a dihydro derivative of D-glucose. Upon treatment with acid it forms the 1,4-anhydro system (107) in 36 percent yield. The preferential formation of a tetrahydrofuranol ring over any other possibility in these acid-catalyzed dehydrations is a well known phenomenon (86).

Treatment of the acetonide of (107) with one mole of methanesulfonyl chloride at low temperature gave the monomesylate (108) contaminated with some dimesylate. In glycosides such as (8), the acidity of the 2-hydroxyl group is enhanced by the inductive effect of the acetal (100). This effect is absent in the present case, but the observed selectivity could conceivably reflect less hindrance of the 2-hydroxyl group. Scission of the derived oxirane with sodio diethyl malonate was non-regioselective, but the desired isomer was converted to the intermediate (**109**) and thence to the target molecule (**110**) by standard methods.

LOURENS and KOEKEMOER (*60*) reported a different approach to the 11-oxaprostaglandins (Scheme 28) in which the 1,4-anhydrohexitol skeleton (**111d**) was created by the reductive desulfurization of a thio-

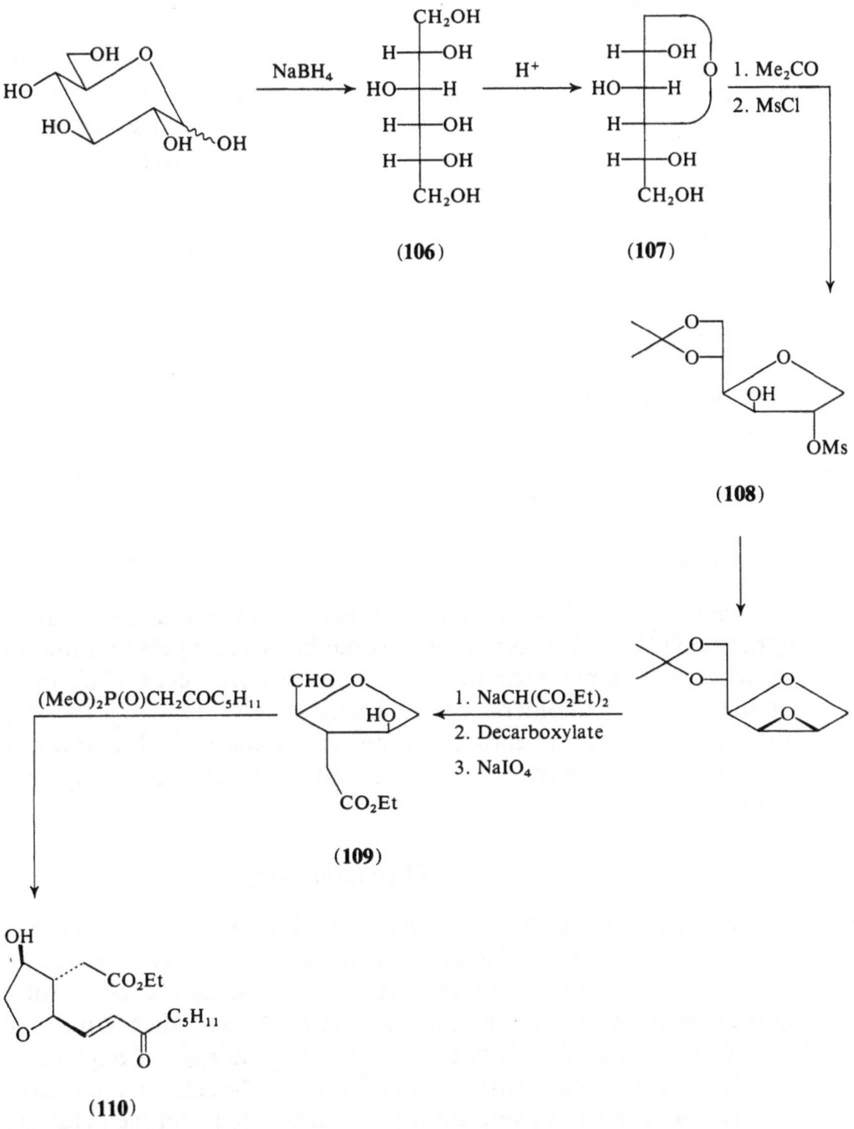

*Scheme 27*

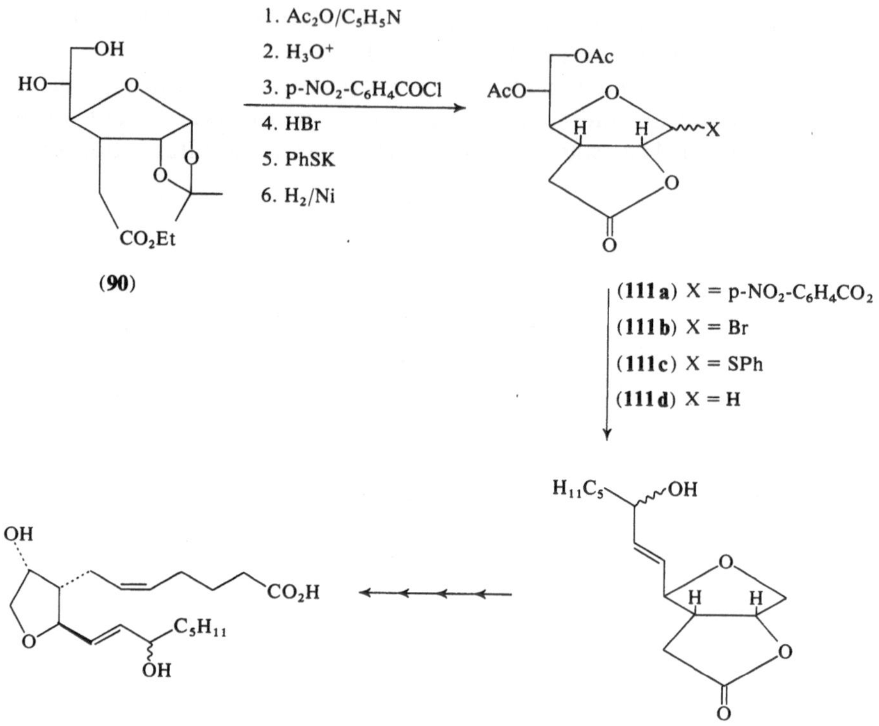

1. $Ac_2O/C_5H_5N$
2. $H_3O^+$
3. $p\text{-}NO_2\text{-}C_6H_4COCl$
4. HBr
5. PhSK
6. $H_2/Ni$

(90)

(111a) $X = p\text{-}NO_2\text{-}C_6H_4CO_2$
(111b) $X = Br$
(111c) $X = SPh$
(111d) $X = H$

*Scheme 28*

glycoside (111c). The route to the latter is ingenious in its use of the tetraester (111a) which is obtained by methods analogous to those depicted in Scheme 23. Upon treatment with hydrobromic acid (111a) reacts only at the activated anomeric centre giving (111b). This sets the stage for the mercaptolysis and subsequent reduction and with this accomplished, the remaining operations were readily carried out using standard techniques.

### f) d-Oxybiotin (115)

An alternative to the foregoing methods in Schemes 27 and 28 for the formation of tetrahydrofuranols is attributed to Defaye and Hildesheim (14) who reported that methanolysis of 1,2-O-alkylidene-5-O-sulfonates affords rearranged acetals such as (114). Ohrui then conceived this transformation as the pivotal reaction for a synthesis of oxybiotin (115).

The starting material was effectively the 3-azido sugar (113) which was readily prepared from the bis-cyclohexylidene ketone (112) [cf. (23), Scheme 1] in four simple steps. Three standard operations afford the key azide (113), the rearrangement of which proceeds in very high yield. The product (114) is well correlated with the target molecule (115) as the subsequent transformations in Scheme 29 indicate.

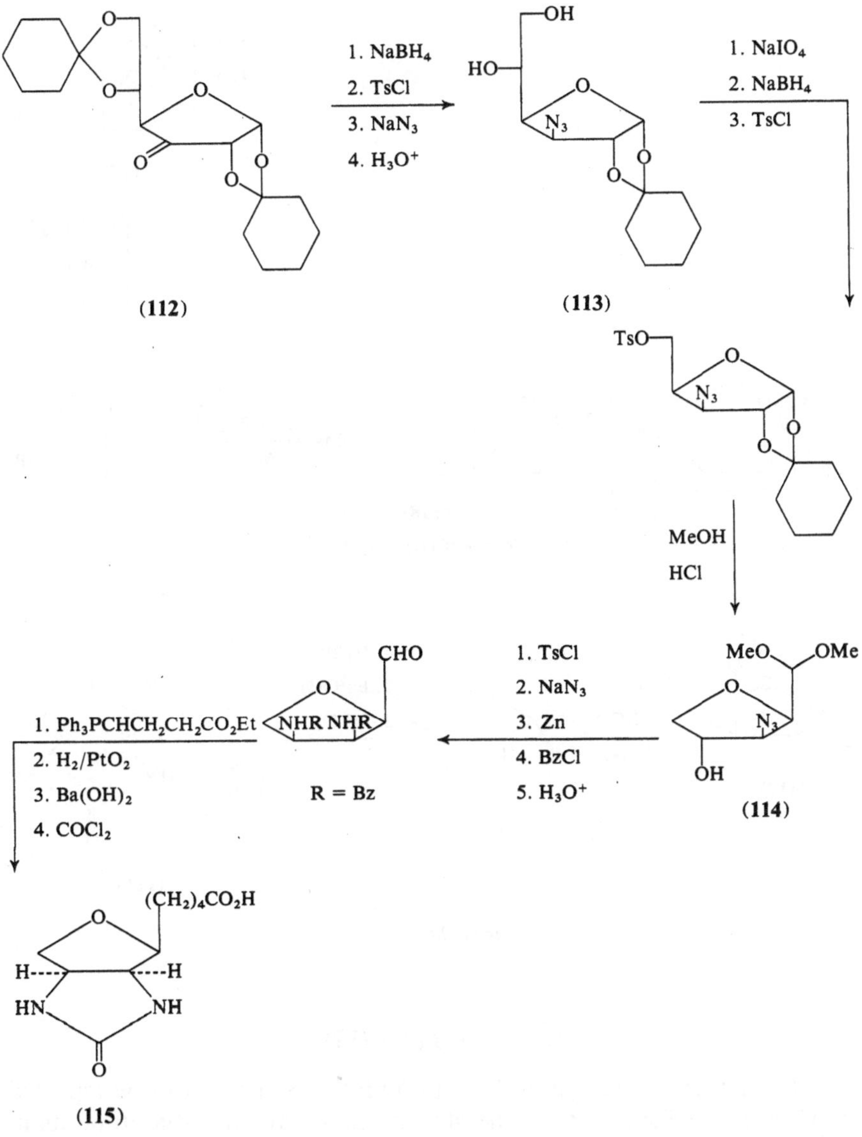

(112)

1. NaBH$_4$
2. TsCl
3. NaN$_3$
4. H$_3$O$^+$

(113)

1. NaIO$_4$
2. NaBH$_4$
3. TsCl

MeOH
HCl

(114)

1. TsCl
2. NaN$_3$
3. Zn
4. BzCl
5. H$_3$O$^+$

R = Bz

1. Ph$_3$PCHCH$_2$CH$_2$CO$_2$Et
2. H$_2$/PtO$_2$
3. Ba(OH)$_2$
4. COCl$_2$

(115)

*Scheme 29*

3*

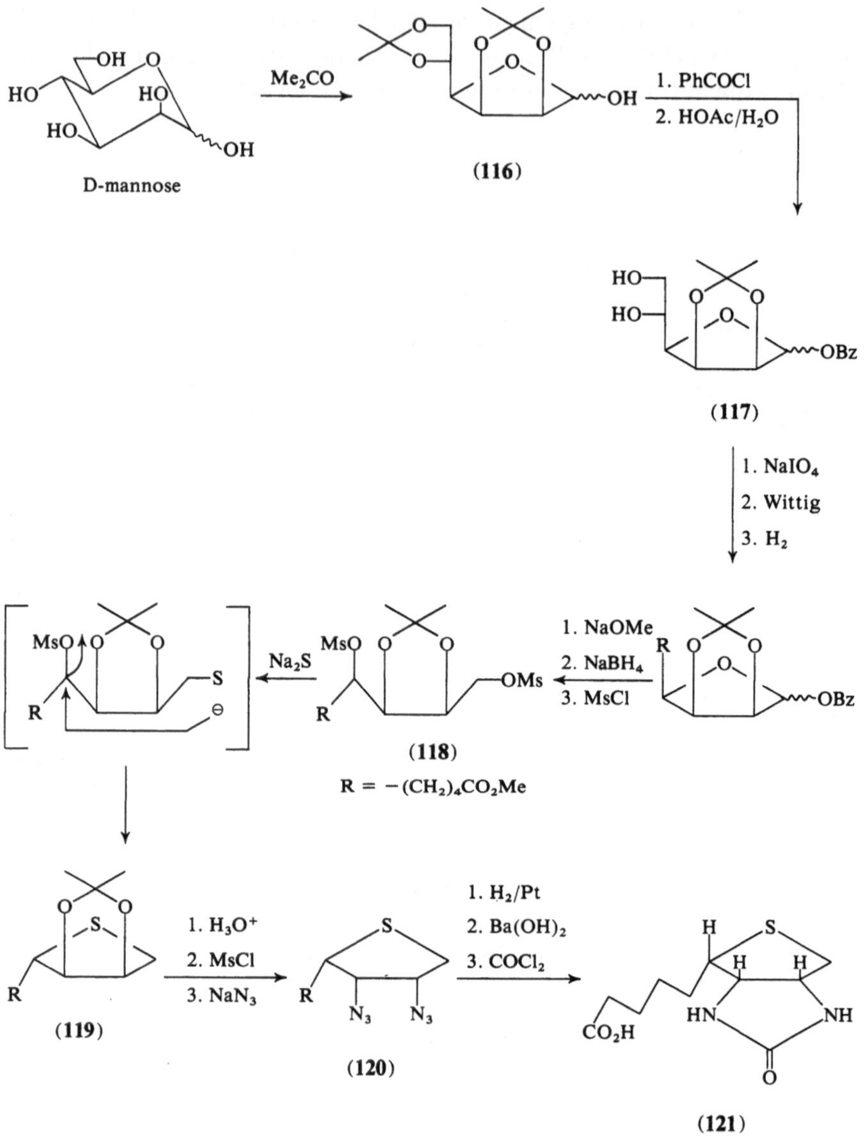

**(116)**

**(117)**

1. NaIO₄
2. Wittig
3. H₂

**(118)**

R = −(CH₂)₄CO₂Me

**(119)** **(120)** **(121)**

*Scheme 30*

## g) (+)-Biotin (121)

The first asymmetric synthesis of (+)-biotin (Scheme 30) was reported by Ohrui and Emoto (68). The starting material was "diacetone mannose" (2,3:5,6-di-O-isopropylidene mannofuranose), **(116)**. Although the selective hydrolysis of the 5,6-isopropylidene group is less facile than in the analogous glucose series (e. g. Scheme 11), the reaction can be carried out as the isolation of **(117)** shows.

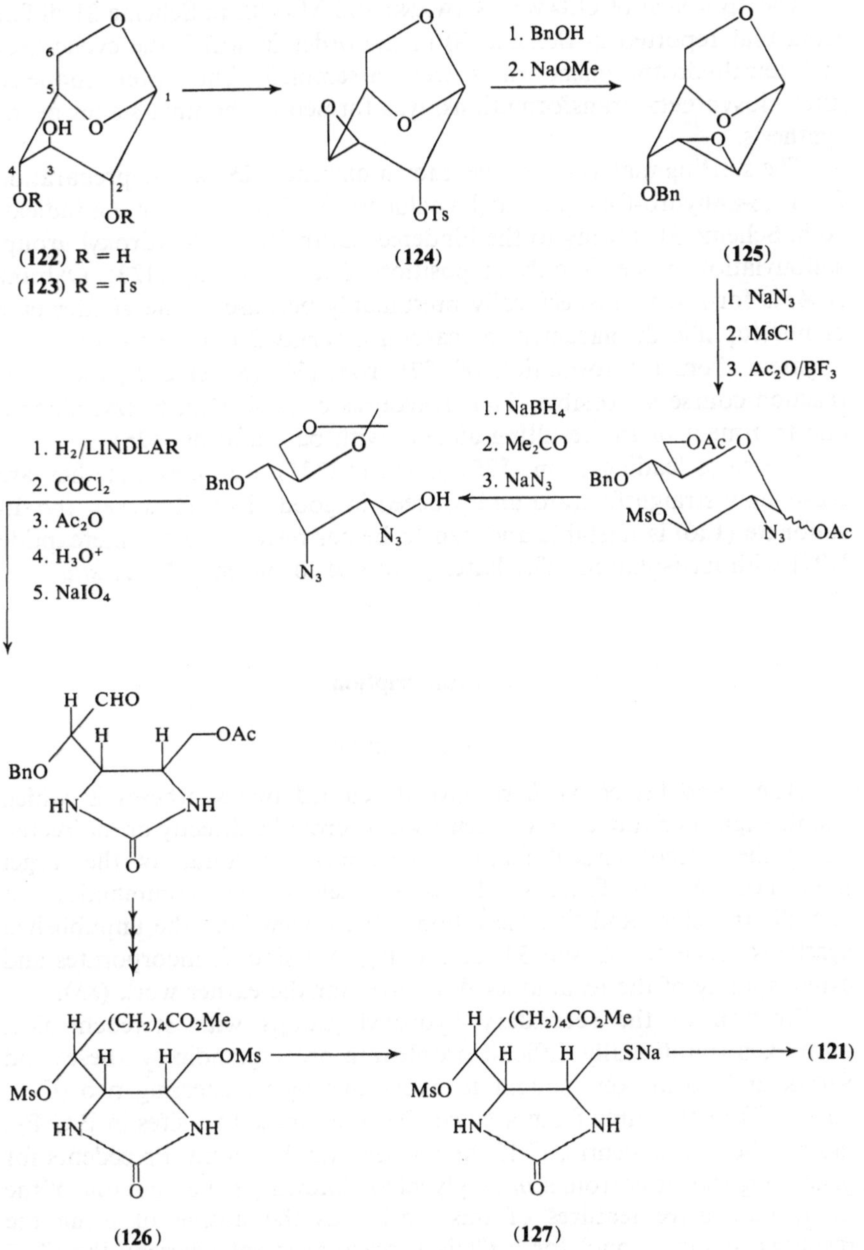
(122) R = H
(123) R = Ts
(124)
1. BnOH
2. NaOMe
(125)
1. NaN3
2. MsCl
3. Ac2O/BF3
1. NaBH4
2. Me2CO
3. NaN3
1. H2/LINDLAR
2. COCl2
3. Ac2O
4. H3O+
5. NaIO4
(126)
(127)
(121)

*Scheme 31*

The succeeding steps leading to the disulfonate (118) are straight-forward. The formation of the tetrahydrothiophene ring (119) and the bis-azidolysis leading to (120) both proceed in ∼ 75 percent yield.

The approach of Ogawa, Kawano and Matsui in Scheme 31 differs from that reported in Scheme 30 in the order in which the cyclic urea and tetrahydrothiophene rings were assembled. The latter, modeled after biosynthetic transformations, was formed in the final stages of the synthesis.

The starting material was the known oxirane (125) whose preparation from 1,6-anhydro-β-D-glucose (levoglucosan) (122) *(vide supra)* is indicated in Scheme 31. Owing to the hindered nature of the 3-hydroxyl group, sulfonylation proceeds only at positions 2 and 4 giving (123). Oxirane (124) is formed regiospecifically presumably because of the greater ease of nucleophilic displacement at carbon 4 *versus* 2 (53). However, as is apparent from the formation of (37) from (36) (Scheme 9), the other reaction course is possible. Thus solvolysis of (124) with benzyl alcohol and treatment of the resulting alcohol with base affords (125).

The steps leading from (125) to biotin (121) are many but they are reasonably straightforward and proceed in good yield. However, the di-sulfonate (126) is unstable and had to be converted into the mercaptide (127) without isolation. The latter gave rise to biotin (121) *in situ*.

## 4. Transcription

### a) Prostaglandins

Stork and his co-workers have developed two syntheses in which simple sugar derivatives have been used to provide, directly or indirectly, all of the stereochemical features, geometric or chiral, of the target molecules (88, 89). These syntheses represent elegant combinations of "acyclic transfer" and "transcription". We review here the unpublished synthesis (Schemes 32 and 33) of P G $F_{2\alpha}$ (89) since it incorporates and extends many of the techniques developed for the earlier work (88).

Creation of the secondary hydroxyl groups with stereochemical purity is a traditionally difficult problem in prostaglandin syntheses and Stork and co-workers sought to avoid this by transferring two (C-11 and C-15) of the three from a sugar. Between these two sites in P G $F_{2\alpha}$ lies another chiral centre, C-12, and a *trans*-double bond. Precedents for generating the latter from a *threo*-glycol are known (16, 92), but one of the many innovative features of this work was the notion of using the resulting allylic alcohol for a Claisen rearrangement whereby the C-12 chiral centre is created by "transcription".

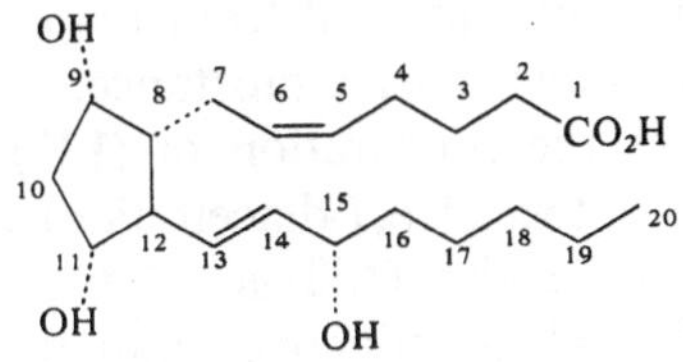

PGF$_{2\alpha}$

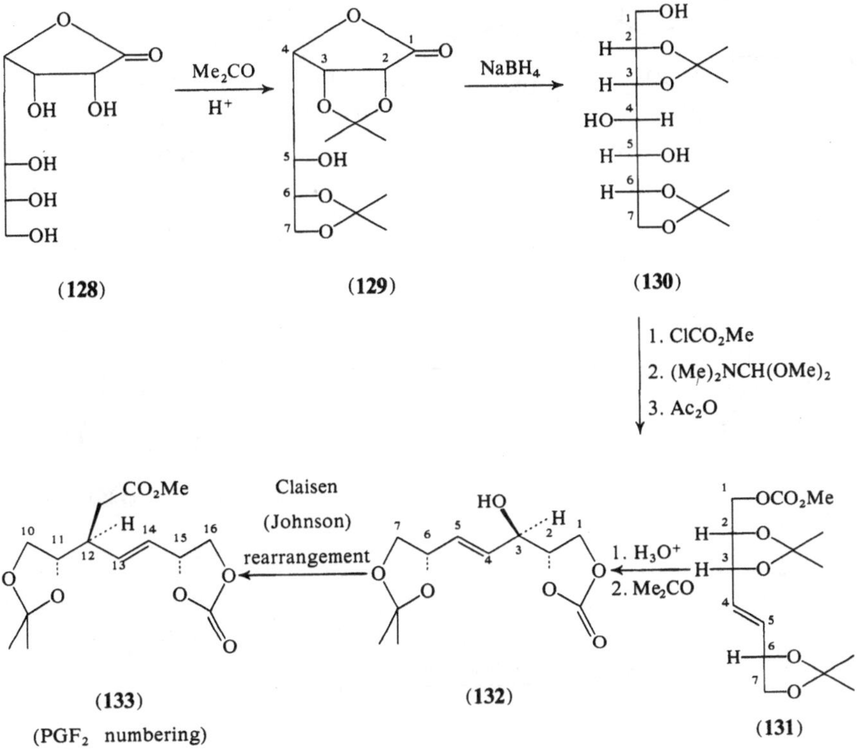
Me$_2$CO
H$^+$
NaBH$_4$
(128)
(129)
(130)
1. ClCO$_2$Me
2. (Me)$_2$NCH(OMe)$_2$
3. Ac$_2$O
Claisen
(Johnson)
rearrangement
1. H$_3$O$^+$
2. Me$_2$CO
(133)
(PGF$_2$ numbering)
(132)
(131)

*Scheme 32*

The required stereochemical implements are all present in the allylic alcohol (**132**) from which (**133**) is derived. The C-8 and C-9 chiral centres of P G F$_{2\alpha}$ are the only ones not represented in (**133**) but the instruments for their stereocontrolled creation are present as will become apparent. The ester (**133**) can therefore be regarded as the chiral lynch-pin for the synthesis.

The progenitor of allylic alcohol (**132**) must be a sugar containing five contiguous secondary hydroxyl groups, and the requirements for producing the natural enantiomer of P G F$_{2\alpha}$ suggest the protected heptitol (**130**).

Gratifyingly, this is readily obtained from D-*glycero*-D-*gulo*-heptono-γ-lactone (**128**) a relatively inexpensive substance.

Scheme 32 outlines the preparation of (**133**). Acetonation of (**128**) gives (**129**) (plus some of the 3,5-6,7-diacetonide) and after reduction with sodium borohydride the heptitol (**130**) is isolated. The primary position is selectively acylated after which reductive elimination is effected (*16*) giving

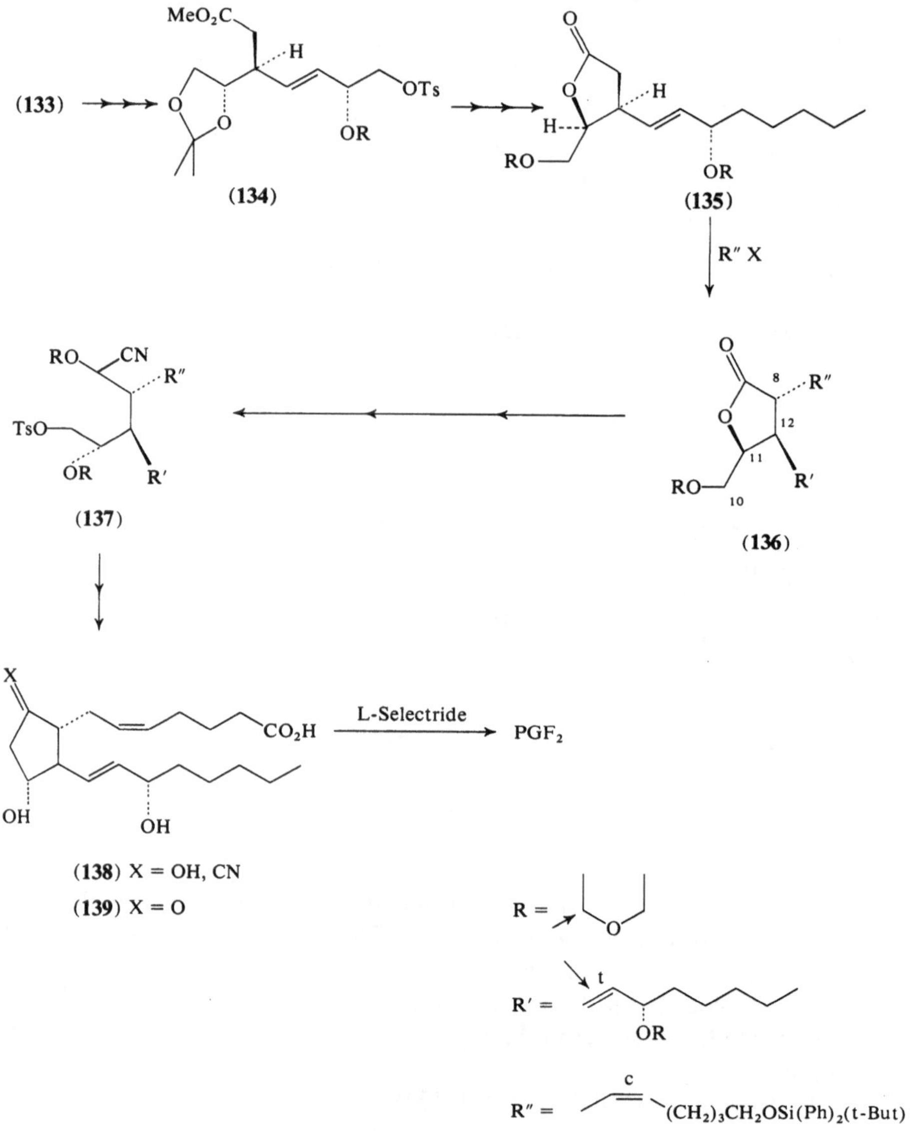

(**133**) ⟶⟶⟶

(**134**)

(**135**)

R″ X

(**137**)

(**136**)

(**138**) X = OH, CN
(**139**) X = O

L-Selectride ⟶ PGF₂

R =

R′ =

R″ =

*Scheme 33*

(131). The choice of methyl carbonate protection for the C-1 hydroxyl group was inspired by the expectation that upon cleavage of the acetonides, a cyclic carbonate would form spontaneously. Reacetonation of the vicinal glycol then affords the allylic alcohol (132) which is required for the Claisen rearrangement. For this rearrangement, the stereochemical outcome at C-12 of (133) was predictable on the basis of the chair-like transition state in which the bulky cyclic carbonate is equatorially oriented (88).

The procedure for converting (133) into P G F$_{2\alpha}$ is summarized in Scheme 33. As in the formation of (93) from (92) (Scheme 23), hydrolysis of the acetonide of (134) leads to the $\gamma$-lactone (135). The latter is of appreciable importance since the subsequent alkylation then gives the

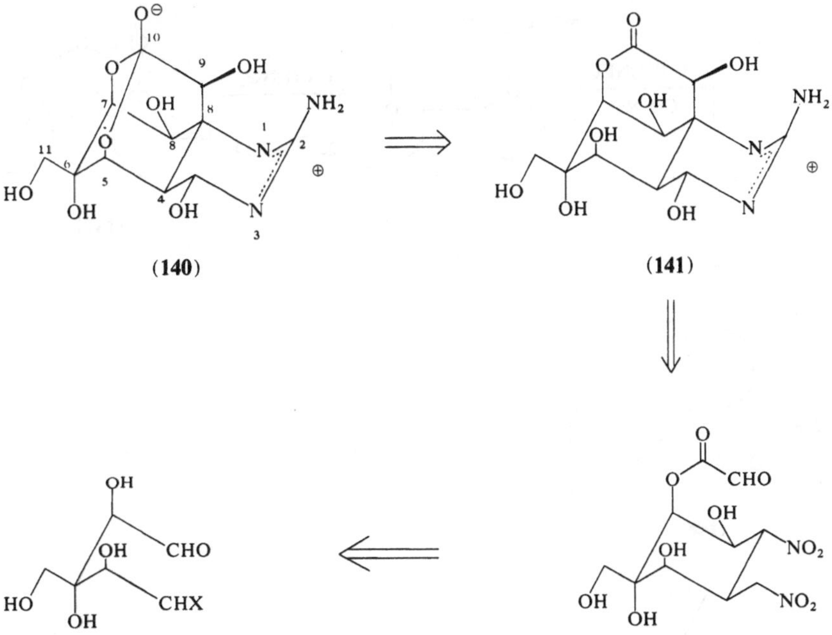

(140)                    (141)

(143) X = O
(144) X = (CH$_2$NO$_2$)$_2$

(142)

*Scheme 34*

thermodynamically favoured *trans*-adduct (136). Thus the stereochemistry at C-8 is directly controlled by that at C-12 which was, in turn, transferred from the C-3-OH of (132). The protected cyanohydrin (137) is prepared by standard methods and the anion derived therefrom is alkylated intramolecularly giving the masked ketone (138). Finally, the unmasked ketone (139) is reduced stereoselectively with L-Selectride.

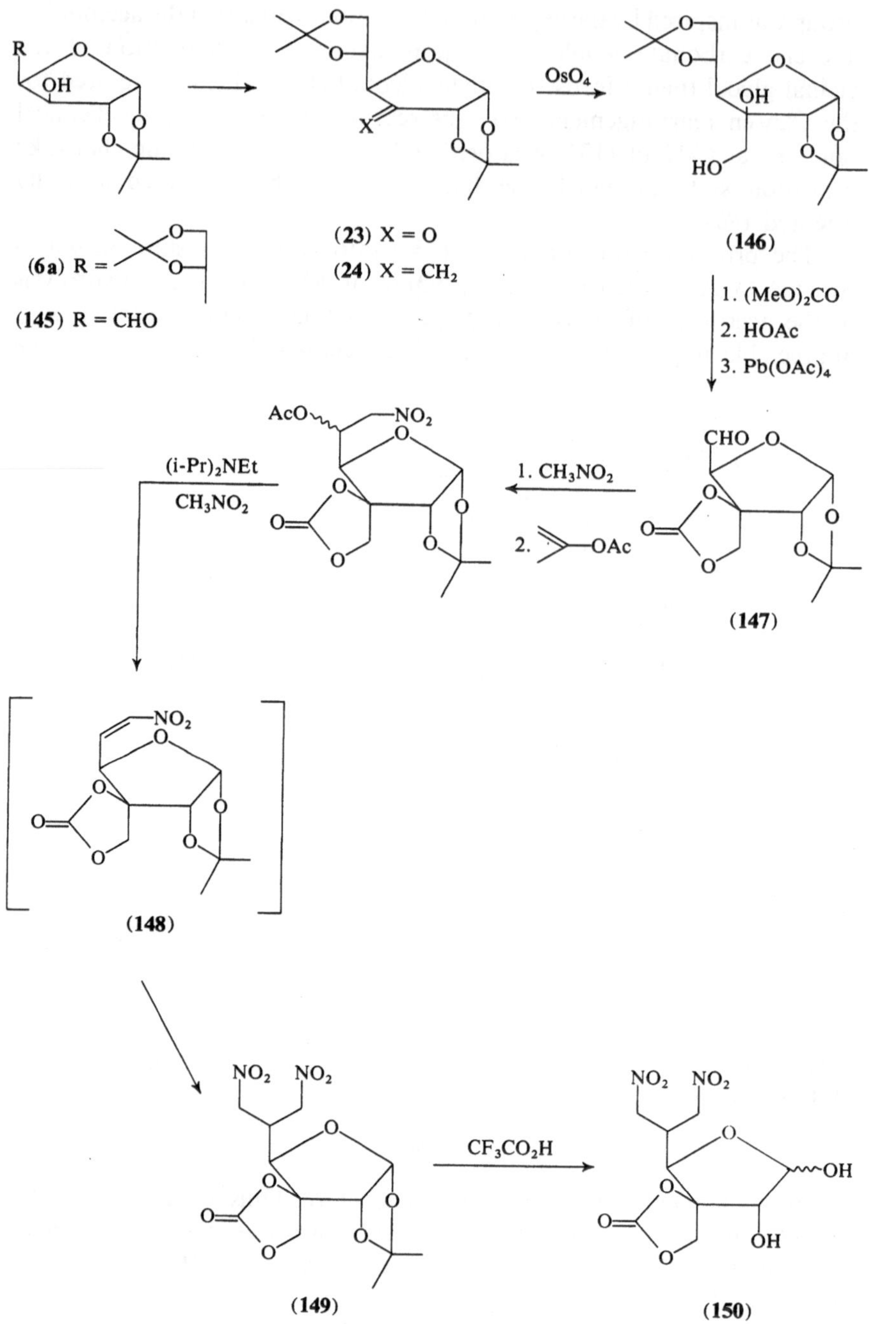
R
OH
(6a) R =
(145) R = CHO
X
(23) X = O
(24) X = $CH_2$
$OsO_4$
OH
HO
(146)
1. $(MeO)_2CO$
2. HOAc
3. $Pb(OAc)_4$
AcO
$NO_2$
$(i\text{-}Pr)_2NEt$
$CH_3NO_2$
1. $CH_3NO_2$
2. OAc
CHO
(147)
$NO_2$
(148)
$NO_2$ $NO_2$
$CF_3CO_2H$
OH
OH
(149)
(150)

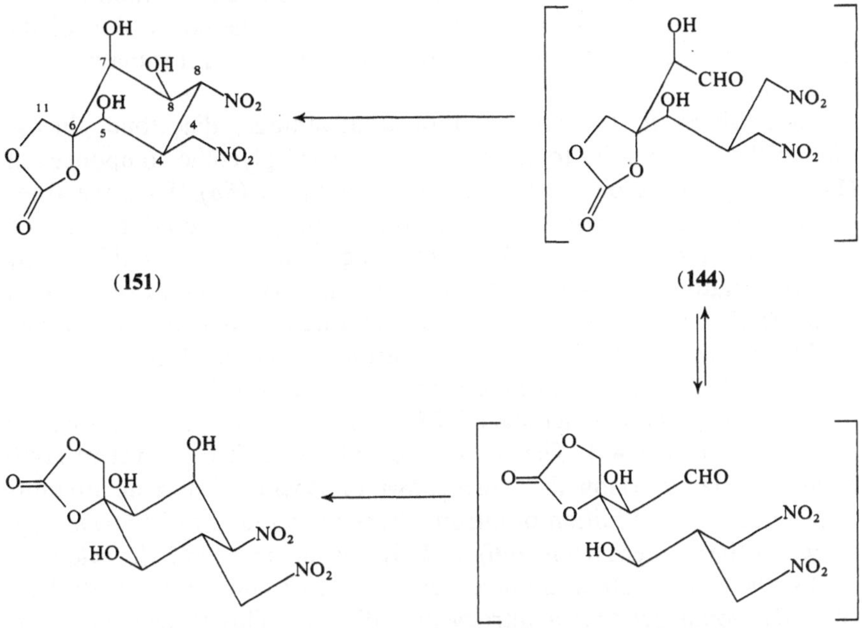

*Scheme 35*

Thus the chiral centres at C-11, C-12 and C-15 are obtained directly or indirectly from the sugar, while formation of the remaining two at C-8 and C-9 is controlled by those already present.

### b) An Approach to Tetrodotoxin

An approach to an asymmetric synthesis of the structurally unique molecule tetrodotoxin (140) constitutes further evidence for the versatility of "diacetone glucose" (6a) as a chiral synthon. Although related studies have been undertaken in YOSHIMURA's (105) laboratory, the work reviewed here is that originated in WOODWARD's laboratory by SITRIN (85) and continued by UPESLACIS (95).

The strategy being pursued is guided by knowledge obtained during the structure elucidation that the hemilactal component of (140) (Scheme 34) exists in equilibrium with α-hydroxylactone forms such as (141) (103). One mode of disconnection of the latter (Scheme 34) leads sequentially to the glyoxylate ester (142) and thence to the meso dialdehyde (143). Actually, employment of (143) would yield a racemic product unless the aldehydic centres were differentiated as in the case of dinitro equivalent

(144) and the acetonide (145) (Scheme 35). In fact extensive studies on the condensation of nitroalkanes with the latter and the conversion of the products derived therefrom into nitrocyclitols provided invaluable guidance (58).

In addition to the aforementioned advantage, diacetone glucose offered the opportunity to create the C-6—C-11 glycollic component of (140) by operating on the free hydroxyl group of (6a). Thus the three contiguous chiral centres, once assembled in (146) could be carried through to (144) without further tampering; and if C-11 should occupy a "pro-equatorial" (144) rather than the "pro-axial" (152) orientation (Scheme 35), cyclisation to a six-membered ring would lead preferentially to the stereoisomer (151). For these reasons diacetone glucose (6a) was seen to "offer some astonishing synthetic advantages" (95).

The synthesis and reactions of (24) (Scheme 7) adequately rationalize the formation of (146). The preparation of the dinitro derivative (149) from (147) had to follow the stages shown in Scheme 35 since all attempts to achieve double addition of nitromethane were found to be unavailing. However the six steps from (146) to (149) were achieved in 21% yield (85).

On the basis of literature precedents (58) it was hoped that hydrolysis of (149) would produce a nitrocyclitol directly. This turned out to be the case although the intervening hemiacetal (150) could be isolated by judicious crystallisation. However the nitrocyclitol was not the desired substance (151) but the stereoisomer (153a).

This adverse course of events could perhaps be avoided by employing the glyoxylate ester (155) (Scheme 36), for which di-condensation on the same nitromethyl group would lead to the formation of two six-membered rings in (156). Accordingly a sequence leading to the crystalline hemiacetal (154) was developed and, as with (150), cyclisation proceeded to the analogous stereoisomer (153b) bearing a cinnamoyl ester at C-7. Ozonolysis was expected to give the glyoxylate (153c) which upon gentle treatment with base could cyclorevert to (155), and thence yield (156). Unfortunately compound (153c) proved elusive, deesterification occurring in situ to give the previously encountered triol (153a).

The formation of (153) rather than (151) indicated that the establishment of correct stereochemistries at three centres (C-2, C-3, C-4) of the synthon (149) was not sufficient to ensure the stereochemistry at C-4a of tetrodotoxin. Probably creation of the latter as a fourth chiral centre at C-5 of the carbohydrate synthon would be advantageous, and if the carbons attached to this centre were of vastly different acidities as in (157) the course of cyclisation would be controllable.

The chirality desired at C-4a is (R), and although most 1,2 or 1,4 additions to trigonal centres at C-5 of sugars give mainly the (S)-adduct, there are circumstances in which this trend is reversed (84). Fortunately

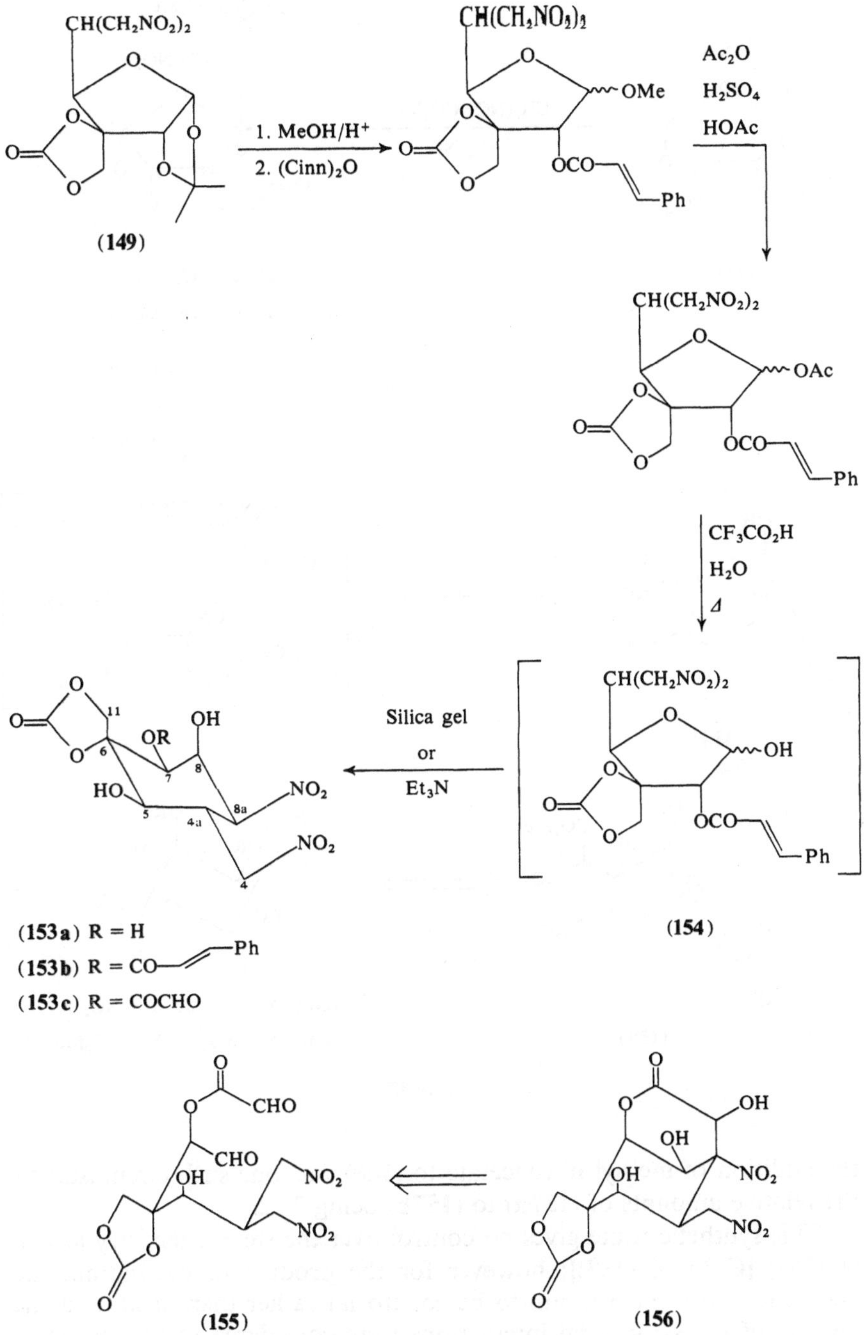
CH(CH2NO2)2
(149)
1. MeOH/H+
2. (Cinn)2O
OMe
OCO
Ph
Ac2O
H2SO4
HOAc
OAc
CF3CO2H
H2O
Δ
OH
Silica gel
or
Et3N
(154)
OR
HO
NO2
(153a) R = H
(153b) R = CO—Ph
(153c) R = COCHO
CHO
(155)
(156)

*Scheme 36*

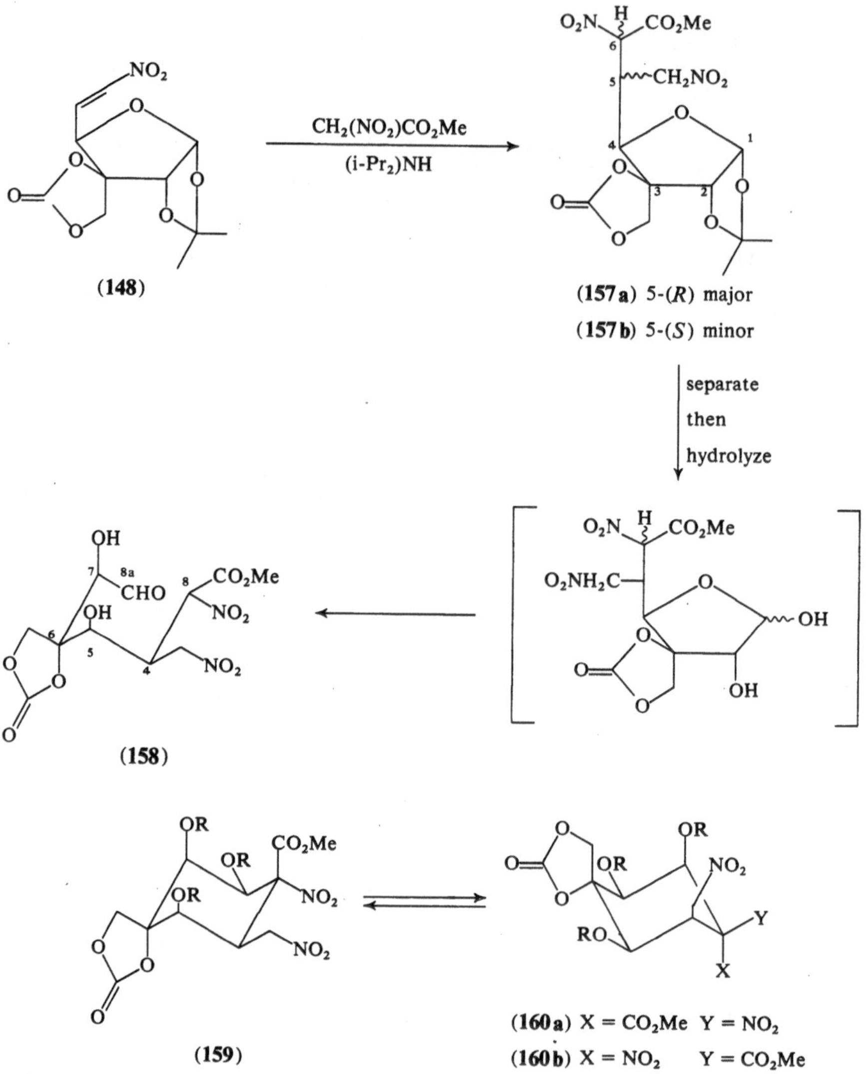

(148)

CH$_2$(NO$_2$)CO$_2$Me

(i-Pr$_2$)NH

O$_2$N—CH—CO$_2$Me (6, H)

$_5$⟳CH$_2$NO$_2$

(157a) 5-(R) major
(157b) 5-(S) minor

separate
then
hydrolyze

OH

$_7$ $^{8a}$ CHO   $_8$ CO$_2$Me

OH       NO$_2$

$_6$   $_5$   $_4$   NO$_2$

(158)

[
O$_2$N—CH—CO$_2$Me (H)

O$_2$NH$_2$C—

⟳OH

OH
]

OR

OR   CO$_2$Me

OR       NO$_2$

NO$_2$

(159)

OR

OR   NO$_2$

RO       Y

X

(160a) X = CO$_2$Me  Y = NO$_2$
(160b) X = NO$_2$     Y = CO$_2$Me

*Scheme 37*

the addition of methyl nitroacetate to (148) was one such circumstance, the relative amounts of (157a) to (157b) being 7 : 2.

This synthetic route gives no control over the stereochemistry at C-6 of (157) [C-8a of (158)]; however for the product of cyclisation, the preference for a nitro group to be equatorial rather than axial, and the severity of the syn-diaxial interactions were considered to favour (159). However the product proved to be a mixture of (160a) and (160b).

The latter is in fact the desired stereoisomer **(159)**—but unfortunately it exists in the undesired conformation!

### c) Chiral Cycloalkanes from Annulated Pyranosides

The examples cited above under "cyclic and acyclic transfer", provide ample evidence for the use of sugar nuclei as stereochemical templates. Rather less well utilized are the latent functionalities which may be unveiled, and one of the most valuable of these is the $\alpha,\beta$-unsaturated carbonyl system. The most readily available is enone **(83)** (Scheme 20); however, the alkylated analogues **(161)** and **(162)** (Scheme 38) have been prepared (38) as have the congeners **(163)**, **(164)** and **(165)** (23).

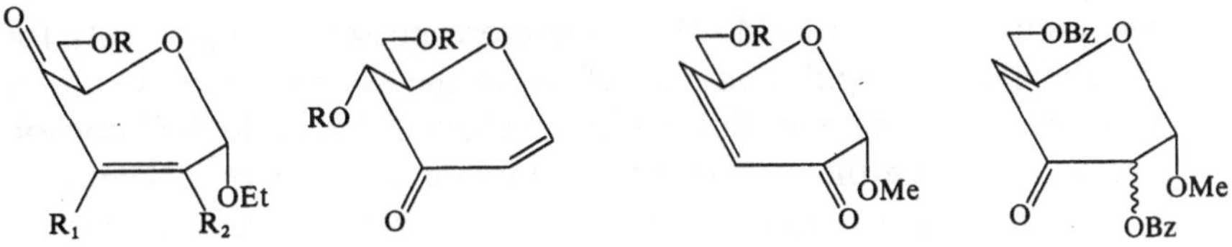

**(83)** $R_1 = R_2 = H$          **(163)**          **(164)**          **(165)**

**(161)** $R_1 = H, R_2 = Me$

**(162)** $R_1 = Me, R_2 = H$

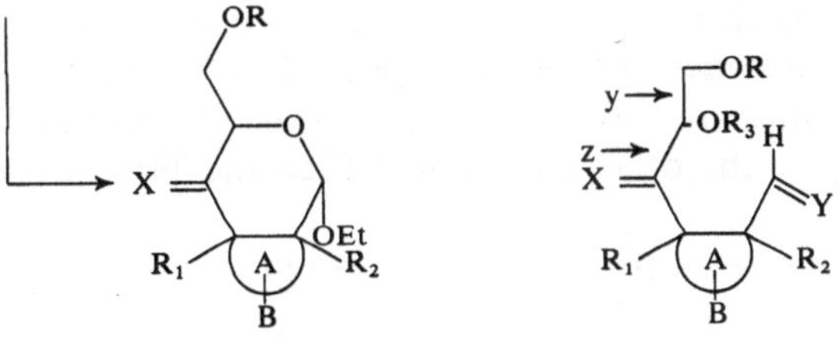

**(166)**                    **(167)**

Scheme 38

Cycloadditions to these enones yield "annulated pyranosides", a number of which have become available in recent years in the authors' laboratory (23). These bicyclic molecules open avenues for "transcription" of stereochemical information from sugars to carbocyclic systems, generalized procedures for which are currently not available.

Enone **(83)** has been our most widely used substrate. The annulation may be stereoselective, in which case only one diastereomer of **(166)** is obtained and the problem is to determine which. If the annulation is non-stereoselective, then diastereomeric forms of **(166)** are given and these should be separable chromatographically. In either case the proof of stereochemistry is greatly facilitated by pmr methods, as summarized in Scheme 4. Thus, it is possible to determine the orientation, and hence chirality, at the ring junctions, C-2 and C-3.

For the case of (166) where $X = O$, $R_1 = H$, C-3 is a potentially epimerisable centre, if not on the bicycle itself, then on the acyclic form (167), derivable by opening of the pyranose ring. Similarly, the anomeric centre of (166) is a protected aldehyde; hence for (167) where $Y = O$ and $R_2 = H$, C-2 is also a potentially epimerisable centre.

Our previous discussion indicates that for the case where C-4 is a tetrahedral centre [i. e. (166): $X = Y, Z$], it should be possible by nuclear magnetic resonance methods to assign the orientation and hence the chiralities of these groups. The same holds for the substituents A and B on the cycloalkyl moiety.

The annulated pyranoside (166) therefore possesses implements for diverse chemical manipulations as well as for the assembly of a wealth of chiral information and once this has been achieved, the cycloalkyl portion may be excised from the carbohydrate template at sites $y$ or $z$. These are locations for possible periodate cleavage which would have to be carried out on the acyclic form (167). However, an alternative procedure for cleaving the bicyclic from (166) $(X = O)$ is described in Scheme 43.

The nature of the functional groups at carbons-1 and -4 (i. e. the $\alpha$ and $\alpha'$ positions of the cycloalkyl moiety) makes these cycloalkyl pyranosides intriguing chiral synthons. Thus in the case where annulation gives only one diastereomer of (166), it is possible to conceive of one set of manipulations, whereby one enantiomer would be produced, and a different set leading to the other enantiomer (Scheme 39; equation i).

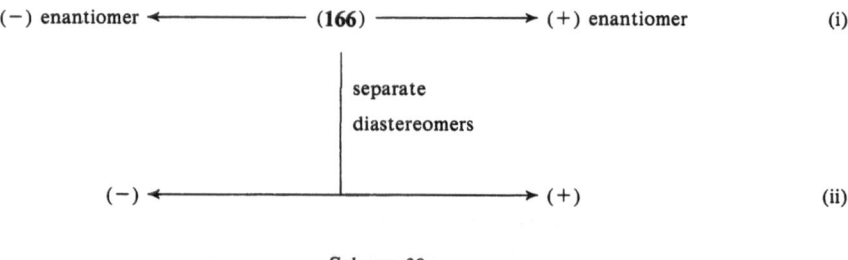

*Scheme 39*

If on the other hand the annulation is non-stereoselective, the resulting diastereomers of (166) are chromatographically separable. Theoretically one of these can be transformed into the $(+)$ enantiomer while the other will afford the $(-)$ enantiomer (Scheme 39, equation ii).

Thus in either case, the annulated pyranoside has the capability for enantioselective transformation.

## d) Cyclopropano Pyranosides

A project directed towards the synthesis of chrysanthemic acid enantiomers illustrates some of our recent work* on cyclopropano-pyranosides. The sequence (Scheme 40) that had worked so well (25) for the preparation of the simple cyclopropyl ketone (169) from the methanol adduct (168) was not adaptable for preparation of the gem-dimethyl analogue (171). The photoaddition of isopropanol to (83) gave an excellent yield of (170), but efforts to convert this into (171) were not encouraging. However, the Wittig cyclopropanation (24) of epoxide (51) gave the ester (172) whose stereochemistry was deduced by two pieces of nmr data; (a) the value $J_{12} < 1$ Hz (76) (see Scheme 4), and (b) a ten percent Nuclear Overhauser Effect between H-1 and the methyl protons.

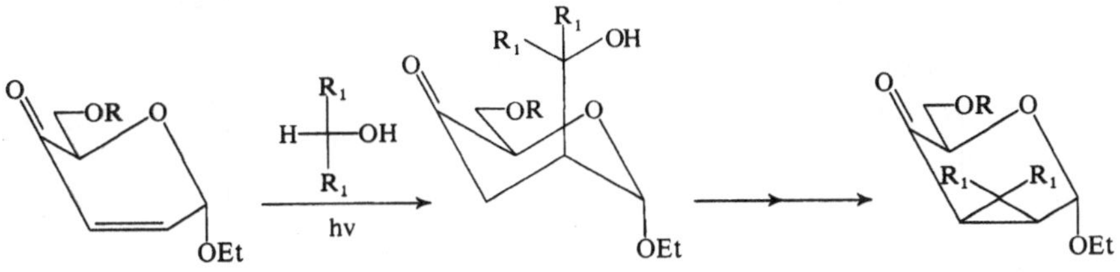

(83)

(168) R₁ = H
(170) R₁ = Me

(169) R₁ = H
(171) R₁ = Me

*Scheme 40*

The ester (172) has been converted to the gem-dimethyl substance (173a) which, in keeping with our earlier studies (76) is hydrolysed to the hydroxyaldehyde (174) upon heating in water. This key intermediate is amenable to transformation into both enantiomers of chrysanthemic acid (177) as summarized in Scheme 41. Thus the Wittig reaction of (174) gives only one geometric isomer whose stereochemistry is judged to be (175). The acetal of this substance has been hydrolysed, and the resulting triol cleaved to the aldehyde (176), which has been converted into the natural enantiomer (+) (177) by using known methodology (19).

The key intermediate (174) has also provided access to the unnatural (−) (177) (Scheme 41) by first epimerizing the aldehyde and then carrying out identical steps to those described above.

It is evident from Scheme 41 that the sequences allow for isotopic labels to be introduced at sites indicated by asterisks, the chiralities of which have been "transcribed" from the carbohydrate template.

---

\* For previous preparations of cyclopropanated pyranosides see reference (23).

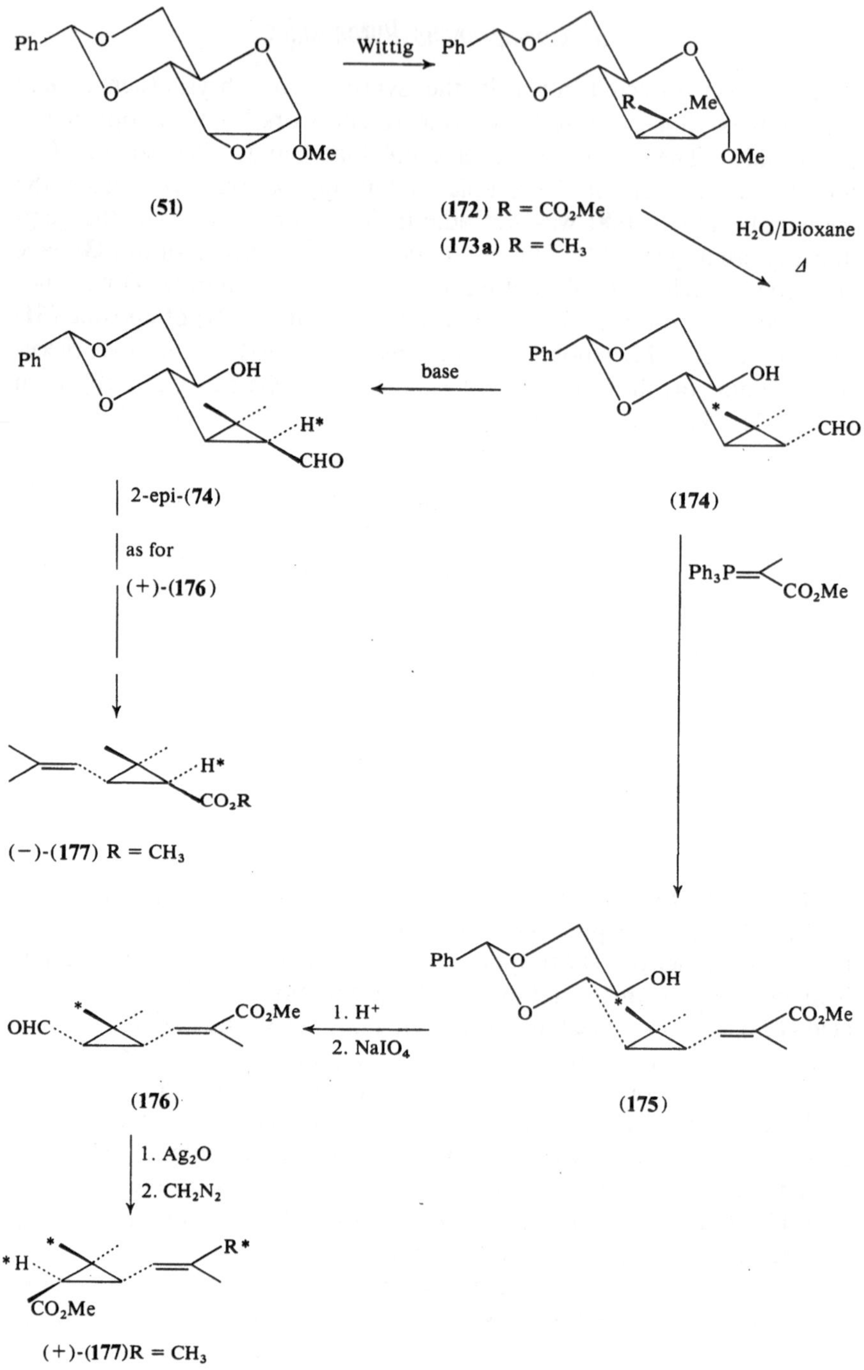

(51)

Wittig

(172) R = CO₂Me
(173a) R = CH₃

H₂O/Dioxane
Δ

2-epi-(74)

base

(174)

Ph₃P=⟨CO₂Me

as for
(+)-(176)

(−)-(177) R = CH₃

OHC

(176)

1. H⁺
2. NaIO₄

(175)

1. Ag₂O
2. CH₂N₂

(+)-(177) R = CH₃

*Scheme 41*

## e) Cyclobutano Pyranosides

The foregoing prospectus for the chrysanthemic acid synthesis represents enantioselective transformations of the first kind (Scheme 39, equation i). The second kind (equation ii) is represented by studies directed toward the (+) and (−) grandisol enantiomers (183).

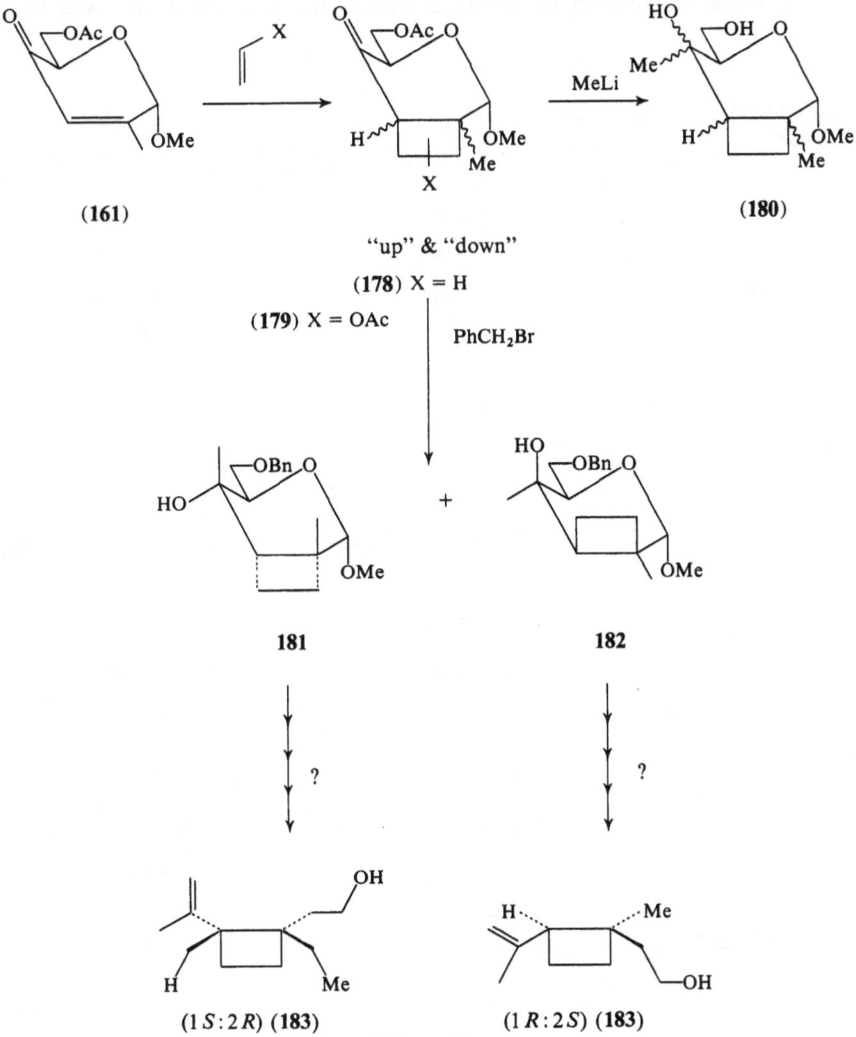

(161)

(180)

"up" & "down"

(178) X = H

(179) X = OAc

PhCH$_2$Br

181                +                182

?                               ?

(1 S : 2 R) (183)               (1 R : 2 S) (183)

*Scheme 42*

Photoannulation of ethylene or vinyl acetate to enone (161) gave mixtures of diastereomers (178) and (179) respectively, which were not well resolved by column chromatography. However further transformations

4*

were carried out, in the expectation that polar derivatives would be more amenable to separation. In fact, the product from reaction with methyl lithium was a mixture of two well resolved substances (180). However, column chromatography was delayed until after benzylation, which occurred at the primary position only.

The decision as to whether the ring was "up" or "down" could not be made by pmr in accordance with Scheme 4, since H-1 was in a

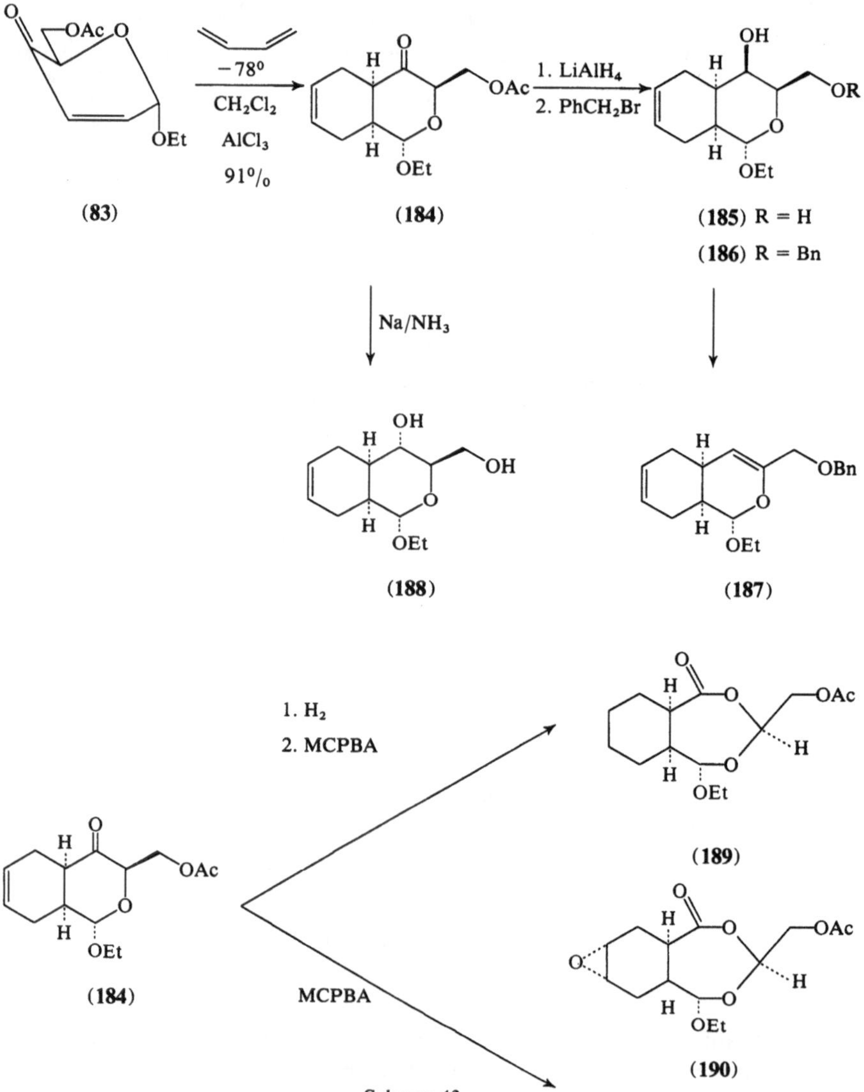

Scheme 43

singlet for both (181) and (182). However, cmr made it possible to completely assign the structures indicated.

The theoretical correlation between these synthons and the enantiomers is indicated in Scheme 42 and is self explanatory.

### f) Cyclohexano Pyranosides

Cyclohexano pyranosides are of paramount importance in view of the omnipresence of the cyclohexyl ring among natural products of various types. Their preparation is therefore of particular interest.

The attempted high temperature addition of butadiene to enone (83) led to massive decomposition. However, the low temperature Lewis acid catalyzed procedure (104) gave an excellent yield of (184) (Scheme 43). The structure of the latter was readily deduced from the pmr spectrum in which H-1 appeared as a doublet $J_{12} = 4.0$ Hz. This value, which is uncharacteristically large for trans-diequatorial protons (see Scheme 4), is usually observed when a trigonal centre is present at C-4 (25). Accordingly reduction with lithium aluminum hydride gives the diol (185) in which $J_{12} < 1$ Hz.

The secondary alcohol in (185) is so hindered that benzylation occurs only at the primary position giving (186) from which the diene (187) is prepared via the sulfonate ester. On the other hand the epimeric diol (188) is obtained in low yield by dissolving metal reduction.

As an alternative to the procedure used in Scheme 41 for excising the cycloalkyl moiety, the Baeyer-Villiger oxidation was applied to the dihydro derivative of (184). Gratifyingly, oxygen-insertion occurs predominantly at the bond of higher electron density to give the bis-acetal (189). With (184) itself, an epoxidized bis-acetal is obtained, whose structure is judged to be (190). The stereochemistry at the oxirane is assigned on the basis of addition from the exo-face of (184). These results confirm our expectation that reactions at the carbonyl and alkene sites of (184) may be carried out with encouraging stereo- and regio-selectivity.

### g) Cyclopentano Furanosides

Attempts to bring about ring-contraction of the cyclohexyl moiety of any of the compounds (184)—(188), thereby yielding cyclopentano pyranosides were unsuccessful. However it was found that upon treatment with dilute ethanolic hydrogen chloride, the pyranoside ring of (185) was rearranged to the furanoside (191). Thus acetonation of the latter and cleavage of the alkene gave the diester (192) from which the Dieckmann product (193) was obtained smoothly.

Our present evidence suggests that (193) is a single diastereomer although the precise structure is at this time unknown. However, deesterification under standard conditions afforded the cyclopentano furanoside (194).

These cyclohexano pyranosides, cyclopentano furanosides, and their transformation products promise to be valuable synthons judging from recent efforts directed towards biomimetic synthesis of camptothecin (196) and its analogues from the congener (195) by Hutchinson and co-workers (61).

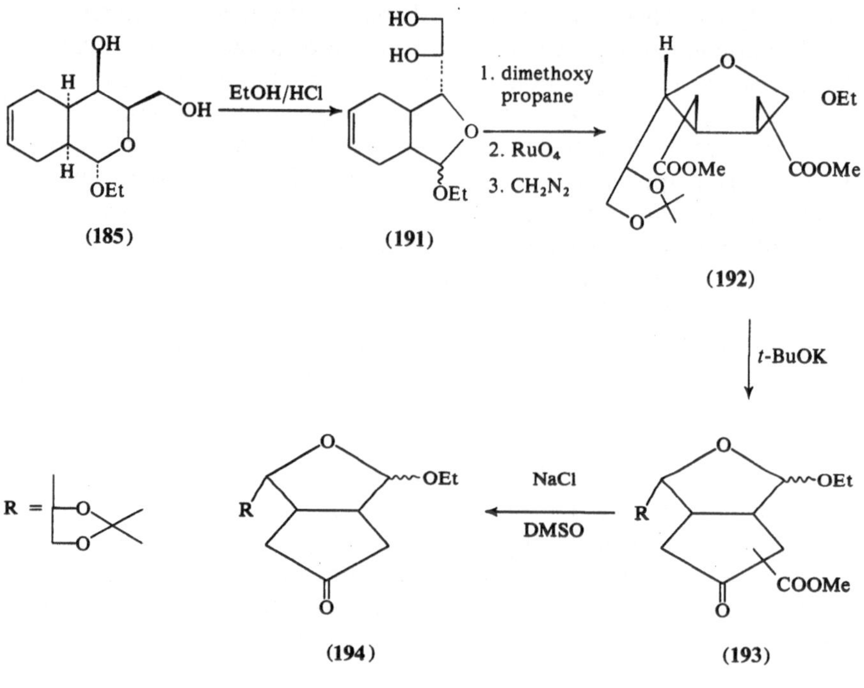

*Scheme 44*

A further prospect for these derivatives originates with the substituted enones (161) and (162) (Scheme 38) which are expected to give the Diels-Alder adducts (194) and (198) respectively. The latter is of particular interest since conversion of the pyranose into a cyclohexane ring (see for example Schemes 35—37), would afford the decalone (199) (*cis* or *trans*) which has the proper absolute stereochemistry for the A and B rings of the majority of terpenes. Furthermore, the functionalization of (199) makes it a synthon of the perhydrophenanthrene (200) and perhydro-anthracene (201) skeletons.

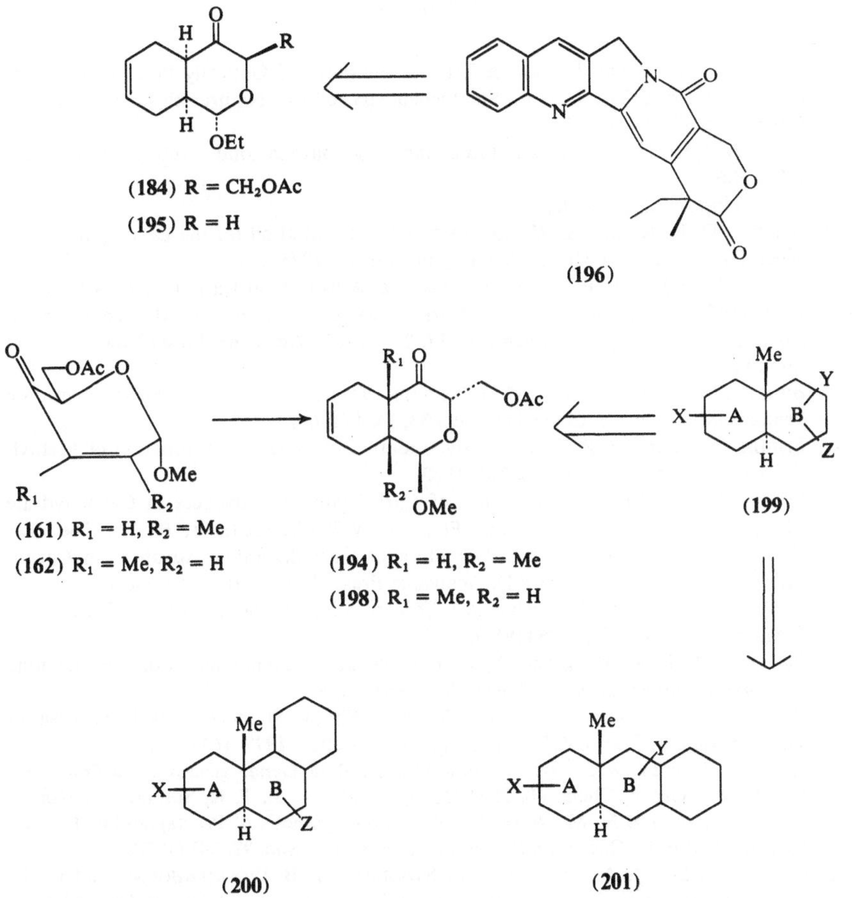

(184) R = CH₂OAc

(195) R = H

(196)

(161) R₁ = H, R₂ = Me

(162) R₁ = Me, R₂ = H

(194) R₁ = H, R₂ = Me

(198) R₁ = Me, R₂ = H

(199)

(200)    (201)

*Scheme 45*

It will be of interest to see how successfully the chiral information can be transmitted from the original sugar to the complex polycyclic systems and studies directed with this objective in mind are in progress in the authors' laboratory.

**Note Added in Proof.** In a recent communication ["Stereospecific Synthesis of (−)-α-Multistriatin from D-glucose" by P.-E. SUM and L. WEILER, Can. J. Chem. **56**, 2700 (1978)] the hydrogenation leading to (**85**) is carried out stereoselectively using an homogeneous catalyst. Other aspects of their synthesis differ significantly from that in Scheme 21.

## References

*1.* Anderson, R. C., and B. Fraser-Reid: A Synthesis of Optically Pure Avenaciolide from D-Glucose. The Correct Stereochemistry of the Natural Product. J. Amer. Chem. Soc. **97**, 3870 (1975).

*2.* — — A Synthesis of Naturally Occurring (−)-Isoavenaciolide. Tetrahedron Letters **1977**, 2865.

*3.* — — Unpublished results.

*4.* Barton, D. H. R., and S. W. McCombie: A New Method for the Deoxygenation of Secondary Alcohols. J. Chem. Soc. Perkin Trans. I, **1975**, 1574.

*5.* Bayne, S., and S. A. Fewster: The Osones. Advances in Carbohydrate Chemistry **11**, 43 (1956), Academic Press: New York; Theander, O.: Dicarbonyl Carbohydrates. Advances in Carbohydrate Chemistry **17**, 233 (1962), Academic Press: New York and London.

*6.* Bhacca, N. S., D. Horton, and H. Paulsen: The Conformation of α-D-Idopyranose Pentaacetate. J. Organ. Chem. (U. S. A.) **33**, 2484 (1968).

*7.* Bishop, C. T., and F. P. Cooper: Glycosidation of Sugars. 1. Formation of Methyl-D-Xylosides. Can. J. Chem. **40**, 224 (1962).

*8.* Bobbit, J. M.: Periodate Oxidation of Carbohydrates. Advances in Carbohydrate Chemistry **11**, 1 (1956), Academic Press: New York; Guthrie, R. D.: The "Dialdehydes" From the Periodate Oxidation of Carbohydrates. Advances in Carbohydrate Chemistry **16**, 105 (1961), Academic Press: New York and London.

*9.* Bonner, W. A.: The Stereochemical Configuration of the 1-C-Phenyl-D-pentitols. J. Amer. Chem. Soc. **73**, 3126 (1951).

*10.* Brookes, D., B. K. Tidd, and W. B. Turner: Avenaciolide, an Antifungal Lactone from *Aspergillus avenaceus*. J. Chem. Soc. **1963**, 5385.

*11.* Corey, E. J., M. Shibasaki, and J. Knolle: Simple Stereocontrolled Synthesis of Thromboxane B₂ from *D*-Glucose. Tetrahedron Letters **1977**, 1625.

*12.* Corey, E. J., and J. W. Suggs: A Method for Catalytic Dehalogenations via Trialkyltin Hydrides. J. Organ. Chem. **40**, 2554 (1975); Corey, E. J., T. K. Schaaf, W. Huber, U. Koelliker, and N. M. Weinshenker: Total Synthesis of Prostaglandins F₂ₐ and E₂ as the Naturally Occurring Forms. J. Amer. Chem. Soc. **92**, 397 (1970).

*13.* Curtis, W. D., D. A. Laidler, J. F. Stoddard, J. B. Wolstenholme, and G. H. Jones: Enantiomeric Differentiation by a Chiral Symmetrical Crown Derived from L-Iditol. Carbohyd. Res. **57**, C 17 (1977); Laidler, C. L., and J. F. Stoddard: Chiral Asymmetrical Crown-Ethers. Carbohyd. Res. **55**, C 1 (1977); Laidler, D. A., and J. F. Stoddard: Stereoselectivity in Complexation of Primary Alkylammonium Cations by the Diastereotopic Faces of Chiral Asymmetric Crowns. Chem. Commun. **1977**, 481.

*14.* Defaye, J., and J. Hildesheim: Synthèses dans la Série des 2,5-Anhydrides des Sucres. Solvolyse d'Esters Sulfoniques. Tetrahedron Letters **1968**, 313.

*15.* David, S., O. Eisenstein, W. J. Hehre, L. Salem, and R. Hoffman: Superjacent Orbital Control. An Interpretation of the Anomeric Effect. J. Amer. Chem. Soc. **95**, 3806 (1973).

*16.* Eastwood, F. W., K. J. Harrington, J. S. Josan, and J. L. Puran: The Conversion of 2-Dimethylamino-1,3-Dioxolanes into Alkenes. Tetrahedron Letters **1970**, 5223.

*17.* Edward, J. T.: Stability of Glycosides to Acid Hydrolysis. Chem. and Ind. **1955**, 1102.

*18.* Eliel, E. L.: Conformational Analysis in Heterocyclic Systems: Recent Results and Applications. Angew. Chem. Intern. **11**, 739 (1972).

*19.* Elliott, M. (ed.): Synthetic Pyrethroids: A Symposium Sponsored by the Division of Pesticide Chemistry at the 172nd Meeting of the American Chemical Society, San Francisco, Calif., Aug. 30—31, 1976; ACS: Wash., D. C., 1977.

20. FERRIER, R. J., and N. VETHAVIYASAR: Unsaturated Carbohydrates. Part XVII. Synthesis of Branched-Chain Sugar Derivatives by Application of the Claisen Rearrangement. J. Chem. Soc. Perkin I 1973, 1791.

21. FISCHER, E.: Über die Verbindungen der Zucker-Arten mit den Mercaptanen. Chem. Ber. 27, 673 (1894).

22. FITZSIMMONS, B., and B. FRASER-REID: Unpublished Results.

23. FRASER-REID, B.: Some Progeny of 2,3-Unsaturated Sugars — They Little Resemble Grandfather Glucose. Accts. Chem. Res. 8, 192 (1975).

24. FRASER-REID, B., and B. J. CARTHY: The Synthesis of Some Cyclopropylcarbonyl Glycopyranosides. Can. J. Chem. 50, 2928 (1972).

25. FRASER-REID, B., N. L. HOLDER, D. R. HICKS, and D. L. WALKER: Synthetic Applications of the Photochemically Induced Addition of Oxycarbinyl Species to Enones. Part I. The Addition of Simple Alcohols. Can. J. Chem. 55, 3978 (1977).

26. FRASER-REID, B., A. McLEAN, E. W. USHERWOOD, and M. YUNKER: Pyranosiduloses. II. The Synthesis and Properties of Some Alkyl 2,3-dideoxy-2-enopyranosid-4-uloses. Can. J. Chem. 48, 2877 (1970).

27. FRASER-REID, B., and A. SHERK: Unpublished results.

28. GORE, N. E., G. T. PEARCE, and R. M. SILVERSTEIN: Relative Stereochemistry of Multistriatin (2,4-Dimethyl)-5-ethyl-6,8-dioxabicyclo [3.2.1] Octane. J. Org. Chem. (U.S.A.) 40, 1705 (1975).

29. HALL, C. R., and T. D. INCH: The Preparation and Absolute Configuration of Some Chiral O,S-Dialkyl Phosphoramidothioates. Tetrahedron Letters 1977, 3761; HALL, C. R., and T. D. INCH: Acid and Base Catalysed Alcoholysis of Some Chiral O,S-Dialkyl Phosphoramidothioates. Tetrahedron Letters 1977, 3765.

30. HANESSIAN, S., P. DEXTRASE, A. FOUGEROUSSE, and Y. GUIDON: Synthese Stereocontrolee Des Precurseurs Chriaux Des 11-Oxaprostaglandines. Tetrahedron Letters 1974, 3983.

31. HANESSIAN, S., and P. LAVALLEE: A Stereospecific, Total Synthesis of Thromboxane $B_2$. Can. J. Chem. 55, 562 (1977).

32. HANESSIAN, S., and G. RANCOURT: Carbohydrates as Chiral Intermediates in Organic Synthesis. Two Functionalized Chemical Precursors Comprising Eight of the Ten Chiral Centers of Erythronolide A. Can. J. Chem. 55, 1111 (1977).

33. HANESSIAN, S. G., G. RANCOURT, and Y. GUINDON: Assembly of the Carbon Skeletal Framework of Erythronolide A. Can. J. Chem. 56, 1843 (1978)

34. HAWORTH, W. N.: The Constitution of Sugars. London: Arnold. 1929.

35. HERNANDEZ, O.: Chiral Synthesis of Thromboxane $B_2$ Intermediates: Tetrahedron Letters 1978, 219.

36. HICKS, D. R., and B. FRASER-REID: Synthesis of One Enantiomer, the Other Enantiomer, and a Mixture of Both Enantiomers of Frontalin from a Derivative of Methyl-α-D-glucopyranoside. J. Chem. Soc. Chem. Commun. 1976, 869.

37. — — Selective Sulphonylation with N-Tosylimidazole. A One-Step Preparation of Methyl 2,3-Anhydro-4,6-O-benzylidene-α-D-mannopyranoside. Synthesis 1974, 203.

38. — — The 2- and 3-C-Methyl Derivatives of Methyl 2,3-Dideoxy-α-D-erythro-hex-2-enopyranosid-4-ulose. Can. J. Chem. 53, 2017 (1975).

39. HOLDER, N. L., and B. FRASER-REID: The Synthesis of Some Alkyl Hex-3-enopyranosiduloses. Can. J. Chem. 51, 3357 (1973).

40. HOLLAND, C. V., D. HORTON, and J. S. JEWELL: The Favored Conformation of Tri-O-acetyl-β-D-Xylopyranosyl Chloride. An All-Axial Tetrasubstituted Six-Membered Ring. J. Org. Chem. (U.S.A.) 32, 1818 (1967).

41. HOUGH, L., and T. J. TAYLOR: 1:1-Diethylsulphonyl Derivatives of L-Rhamnose and Their Conversion into 5-Deoxy-L-Arabinose. J. Chem. Soc. 1955, 3544.

42. Inch, T. D.: The Use of Carbohydrates in the Synthesis and Configurational Assignments of Optically Active, Non-carbohydrate Compounds. Advances in Carbohydrate Chemistry and Biochemistry 27, 191 (1972), Academic Press: New York and London.

43. Jeffrey, G. A., J. A. Pople, and L. Radom: The Application of *ab initio* Molecular Orbital Theory to the Anomeric Effect. A Comparison of Theoretical Predictions and Experimental Data on Conformations and Bond Lengths in Some Pyranoses and Methyl Pyranosides. Carbohydr. Res. 25, 117 (1972).

44. Jones, J. K. N., and J. L. Thompson: A Synthesis of 5,6-Dideoxy-D-Xylohexose (5-Deoxy-5-C-Methyl-D-Xylose). Can. J. Chem. 35, 955 (1957).

45. Karplus, M.: Contact Electron-Spin Coupling of Nuclear Magnetic Moments. J. Chem. Phys. 30, 11 (1959); Karplus, M.: Vicinal Proton Coupling in Nuclear Magnetic Resonance. J. Amer. Chem. Soc. 85, 2870 (1963).

46. Kato, M., M. Kageyama, R. Tanaka, K. Kawahara, and A. Yoshikoshi: Synthetic Study of (±)-Canadensolide and Related Dilactones. Double Lactonization of Unsaturated Dicarboxylic Acids via Acyl Hypoiodite Intermediates. J. Organ. Chem. (U.S.A.) 40, 1932 (1975).

47. Kinzer, G. W., A. F. Fetiman, Jr., T. F. Page, Jr., R. L. Foltz, J. P. Vite, and G. B. Pitman: Bark Beetle Attractants: Identification, Synthesis and Field Bioassay of a New Compound Isolated from *Dendroctonus*. Nature (London) 221, 477 (1969).

48. Kocienski, P. J., and R. W. Ostrow: A Stereoselective Total Synthesis of *exo-* and *endo*-Brevicomin. J. Organ. Chem. (U.S.A.) 41, 398 (1976).

49. Konotshita, M., and S. Mariyama: Synthesis of (2R,3S,4S)-4-Amino-3-hydroxy-2-methyl-5(3-pyridyl)-pentanoic Acid Present in Antibiotic Pyridomycin. Bull. Soc. Chem. Japan. 48, 2081 (1975).

50. Lemieux, R. U., L. Anderson, and A. H. Conner: The Mutarotation of 2-Deoxy-β-D-*erythro*-Pentose ("2-Deoxy-β-D-Ribose). Conformations, Kinetics and Equilibria. Carbohydr. Res. 20, 59 (1971).

51. Lemieux, R. U., and J. Howard: The Absolute Configuration of Dextro-I-Deuterioethanol. Can. J. Chem. 41, 308 (1963).

52. Lemieux, R. U., and S. Koto: The Conformational Properties of Glycosidic Linkages. Tetrahedron 30, 1933 (1974); Bailey, W. F., and E. L. Eliel: Conformational Analysis. XXIX. 2-Substituted and 2,2-Disubstituted 1,3-Dioxanes. The Generalized and Reverse Anomeric Effects. J. Amer. Chem. Soc. 96, 1798 (1974); Anet, F. A. L., and I. Yavari: Generalized Anomeric Effect and Barrier to Internal Rotation about the Oxygen-Methylene Bond in Chloromethyl Ether. J. Amer. Chem. Soc. 99, 6752 (1977).

53. Lemieux, R. U., R. K. Kullnig, H. J. Bernstein, and W. G. Schneider: Configurational Effects in the Proton Magnetic Resonance Spectra of Acetylated Carbohydrates. J. Amer. Chem. Soc. 79, 1005 (1957).

54. — — — — Configurational Effects on the Proton Magnetic Resonance Spectra of Six-membered Ring Compounds. J. Amer. Chem. Soc. 80, 6098 (1958).

55. Lemieux, R. U., J. D. Stevens, and R. R. Fraser: Observations on the Karplus Curve in Relation to the Conformation of the 1,3-Dioxolane Ring. Can. J. Chem. 40, 1955 (1962).

56. Lesage, S., and A. S. Perlin: Synthesis and Stereochemistry of 2-Pyrone Derivatives Related to Some Fungal Metabolites. Can. J. Chem. 56, 2889 (1978).

57. Levene, P. A., A. L. Raymond, and R. T. Dillon: Glucoside Formation in the Commoner Monoses. J. Biol. Chem. 95, 699 (1932).

58. Lichtenthaler, F. W.: Konfiguration der bei Cyclisierung von 6-Nitro-D-glucose und -L-idose gebildeten Desoxynitroinosite und ihre Isomerisierungen mit Alkali. Chem. Ber. 94, 3071 (1961); Grosheintz, J. M., and H. O. L. Fischer: Cyclization

of 6-Nitrodesoxyaldohexoses to Nitrodeoxyinositols. J. Amer. Chem. Soc. **70**, 1479 (1948); KOVAR, J., and H. H. BAER: Cyclisations of Dialdehydes with Nitromethane. XV. Synthesis of Four Stereoisomeric Deoxynitroinositol Monomethyl Ethers. Can. J. Chem. **51**, 1801 (1973).

*59.* LOS, J. M., L. B. SIMPSON, and K. WIESNER: The Kinetics of Mutarotation of D-Glucose with Consideration of an Intermediate Free-aldehyde Form. J. Amer. Chem. Soc. **78**, 1564 (1956); LOS, J. M., and K. WIESNER: A Polarographic Investigation of the Mechanism of Mutarotation of D-Glucose. J. Amer. Chem. Soc. **75**, 6346 (1956).

*60.* LOURENS, G. J., and J. M. KOEKEMOER: The Novel Stereospecific Synthesis of 11-Oxaprostaglandin F₂ . Tetrahedron Letters **1975**, 3719.

*61.* MATTES, K. C., M. T. HSIA, C. R. HUTCHINSON, and S. A. SISK: A Useful Synthon for Elaboration Into Iridoids and Alkaloids. Tetrahedron Letters **1977**, 3541.

*62.* MILLS, J. A.: The Stereochemistry of Cyclic Derivatives of Carbohydrates. Advan. Carbohydr. Chem. **10**, 1 (1955).

*63.* NELSON, N. A., and R. W. JACKSON: Total Synthesis of Thromboxane B₂. Tetrahedron Letters **1976**, 3275; KELLY, R. C., I. SCHLETTER, and S. J. STEIN: Synthesis of Thromboxane B₂. Tetrahedron Letters **1976**, 3279.

*64.* NESS, R. K., and H. G. FLETCHER, JR.: 2-Deoxy-D-*erythro*-pentose. X. Synthesis of 1,4-Anhydro-3,5-di-O-benzoyl-2-deoxy-D-*erythro*-pentose-1-enol. Derivatives of a Furanose-related Glycal. J. Organ. Chem. (U.S.A.) **28**, 435 (1963); HAGA, M., and R. K. NESS: Preparation and Properties of 3,5-Di-O-p-anisoyl-1,2-dideoxy-D-*erythro*-pentofuranose-1-ene. Various p-Anisoylated Derivatives of D-Ribofuranose. J. Organ. Chem. (U.S.A.) **30**, 158 (1965).

*65.* OGAWA, T., T. KAWANO, and M. MATSUI: A Biomimetic Synthesis of (+) biotin from D-Glucose. Carbohydr. Res. **57**, C 31 (1970).

*66.* OHRUI, H., and S. EMOTO: A Synthesis of (S)-(−)-Frontalin from D-Glucose. Agric. Biol. Chem. **40**, 2267 (1967).

*67.* — — Stereoselective Synthesis of (+) and (−) Avenaciolide from D-Glucose. The Correct Absolute Configuration of Natural Avenaciolide. Tetrahedron Letters **1975**, 3657.

*68.* — — Stereospecific Synthesis of (+)-Biotin. Tetrahedron Letters **1975**, 2765.

*69.* OHRUI, H., H. KUZUHARA, and S. EMOTO: Synthesis with Azido Sugars. Part III. Preparation of d-Oxybiotin. Agric. Biol. Chem. **34**, 375 (1970).

*70.* PAULSEN, H., and H. BEHRE: Umwandlung von 1,2:5,6-Di-O-isopropyliden-3-desoxy-α-D-glucose-3-en in 1,2:5,6-Di-O-isopropyliden-α-D-galactofuranose durch selektive Hydroborierung. Carbohydr. Res. **2**, 80 (1966).

*71.* PAULSEN, H., W.-P. TRAUTWEIR, F. G. ESPINOSA, and K. HEYNS: Einfache Synthese von D-Idose aus D-Glucose durch mehrfache Acetoxonium-Ion-Umlagerungen. Darstellung eines stabilen Acetoxonium-Salzes der Tetracetyl-Idose. Chem. Ber. **100**, 2822 (1967).

*72.* PEARCE, G. T., W. E. GORE, R. M. SILVERSTEIN, J. W. PEACOCK, R. A. CUTHBERG, G. N. LANIER, and J. B. SIMEONE: Chemical Attractants for the Smaller European Elm Bark Beetle *Scolytus multistriatus* (Coleoptera: Scolytidae). J. Chem. Ecol. **1**, 115 (1975).

*73.* PEAT, S.: The Chemistry of Anhydro Sugars. Advances in Carbohydrate Chemistry **2**, 38 (1946). New York: Academic Press.

*74.* PERLIN, A. S.: Carbon-13 N.M.R. Spectroscopy of Carbohydrates. International Review of Science (ASPINALL, G. O., ed.). Series 2, **7**, 1 (1976).

*75.* PLAUMANN, D. E.: A Stereoselective synthesis of (1S:2R:4S;5R)-2,4-Dimethyl-5-Ethyl-6,8-Dioxabicyclo (3.2.1) Octane, α-Multistriatin. M. Sc. Thesis, University of Waterloo, 1977.

76. Radatus, B., and B. Fraser-Reid: Cyclopropylcarbinyl-oxo-carbonium Ions, Part V. Synthesis and Chemistry of Some Cyclopropyl Glucopyranosides. Can. J. Chem. **50,** 2909 (1972).

77. Radom, L., W. J. Hehre, and J. A. Pople: Molecular Orbital Theory of the Electronic Structure of Organic Compounds. XIII. Fournier Component Analysis of Internal Rotation Potential Functions in Saturated Molecules. J. Amer. Chem. Soc. **94,** 2371 (1972).

78. Reeves, R. E.: Cuprammonium-Glycoside Complexes. Advances in Carbohydrate Chemistry **6,** 108 (1951). New York: Academic Press.

79. — The Optical Rotation of Cellulose and Glycosides in Cuprammonium Hydroxide Solution. Science **99,** 148 (1944); Reeves, R. E.: The Optical Activity of the Copper Complexes of Polysaccharides and Substituted Methyl Glucosides. J. Biol. Chem. **154,** 49 (1944).

80. Richardson, A. C.: Nucleophilic Replacement Reactions of Sulphonates. Part VI. A Summary of Steric and Polar Factors. Carbohydr. Res. **10,** 395 (1969).

81. Richtmyer, N. K.: D-*altro*-Heptulose (Sedoheptulose). Methods in Carbohydrate Chemistry I, 167. New York and London: Academic Press.

82. Rosenthal, A., and L. B. Nguyen: New Route to Branch-Chain Sugars by Application of Modified Wittig Reaction to Ketoses. Tetrahedron Letters **1967,** 2393.

83. Schwarz, J. C. P.: Rules for Confirmation Nomenclature for Five- and Six-membered Rings in Monosaccharides and their Derivatives. Chem. Commun. **1973,** 505.

84. Shafizadeh, F., and M. L. Wolfrom: Synthesis of L-Iduronic Acid and an Improved Production of D-Glucose-6-C′. J. Amer. Chem. Soc. **77,** 2568 (1955); Whistler, R. L., and R. E. Pyler: Synthesis of 6-Amino-6-deoxy-5-thio-D-Glucopyranose. Carbohydr. Res. **12,** 201 (1970).

85. Sitrin, R. D.: Studies on the Synthesis of Tetrodotoxin. Ph. D. Thesis, Harvard University, 1972.

86. Soltzberg, S., R. M. Goepp, Jr., and W. Freudenberg: Hexitol Anhydrides, Synthesis and Structure of Arlitan, the 1-4 Monoanhydride of Sorbitol. J. Amer. Chem. Soc. **68,** 919 (1946).

87. Stevens, J. D., and H. G. Fletcher, Jr.: The Proton Magnetic Resonance Spectra of Pentofuranose Derivatives. J. Org. Chem. **33,** 1799 (1968); Lemieux, R. U., and R. Nagarajan: The Configuration and Conformation of "Di-D-Fructose Anhydride 1". Can. J. Chem. **42,** 1270 (1964).

88. Stork, G., and S. Raucher: Chiral Synthesis of Prostaglandins From Carbohydrates. Synthesis of (+)-15-(−)-Prostaglandin A₂. J. Amer. Chem. Soc. **98,** 1583 (1976).

89. Stork, G., T. Takahashi, T. Suzuki, and I. Kawamoto: Unpublished results.

90. Taylor, E. C., and P. A. Jacobi: Pteridines. XXXVII. A Total Synthesis of L-*erythro*-Biopterin and Some Related 6-(Polyhydroxyalkyl) pterins. J. Amer. Chem. Soc. **98,** 2301 (1976).

91. Theander, O.: 1,2 : 5,6-Di-O-isopropylidene Derivatives of D-Glucohexodialdose and D-*ribo*-hexos-3-ulose. Acta Chem. Scand. **18,** 2209 (1964).

92. Tipson, R. S., and A. Cohen: Action of Zinc Dust and Sodium Iodide in N,N-dimethyl-formanide on Contiguous, Secondary Sulfonyloxy Groups: A Simple Method for Introducing Nonterminal Unsaturation. Carbohydr. Res. **1,** 338 (1965).

93. Toms, B. A.: L-Galactose as a Component of a Polysaccharide of Animal Origin. Nature **146,** 559 (1940).

94. Triggle, D. J., and B. Belleau: Studies on the Chemical Basis for Cholinomimetic and Cholinolytic Activity. Part 1. The Synthesis and Configuration of Quaternary Salts in the 1,3-Dioxolane and Oxazoline Series. Can. J. Chem. **40,** 1201 (1962).

95. Upeslacis, J.: A Carbohydrate Approach to Tetrodotoxin. Ph. D. Thesis, Harvard University, 1975.

*96.* VICKERY, H. B.: Rules of Carbohydrate Nomenclature. J. Organ. Chem. (U.S.A.) **28,** 281 (1963).

*97.* WARD, R. B.: 1,6-Anhydro-β-D-Glucopyranose (Levoglucosan). Methods in Carbohydrate Chemistry II, 394 (1963). New York and London: Academic Press.

*98.* WIGGINS, L. F.: Ethylene Oxide-type Anhydrosugars. Methods in Carbohydrate Chemistry **2,** 188 (1963).

*99.* WILBUR, D. J., C. WILLIAMS, and A. ALLERHAND: Detection of the Furanose Anomers of D-Mannose in Aqueous Solution. Application of Carbon-13 Nuclear Magnetic Resonance Spectroscopy at 68 MHz. J. Amer. Chem. Soc. **99,** 5450 (1977).

*100.* WILLIAMS, J. M., and (in part) A. C. RICHARDSON: Selective Acylation of Pyranosides-I Benzolation of Methyl α-D-Glycopyranosides of Mannose, Glucose and Galactose. Tetrahedron **23,** 1369 (1967).

*101.* WOLFE, S., A. RAUK, L. M. TEL, and T. G. CSIZMADIA: A Theoretical Study of the Edward-Lemieux Effect (The Anomeric Effect). The Stereochemical Requirements of Adjacent Electron Pairs and Polar Bonds. J. Chem. Soc. **B 1971,** 136.

*102.* WOLFROM, M. L., R. U. LEMIEUX, and S. M. OLIN: Configurational Correlation of L-(levo)-Glyceraldehyde With Natural (dextro)-Alanine by a direct Chemical Method. J. Amer. Chem. Soc. **71,** 2780 (1949).

*103.* WOODWARD, R. B.: The Structure of Tetrodotoxin. Pure and Applied Chem. **9,** 49 (1964).

*104.* YATES, P., and P. EATON: Acceleration of the Diels-Alder Reaction by Aluminium Chloride. J. Amer. Chem. Soc. **82,** 4436 (1960).

*105.* YOSHIMURA, J., K. KOBAYASHI, K. SATO, and M. FUNABASHI: On the Configuration of Branched-Chain Derivatives of 1.2 : 5.6-Di-O-isopropylidene-α-D-*ribo*-hexofuranose-3-ulose. Bull. Soc. Chem. Japan. **45,** 1806 (1972).

*106.* YUNKER, M. B., D. E. PLAUMANN, and B. FRASER-REID: The Stereochemistry of Conjugate Addition of Lithium Dialkyl Cuprate Reagents to Some Carbohydrate α-Enones. Can. J. Chem. **55,** 4002 (1977).

*(Received May 30, 1978)*

# Recent Advances in the Biology and Chemistry
# of Vitamin D

By H. Jones and G. H. Rasmusson,
Merck Institute for Therapeutic Research, Rahway, New Jersey, U.S.A.

With 3 Figures

**Contents**

# I. Biology and Biochemistry of Vitamin D

## 1. Introduction

The D-vitamins are a group of fat soluble materials with high anti-rachitic activity. Rickets is a disease of infancy caused by faulty calcium hydroxy apatite deposition in the growing bone. It was Palm (*140*) in 1890 who found that the disease responded favorably to irradiation of patient's food or skin by sunlight or ultraviolet light. Mellanby (*113*) in 1919 first reported the presence of an antirachitic factor in cod-liver oil. This discovery encouraged research on the isolation of the active principle in fish-liver oil, but the low natural concentration of the vitamin made isolation very difficult.

Early investigators showed that superficially purified cholesterol and activatable foods contained a provitamin, or precursor, which was converted into the vitamin on irradiation. Ultimately, two vitamin D precursors were identified. Provitamin $D_2$ was found to be ergosterol (**1a**). Windhaus (*194*) and Askew (*6*) independently isolated pure crystalline vitamin $D_2$ (**2a**) *via* its 3,5-dinitrobenzoate from the complex and unstable irradiation products of ergosterol. Provitamin $D_3$ was more difficult to identify, but it was found by Windhaus (*195*) to be 7-dehydro-cholesterol (**1b**). On irradiation crystalline vitamin $D_3$ (**2b**) was obtained, which was identical with that isolated from tuna-liver oil by Brock-mann (*29*). Vitamins $D_4$ and $D_5$ have been found in nature, each one differing from the other and the ones mentioned above in the constitution of the side chain.

The provitamins are more abundant in nature than the pure vitamins. Ergosterol is found in plants and yeast. 7-Dehydrocholesterol is the provitamin of man; it is also found in animal skins. The most concentrated source, however, is in the snail *Baccinum undatum,* which contains about 27 percent 7-dehydrocholesterol in its dry weight. The D-vitamins themselves are not very widely distributed in nature. The richest sources are fish-liver oils, milk, and bird eggs. In consequence, the main source of the vitamins $D_2$ and $D_3$ is by irradiation of the corresponding provitamins. Although vitamins $D_2$ and $D_3$ have the same activity in man and New World monkeys, vitamin $D_3$ is man's natural vitamin. However, in birds and Old World monkeys vitamin $D_3$ is the natural vitamin and vitamin $D_2$ is virturally inactive. Vitamin $D_2$ was first purified and degraded by WINDHAUS (*196*) and HEILBRON (*73*). The structure was assigned in 1936.

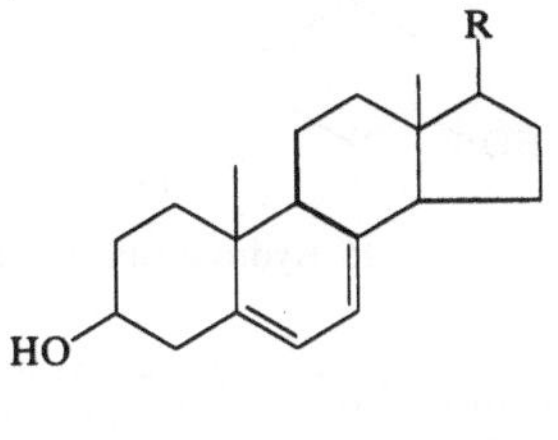

R

R

(1)

HO

HO····

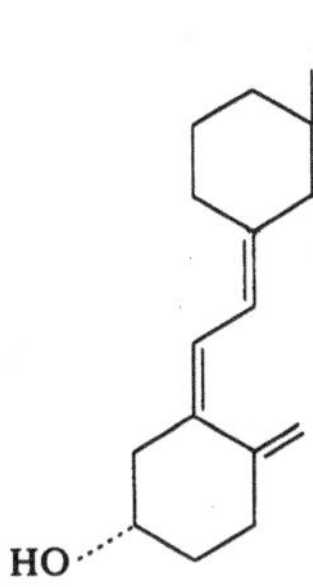

(2)

(a) R = $C_9H_{17}$ = 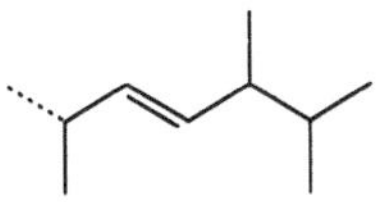    $D_2$

(b) R = $C_8H_{17}$ = 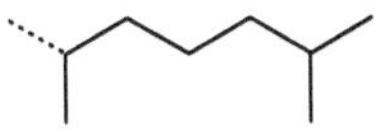    $D_3$

Synthetic work in the vitamin D area attracted a large number of organic chemists during the late 1950's and early 1960's. This work culminated in the photochemical synthesis of vitamin $D_3$ coincidentally by LYTHGOE (*64*) in England and INHOFFEN (*83*) in Germany. New interest has developed recently as a result of the isolation and characterization of the circulating forms of vitamin $D_3$ — namely, 25-hydroxyvitamin $D_3$ and $1\alpha$,25-hydroxyvitamin $D_3$. The characterization of these metabolites was carried out independently in three laboratories, by NORMAN (*133*) at the University of California, Riverside, KODICEK (*94*) in Cambridge, and DeLUCA (*77*) at the University of Wisconsin.

The last decade has seen a startling advance in our understanding of the biochemical mechanism of action of vitamin D. Ten years ago little was known concerning the absorption or tissue localization of this classical nutritional substance, nor was there available in the literature any very clear postulation of its mode of action. Some workers preferred to think of vitamin D as a cofactor of some unidentified enzyme (*37*). It is now clear, however, that vitamin D acts in a manner analogous to

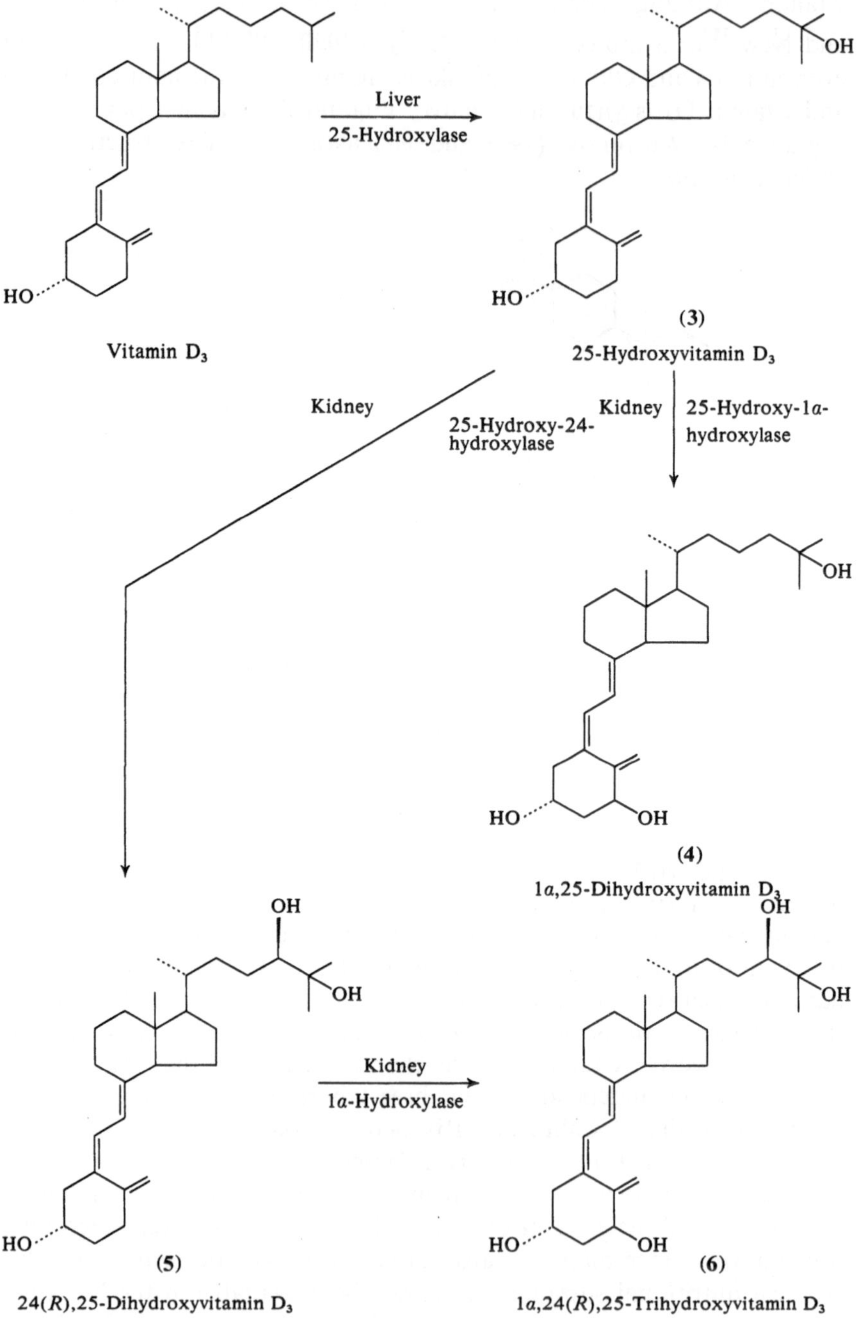

Fig. 1. Biological conversion of vitamin $D_3$ to specific metabolites

that of other classical steroid hormones. However, its exact mechanism of action still remains unresolved nearly 56 years after the identification of a "vitamin-D-like" factor by HESS, GUTTMAN, and McCOLLUM (72, 110). It is clear, however, that vitamin D action is regulated primarily by its conversion to metabolites with specific site activities. The most studied of these transformations are shown in Figure 1.

### 2. 25-Hydroxyvitamin $D_3$ and $1\alpha,25$-Dihydroxyvitamin $D_3$

The early investigators in the 1930's and 1940's attempted to follow the metabolic fate of vitamin $D_3$ in animals. Unfortunately, their efforts were hampered by the small natural concentration of vitamin D and its metabolites. KODICEK (93) in 1955 synthesized radiolabeled vitamin $D_2$. However, the material was of such low specific activity that it was of little value in characterizing the metabolism of vitamin D itself.

NORMAN and DeLUCA (128) in 1963 tritiated vitamins $D_2$ and $D_3$. Their compounds had 1000 times the specific activity of the material prepared by KODICEK. In 1964 they reported (129) isolation of a modified vitamin $D_3$ metabolite which retained the activity of the parent molecule, but was chromatographically distinguishable. MORII (116) in 1967 showed that this metabolite could raise serum calcium, cure rickets, and had a shorter onset of action than the parent vitamin $D_3$. DeLUCA (17) in 1968 showed by isolation and mass spectrum that this metabolite of vitamin $D_3$ was 25-hydroxyvitamin $D_3$ (3). They initially considered this to be the final active form of the vitamin, but in 1968 HAUSSLER (69) suggested that there might be yet another vitamin D metabolite of major biological significance. It was subsequently shown that this material was more potent than 25-hydroxyvitamin $D_3$ and was present in particularly high concentration in the chick intestine. LAWSON et al. (102, 103) in 1969 suggested that vitamin D metabolism involved introduction of yet another oxygen function into the vitamin $D_3$ molecule and showed that the product was identical with the material found by HAUSSLER the year before. Also in 1969 DeLUCA (151) detected by silicic acid chromatography a metabolite more potent than 25-hydroxyvitamin $D_3$, which was subsequently shown to be identical with the potent metabolite of HAUSSLER (68) and LAWSON (102, 103). Coincidentally, KODICEK (94) in 1970, HAUSSLER (67) in 1971, and NORMAN (124) in 1971 showed that this more potent metabolite of vitamin $D_3$ and 25-hydroxyvitamin $D_3$ was also more active than vitamin $D_3$ and increased intestinal absorption of calcium. KODICEK (56) in 1970 demonstrated that the kidney was the unique site of bio-synthesis of this more active vitamin $D_3$ metabolite. Almost immediately the identity of this metabolite was independently and conclusively shown by the three groups (*77, 101, 130*) to be 1,25-dihydroxyvitamin $D_3$ (**4**) by mass spectrometry, U.V. absorption spectrometry and chemical manipulation.

It now appears (Fig. 1) that after vitamin $D_3$ has been absorbed from the gut it finds its way *via* the blood stream to the liver and is hydroxylated in the 25-position by a 25-hydroxylase enzyme. This enzyme is not specifically found in the liver but in several species occurs at many other sites — *i. e.*, in the intestine and kidney (*66*). The 25-hydroxyvitamin $D_3$ then proceeds to the kidney where it is metabolized to 1α,25-dihydroxy-vitamin $D_3$. It is then carried by the blood stream to the intestinal mucosal lining cells where it initiates calcium absorption (see Section 3).

Both the synthetic analog, 1α-hydroxyvitamin $D_3$ (**7**), and 1α,25-hydroxyvitamin $D_3$ are potent stimulators of intestinal calcium absorption in organ culture (*37, 58*) in rats and chickens (*26, 36, 69, 145, 146, 198*) fed vitamin $D_3$ and calcium deficient diets. The ratio of activities of 1α-hydroxy- and 1α,25-dihydroxyvitamin $D_3$ compared to the activity of vitamin $D_3$ itself reaches 1000 in parathyroidectomixed-thyroidectomized animals (*146*). However, what these ratios mean as regards the therapeutic value of 1α-hydroxyvitamin $D_3$ or 1α,25-dihydroxyvitamin $D_3$ is not easy to assess. In clinical situations of impaired calcium utilization, there normally would be an adequate supply of calcium, phosphorous, and vitamin $D_3$ in the diet. However, in certain metabolic dysfunctions (see Section 6) both 1α- and 1α,25-dihydroxyvitamin $D_3$ have been clearly demonstrated to have therapeutic value.

Both 1α-hydroxyvitamin $D_3$ and 1α,25-dihydroxyvitamin $D_3$ increase bone formation and bone resorption *in vitro* and in the rachitic animal (*100, 153, 183, 188, 197*).

It is also clear that the introduction of the 1α-hydroxy function is an absolutely essential metabolic event for calcium absorption. Nephrectomized vitamin D-deficient animals show no increase in intestinal calcium transport or bone calcium mobilization in response to physiological doses of vitamin $D_3$ or 25-hydroxyvitamin $D_3$. On the other hand these animals respond efficiently to 1α,25-dihydroxyvitamin $D_3$.

## 3. Calcium

The calcium homeostatic system, originally attributed exclusively to the parathyroid hormone, now includes the renal hormone 1α,25-di-hydroxyvitamin $D_3$. Calcium is the fifth most abundant element in vertebrate systems and is the most important structural component of the body (*45, 72*). Calcium is also required for the following purposes:

*References, pp. 111—121*

1. Contraction of skeletal and cardiac muscle.

2. Nerve conduction; decreased calcium increases irritability and may produce tetany.

3. Maintenance of membrane permeability; decreased calcium increases permeability; which in turn affects muscle contraction and nerve transmission.

4. Blood coagulation; presence required for conversion of prothrombin to thrombin.

All these processes are dependent on an adequate supply of calcium from the diet. It is the integrated action of parathyroid hormone and vitamin $D_3$ which supplies the required amounts of calcium for all these processes *via* a smoothly operating calcium homeostatic mechanism. When serum calcium concentration falls below 10 mg/100 ml, the sequence of events shown in Fig. 2 takes place.

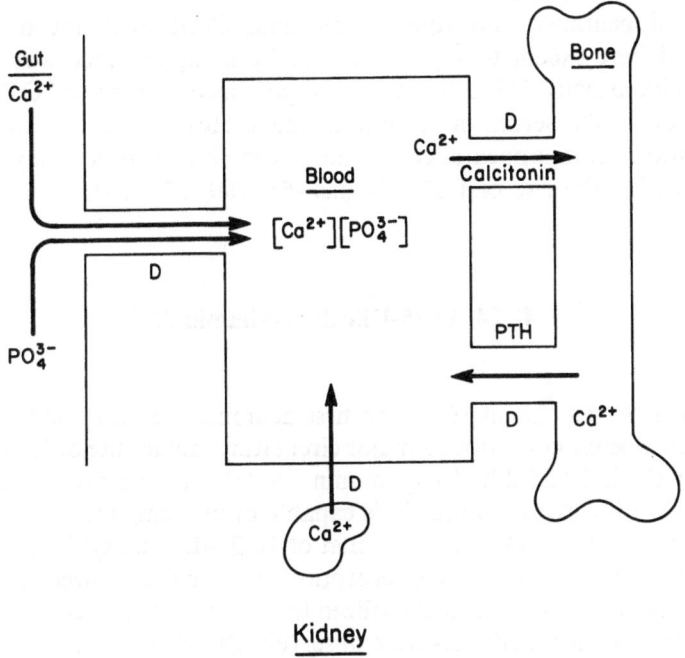

Fig. 2. Calcium transport

The parathyroid glands are stimulated to secrete parathyroid hormone. The latter proceeds directly to two sites. One site is bone, where the hormone mobilizes calcium, provided that there is a form of vitamin $D_3$ present. The second site is the kidney, where the hormone stimulates in some way the synthesis of 25-hydroxy-1α-hydroxylase which metabolizes the already circulating 25-hydroxyvitamin $D_3$ to $1\alpha,25$-dihydroxy-vitamin $D_3$. This renal hormone then proceeds also to two sites. In the bone it stimulates the mobilization of calcium. This calcium then restores the serum calcium to normal level, shutting off parathyroid hormone secretion and completing the feedback loop.

$1\alpha,25$-Dihydroxyvitamin $D_3$ also proceeds to the intestine, where its mechanism of action is similar to that of all steroid hormones. $1\alpha,25$-Dihydroxyvitamin $D_3$ finds its way into the cytoplasm of the intestinal cell and is bound to some specific receptor protein. It is translocated into the cell nucleus, where it is bound to nuclear chromatin and initiates the synthesis of mRNA and in the cytoplasm, a calcium binding protein, which is subsequently secreted. This process stimulates the intestinal absorption of calcium ions, which is the alternative pathway by which the serum calcium concentration may be increased to the optimum value of around 10 mg/100 ml.

The appearance and role of intestinal CaBP will not be further discussed here except to say that originally its appearance was thought to coincide exactly (48, 50) with the appearance and rise of $Ca^{++}$ concentration in the serum of rachitic animals. However, this is now open to question, as its appearance has now been shown to be delayed up to 2 hours after the rise of $Ca^{++}$ begins (63, 100, 177, 191).

## 4. 24(R),25-Dihydroxyvitamin $D_3$

It was Boyle (25) in 1971 who first demonstrated that under normal or hypercalcemic conditions a major circulating metabolite of 25-hydroxyvitamin $D_3$ is 24,25-dihydroxyvitamin $D_3$ (5). He claimed in 1973 (27) that 24,25-dihydroxyvitamin $D_3$ is capable of inducing intestinal calcium transport at dose levels similar to that of $1\alpha,25$-dihydroxyvitamin $D_3$ but has little ability to cause bone resorption and further showed that 24,25-dihydroxyvitamin $D_3$ was metabolized to a more potent metabolite in the kidney before these activities were observed. DeLuca (76) in 1973 showed that in chickens this more potent metabolite was $1\alpha,24,25$-trihydroxy-vitamin $D_3$ (6), although the configuration at the 24-position was not defined at this time, and that in chickens (6) was only 60 percent as active as vitamin $D_3$ in curing rickets and was less active than $1\alpha,25$-dihydroxy-vitamin $D_3$ in stimulating and sustaining intestinal calcium transport and bone resorption. He further indicated that this mixture of 24-isomers had preferential action on the intestine and did not cause bone resorption as much as was expected. More recently, Uskokovic (143) has chemically synthesized both 24(R)- and 24(S),25-dihydroxyvitamin $D_3$ and IKEKAWA (81, 184) prepared $1\alpha$,24(R),25- and $1\alpha$,24(S),25-trihydroxyvitamin $D_3$. The natural metabolites were found to have the 24(R)-configuration.

Some conflicting statements have been made concerning the relative potencies of compounds possessing of the 24(R)- and 24(S)-configuration in the vitamin $D_3$ side chains. DELUCA (185) claims that both synthetically derived 24-hydroxyvitamin $D_3$ isomers stimulate intestinal calcium transport almost equally well in the rat and has shown that the 24(S)-isomer has little or no activity in bone calcium resorption or in bone formation, whereas the 24(R)-isomer is almost as active as 25-hydroxyvitamin $D_3$ in these in vivo tests. However, ATKINS (7) claims that both diastereo-isomers of 24,25-dihydroxyvitamin $D_3$ are potent stimulators of bone resorption in tissue culture and there is no significant difference between the (R)- and (S)-forms. BORIS (24) recently also showed that both 24(R)- and 24(S)-isomers of $1\alpha$,24,25-trihydroxyvitamin $D_3$ promote bone mineralization almost equally well in cockerels. Whether the differentia-tion of the 24(R)- and 24(S)-diastereoisomers in bone formation and resorption occurs only in the absence of the $1\alpha$- and the 25-hydroxy substituents is not known at this time, but it seems unlikely.

NORMAN (152) in 1977 reported that $1\alpha$,24(R),25-trihydroxyvitamin $D_3$ is less potent than $1\alpha$,25-dihydroxyvitamin $D_3$ in the rachitic chick in terms of its ability to stimulate intestinal calcium absorption, mobilize bone calcium, and induce intestinal calcium binding protein. DELUCA (27) had shown earlier that 24,25-dihydroxyvitamin $D_3$ must be converted to $1\alpha$,24,25-trihydroxyvitamin $D_3$ in the kidney in order to have biological activity. GRAY (62) in 1974 had demonstrated presence of a peak in the analysis of the plasma of normal and nephrectomized humans given $^3$H-labeled 25-hydroxyvitamin $D_3$, which was chromatographically similar to authentic $1\alpha$,24,25-trihydroxyvitamin $D_3$. DELUCA (182) has recently shown that the renal 24-hydroxylase enzyme requires presence of a hydroxyl group be on the 25-carbon of the vitamin $D_3$ molecule before further hydroxylation can take place. The 24(R)-hydroxylase is, how-ever, not totally isolated in the kidney as the $1\alpha$-hydroxylase appears to be. DELUCA (182) finds that nephrectomized animals also metabolize $1\alpha$,25-dihydroxyvitamin $D_3$ to $1\alpha$,24(R),25-trihydroxyvitamin $D_3$.

As can be seen from the above work, the role of $1\alpha$,24(R),25-tri-hydroxyvitamin $D_3$ is becoming more ubiquitous. Originally, synthesis of this metabolite was thought to be a mechanism employed by the body for ridding itself of circulating levels of $1\alpha$,25-dihydroxyvitamin $D_3$ after the successful attainment of the normal concentration of calcium in the serum (see Fig. 3). It now seems, however, that this trihydroxy metabolite of vitamin $D_3$ is becoming more influential in the overall conservation of calcium homeostasis.

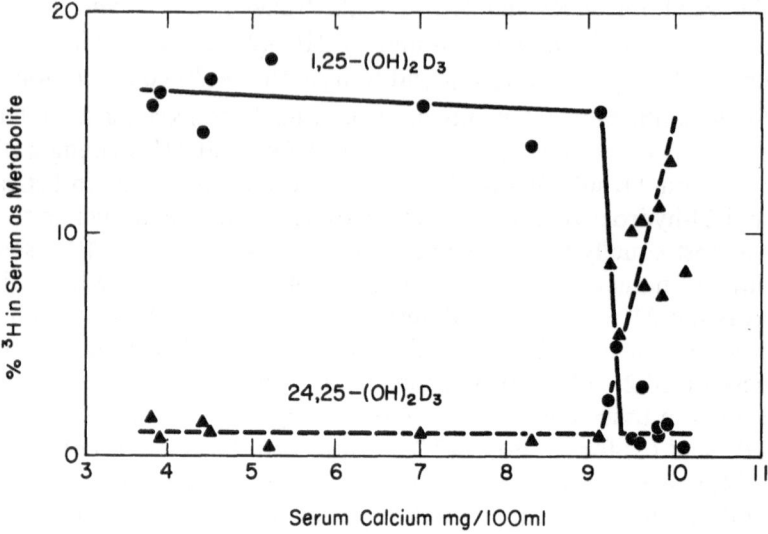

Fig. 3. Relation of serum calcium of rats on a normal phosphorus diet to production of either 1,25-(OH)D or 24,25-(OH)D from 25-OH-D. (Reproduced from Omdahl and DeLuca with the permission of the publisher; rats fed a variety of dietary-calcium levels with and without vitamin D were examined for their ability to convert H-25-OH-D to the two metabolites.)

Clinically, there may indeed be an exciting role for a compound with differential effects on bone. It is now generally considered that the slow onset of osteoporosis after maturity is due to a disequilibrium of bone resorption over bone formation. A compound which exerts a differential effect on bone formation above that of bone resorption could have great value in the treatment of osteoporosis. It remains to be seen whether 24(R),25-dihydroxy- or 1α,24(S),25-trihydroxyvitamin $D_3$ is such a compound.

## 5. 25,26-Dihydroxyvitamin $D_3$

25,26-Dihydroxycholesterol has been shown by Suda (*180*) to be present in pigs given large doses of vitamin $D_3$. At present the site or mechanism of its production is not clear. However, it does stimulate intestinal calcium absorption in the rat and it is claimed that this metabolite does not effect bone resorption. In 1975 LAM (98) found that this compound was less active in nephrectomized animals and suggested that it requires further metabolism in the kidney, possibly to 1α,25,26-trihydroxyvitamin $D_3$. BORDIER (115) in 1976 has shown, however, that 25,26-dihydroxyvitamin $D_3$ is just as active as 24,25-dihydroxyvitamin $D_3$ in bone resorption in vitamin $D_3$-calcium-deficient rats. Neither of these compounds has any effect on either calcium absorption or bone resorption in rats fed on normal calcium-containing diets.

The introduction of the 26-hydroxy function also introduces chirality at C-25, but the configuration of the naturally-occurring 25-hydroxyl derivative is not known. KODICEK (155) has synthesized 25ε,26-dihydroxyvitamin $D_3$ and has shown that while this material stimulates intestinal absorption of calcium, it has little or no effect on the healing of rickets in rats.

## 6. Clinical Aspects

The advances in our understanding of vitamin $D_3$ metabolism which have been described in the previous sections have promoted the rapid use of some of these compounds in the clinic. Both 1α-hydroxyvitamin $D_3$ and 1α,25-dihydroxyvitamin $D_3$ have been studied in several disease states, such as chronic renal failure (28, 32, 33, 51, 71, 115, 144), hypoparathyroidism (74, 159), osteomalacia (23, 87), osteoporosis (106), and neonatal hypocalcemia (47, 96).

In chronic renal disease, the kidney fails to produce enough or any of the 25-hydroxy-1α-hydroxylase enzyme that converts 25-hydroxyvitamin $D_3$ to the circulating hormonal form, 1α,25-dihydroxyvitamin $D_3$. Biochemical and bone abnormalities have been corrected with doses of 1α-hydroxy- or 1α,25-dihydroxyvitamin $D_3$ of 1 µg daily.

In osteomalacia associated with malabsorption problems and nutritional osteomalacia, small doses of 1α-hydroxyvitamin $D_3$ and 1α,25-dihydroxyvitamin $D_3$ are efficacious. Clinical response is impressive even during the first week with doses as low as 0.25 µg daily.

Hypoparathyroidism has long been recognized as a disease which is difficult to manage with large doses of vitamin D and calcium supplements. Because the full effects of vitamin D are slow in both onset and reversal, considerable time may elapse before the action of a particular dose is established or the harmful effects of inadvertent overdose are relieved. An 0.25—1.0 µg daily dose of 1α,25-dihydroxyvitamin $D_3$ corrects the hypocalcemia of parathyroid insufficiency, and hypercalcemia resulting from overdose decreases quickly upon withdrawal of therapy.

Preliminary studies in patients with biliary cirrhosis of the liver have disclosed abnormally low values of serum 25-hydroxyvitamin $D_3$ (*190*). Neither oral nor parenteral vitamin D increases levels of 25-hydroxyvitamin $D_3$ (*105*). However, it is not clear whether the low values of 25-hydroxyvitamin D result from impaired formation of 25-hydroxyvitamin D or from impaired dietary intake of vitamin D. Rosenberg (*190*) in 1976 claimed that his group had successfully treated primary biliary cirrhosis with 25-hydroxyvitamin $D_3$. The daily dose ranged from 100—200 µg; and although this is extremely high, it seems to indicate that impaired hepatic hydroxylation of vitamin D, rather than malabsorption, is the major factor in the failure of conventional vitamin D therapy in this disease. 25-Hydroxyvitamin $D_3$ has also been used by Finberg (*16*) to treat children with rickets associated with extrahepatic biliary atresia.

While the etiology of osteoporosis remains obscure, it does appear that bone loss is caused by an increase in bone resorption which after the age of 45 is most pronounced in the cortex. It is considered that bone loss is unlikely to be caused by a decrease in bone formation. Small doses of $1\alpha,25$-dihydroxyvitamin $D_3$ have been given to a group of osteoporotic patients in a long-term study in the hope that not only will intestinal calcium absorption be increased, but that bone formation will be increased in relation to bone resorption. At 0.25 µg there seems to be evidence of a state of positive calcium balance, but it is as yet too early to give any statistically significant answers. The need for long-term studies on the utility of any therapy for osteoporosis is also put forward in the criticism of Lund's (*106*) evaluation of $1\alpha$-hydroxyvitamin $D_3$ over a period of 3—4 months. It is only recently becoming apparent that therapy used to prevent bone loss or increase bone formation may indeed be leading to more rapid bone loss than would occur normally.

## II. Partial and Total Synthesis

### 1. Introduction

The isolation of the active metabolites of vitamin D from biological tissues is extremely laborious and affords only small quantities of material. Development of synthetic approaches to these materials was thus necessary to provide them and related analogs in adequate quantity for elucidation of their mode of action. The new chemistry required was that in which new functionalities, usually hydroxyl groups, could be introduced into relatively non-activated positions in the A-ring and side chain of the parent vitamin D molecule.

Direct chemical modification of this structure seemed unpromising as the seco-steroid triene system is quite unstable and prone to decomposition. The flexible nature of the A-ring and side chain also make it difficult to control or even determine the stereochemistry of groups which might be directly introduced into these portions of the molecule. For example, direct allylic oxidation of vitamin $D_3$ with selenium dioxide has resulted in a mixture of the four 1-hydroxycholecalciferols (**7—10**), both C-1 epimers with double bond geometry of $5(E)$ and $5(Z)$, being formed in a combined yield of less than 5% (*139b, 148*). Most approaches therefore utilize the more rigid structure of precursor sterols where ample precedent provides assurance that functionalities can be introduced in high yield and with stereochemical selectivity. In these approaches the generation of the vitamin D framework is left to the end of the syntheses and is performed by nature's method, *i. e.,* the photolyses and subsequent thermolysis of the required provitamin. Similarly, in those total syntheses which have been reported, the generation of the triene

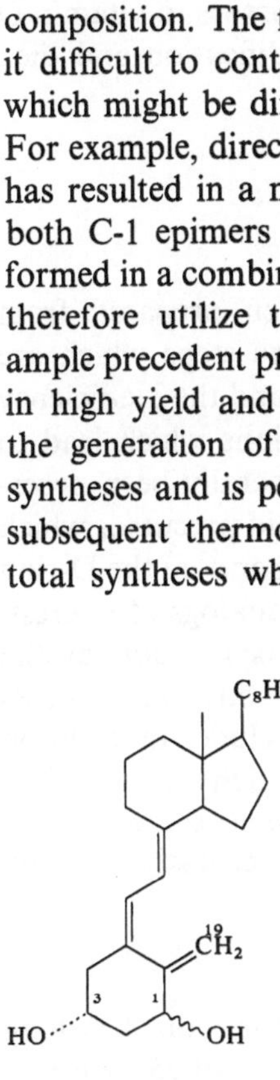

(7) 1*a*-OH
(8) 1*β*-OH

(9) 1*a*-OH
(10) 1*β*-OH

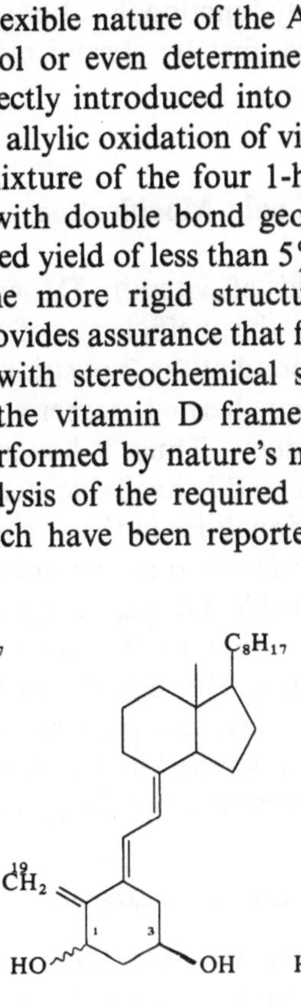

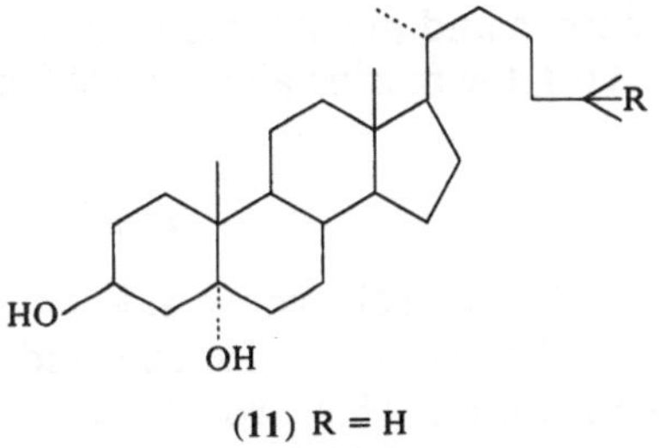

(11) R = H
(12) R = OH

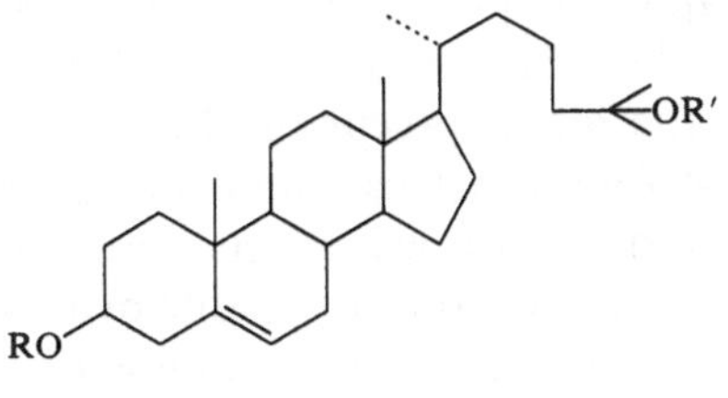

(13) R = R' = COCH₃
(14) R = R' = H

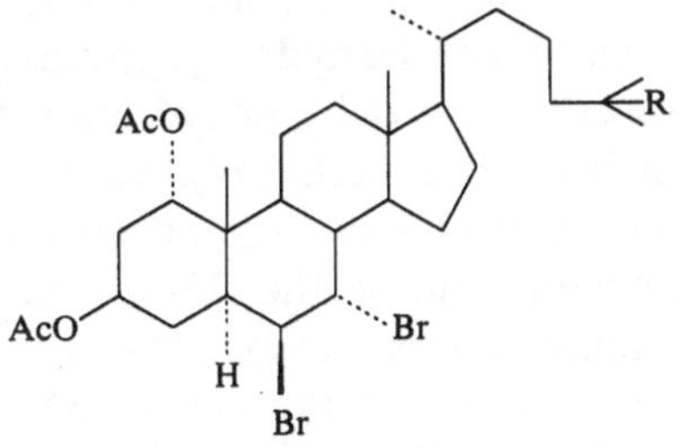

(15) R = H
(16) R = OH

system was completed only after the required functionalities had been attached to the precursor fragments. The topic has been discussed in several reviews including a chapter in an earlier volume of this series (*60, 82, 168*). The discussion below will describe methods which have been used to prepare the properly functionalized precursors to vitamin D and its metabolites and analogs, first by steroid modification and then by total synthesis.

## 2. Side Chain Modification

In general the overall yield of vitamin $D_3$ when prepared from cholesterol is quite low due to the inefficiency of the steps which are required to introduce a double bond at the 7-position and the inefficiency of the subsequent photolytic and thermal isomerizations which lead to significant quantities of by-products. Thus, it has generally been necessary to start with a quantity of a readily available steroid precursor which already possesses requisite functionalities in the A-ring and on the D-ring or side chain for further elaboration into the vitamin analogs of interest. Invariably these precursor steroids possess a hydroxyl group at the 3-position useful for the elaboration of the A-ring, but may possess differently constituted side chains. The precursors include, in addition to cholesterol, the mammalian hormone precursors, dehydroisoandrosterone and pregnenolone, as well as the bile acids, the yeast sterol ergosterol, the marine algae sterol fucosterol and the plant sterol stigmasterol.

### a) 25-Hydroxycholesterols

Cholesterol, whose side chain has no activating function, has long been known to oxidize on standing to give small amounts of 25-hydroxycholesterol. This process can be enhanced as Mazur has found that irradiation of 5α-hydroxycholestanol (**11**) in peracetic acid gives in 50% conversion newly hydroxylated products of which the corresponding 25-hydroxy compound (**12**) is the major component (*157*). Selective acylation of the hydroxyl groups at C-3 and C-25 allowed the C-5 hydroxyl to be removed by dehydration to give the diacyl precursor (**13**) of 25-hydroxycholesterol (**14**). Mazur has also found that passing ozone over the saturated cholestane (**15**) adsorbed on silica gel effects selective introduction of the 25-hydroxyl group to give the 1α,25-dihydroxyvitamin $D_3$ precursor (**16**) in fair yield (51%) but only in modest preparative conversion (11%) (*35*). As such these methods represent convenient short cuts for obtaining small quantities of these valuable intermediates.

Construction of a cholesterol side chain from a C-19 steroid such as dehydroepiandrosterone (17) requires that the adjacent centers, C-17 and C-20, be introduced stereoselectively to give material of natural configuration. Fortunately, the β-orientation of groups at C-17 is favored and hydrogenation of the Δ17,20 double bond affords the natural orientation at this center. However, control of stereochemistry at C-20 is more difficult to manage as hydrogenation of the more easily obtained Δ17,20-E double bond gives predominantly the unnatural isomer and thermodynamics do not favor the desired stereochemistry. To circumvent these difficulties WICHA and BAL (192, 193) have prepared ester (19) by Reformatsky reaction of ethyl bromoacetate with dehydroepiandrosterone acetate (18) followed by dehydration, hydrogenation and protecting group exchange. Alkylation of (19) with the appropriate bromoketal afforded only one isomer, (20), which on LiAlH₄ reduction, tosylation and further hydride reduction afforded the ketal (21) which has the appropriate stereochemistry at C-17 and C-20. Further reactions gave 25-hydroxycholesterol in 42% overall yield from dehydroepiandrosterone acetate.

In studies which may have a bearing on the preparation of other functionalized side chain analogs, TROST and VERHEOVEN have shown

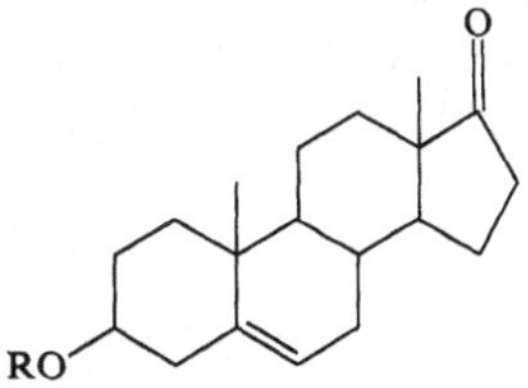

(17) R = H
(18) R = COCH₃

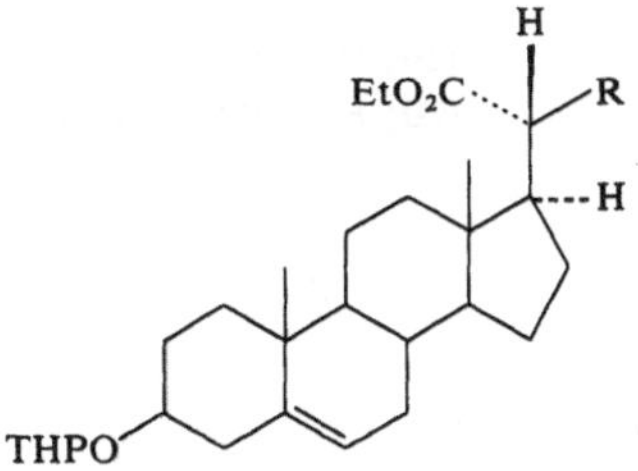

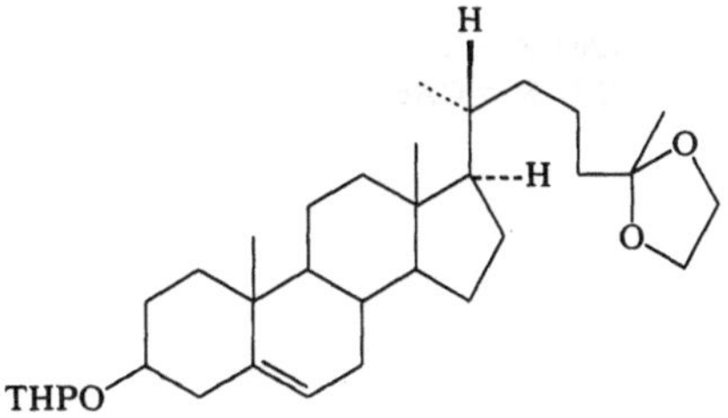

(19) R = H
(20) R = (CH₂)₃−C−CH₃

(21)

that a C-17 ketone (22) can be converted in several steps in high overall yield to the corresponding steroid (25) containing the natural configuration at C-20 (187). The key step in these transformations is the palladium catalyzed displacement of the allylic acetate (24) by malonate anion with retention of configuration.

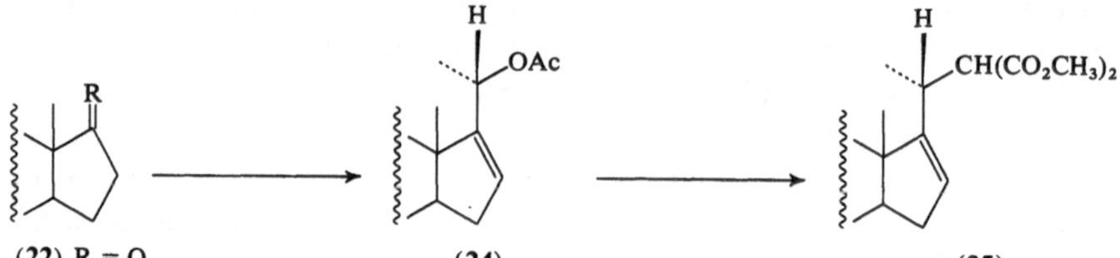

(22) R = O          (24)                    (25)
(23) R = CHCH₃

Conversion of pregnenolone, (26), to 25-hydroxycholesterol has been described by chemists from Hoffman-LaRoche (126). Reaction of the acetate (27) with vinyl magnesium chloride gave a single allylic alcohol which reacted with acetoacetic ester or diketene under conditions of the Carroll rearrangement to give a mixture of the *cis* and *trans* ketones (28) and (29). Regioselective hydrogenation of the Δ20,22 double bond resulted in a mixture of C-20 isomers with the natural 20(R) isomer (31) predominating. Each pure geometric isomer was reduced with poor selectivity. Reaction of (31) with methyl magnesium chloride gave 25-hydroxycholesterol in a yield of about 25% from (27). The ketal (30) has

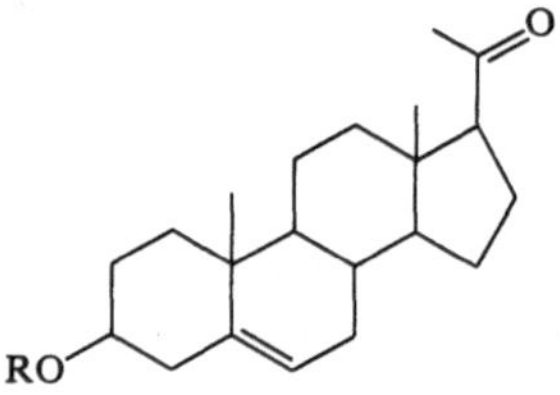

(26) R = H          (28) 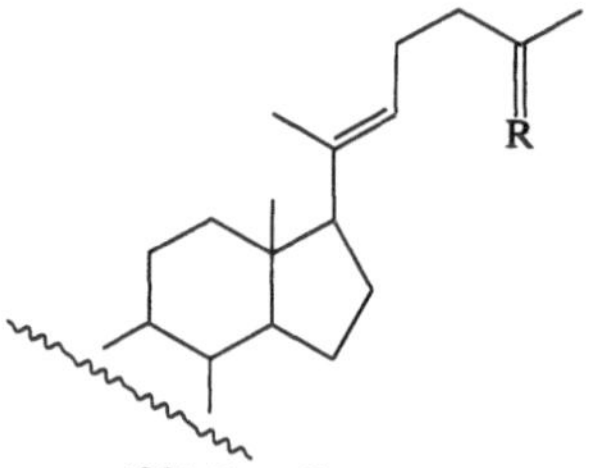        (29) R = O
(27) R = OAc                    (30) R = ethylenedioxy

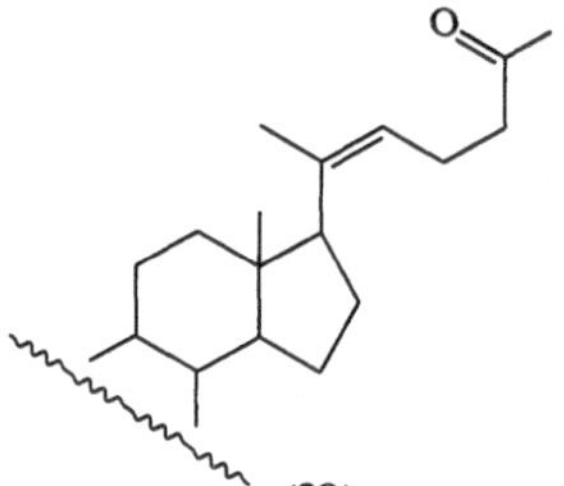

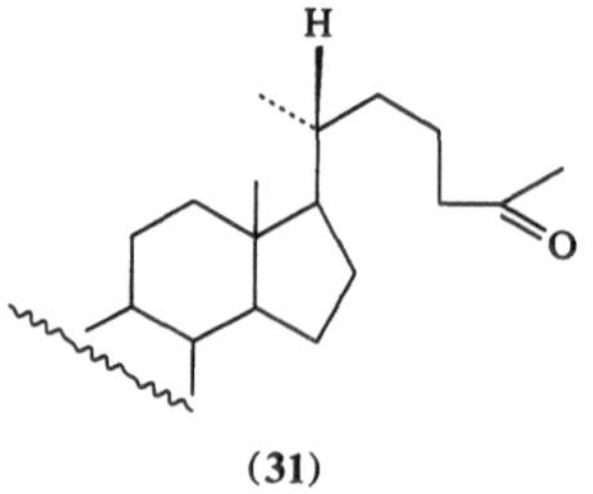

(31)

been obtained *via* a Wittig reaction on pregnenolone with high geometrical purity. Hydrogenation of the ketone (29) prepared from this material gave the product with the desired 20($R$) stereochemistry in 90% yield (*111, 189*). Substitution of an ethyl group at C-24 results in an increased percentage of the 20($S$) isomer being formed in this hydrogenation (*112*).

The bile acid derivative, 3β-acetoxychol-5-en-24-oic acid (32), has been converted in two steps in high yield to 25-hydroxycholesterol (*80*). The homocholenic acid derivative (33) prepared by a photochemical Wolff rearrangement in this transformation served as a key early intermediate in the first chemical synthesis of 1α,25-dihydroxycholecalciferol (*171*). Lithocholic acid, likewise, has been converted to an intermediate (16) used for the preparation of the dihydroxyvitamin (35). Hyodeoxycholic acid (34) also shows promise of being a desirable starting material for 25-hydroxycholesterol (*132*).

Fucosterol (35), a sterol which is potentially available in large quantity from the harvest of kelp, has been used by Japanese chemists as an entry into the preparation of side chain hydroxylated vitamin D$_3$ derivatives (*120, 121, 181*). The acetate of fucosterol (36) may be

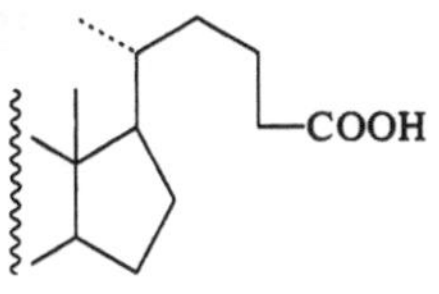

(33)

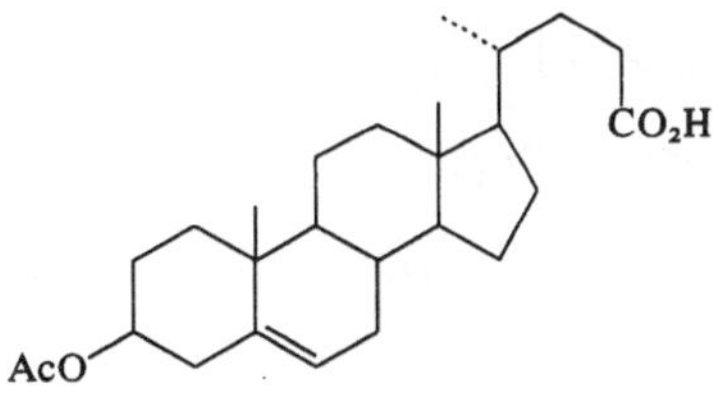

(32)

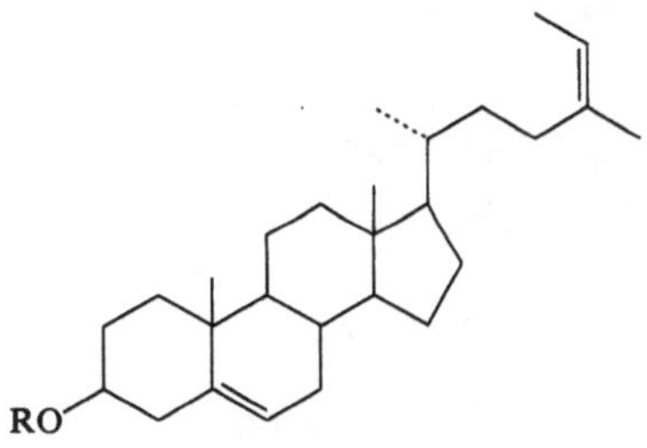

(35) R = H
(36) R = Ac

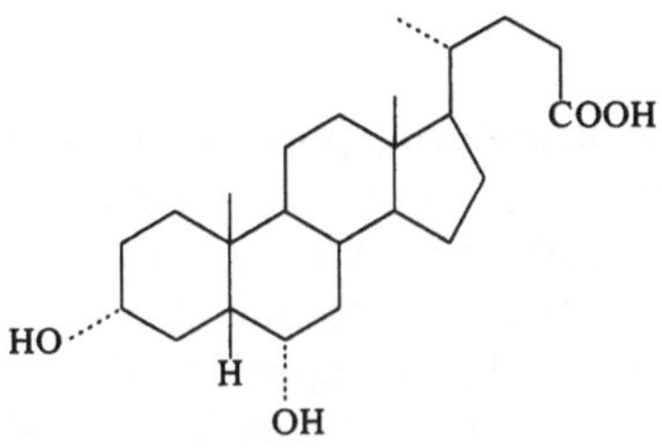

(34)

selectively ozonized or epoxidized at the Δ24,28 double bond to give respectively, 24-keto- (37) or 24,28-epoxyfucosterol (38). Reduction of (37) with sodium borohydride followed by dehydration with phosphorus pentoxide afforded desmosterol acetate (39) in 85% overall yield from (36). Treatment of the epoxide (38) with Lewis acids gave (39) directly, albeit in lower yield. Oxymercuration-demercuration of (39) afforded 25-hydroxycholesterol (14) in high yield. Other less efficient methods of effecting the conversion of desmosterol acetate (39) to (14) include formation of the 24,25-epoxide with subsequent lithium aluminum hydride reduction or by photochemical oxidation to give 3,25-dihydroxy-cholesta-5,23-diene-3-acetate (42) followed by selective hydrogenation of the side chain double bond.

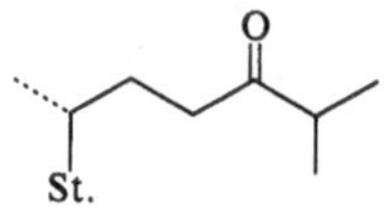

(37)

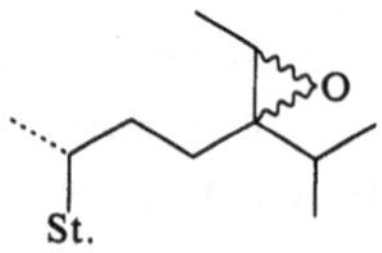

(38)

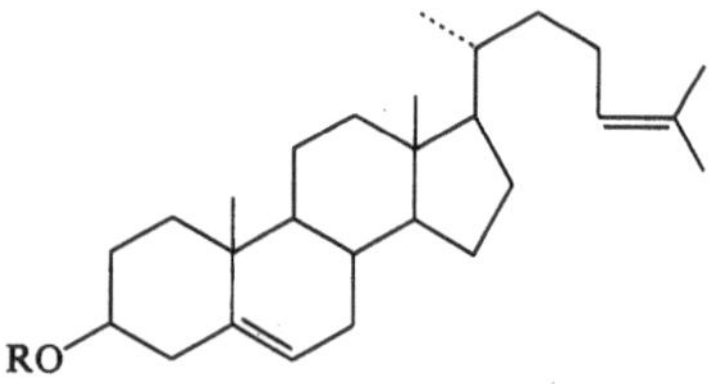

(39) R = Ac
(40) R = H
(41) R = C$_6$H$_5$CO

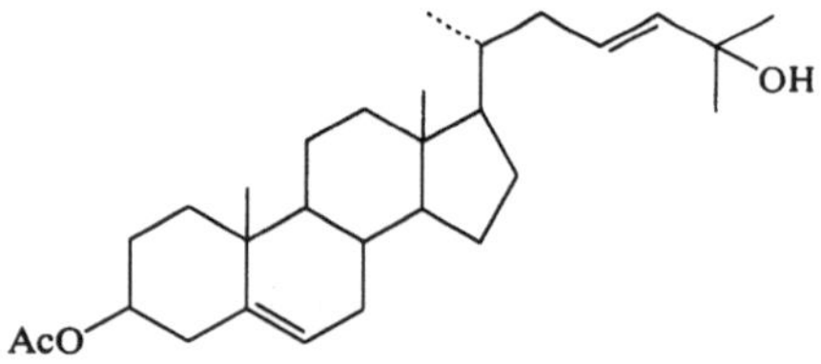

(42)

It would seem that ergosterol, provitamin D$_2$ (1a), would serve as an ideal precursor for side chain oxygenated vitamin D derivatives because of its ready availability from yeast and its preexisting 5,7-diene system. However, the reactivity of the diene moiety in its free or protected form *(vide infra)* has complicated the selective manipulation of the side chain. EYLEY and WILLIAMS have studied the reactions of the aldehyde (43) obtained in modest yield by selective ozonization of the corresponding ring-B protected ergosterol acetate (*53, 54*). This aldehyde reacts with 4-methyl-4-butenyl magnesium chloride to give the C-22 alcohol (**44**). Elimination of the C-22 hydroxyl group by sodium borohydride reduction of the corresponding mesylate in dimethyl sulfoxide gave the desired (**45**) in 79% yield overall from the aldehyde (**43**). The hydration of (**45**) by the process of oxymercuration-demercuration was followed by lithium hydride deblocking of the diene protecting group to give 25-hydroxy provitamin D₃ (**48**). The aldehyde (**43**) reacts with an enolate or with Wittig reagents to give the corresponding C-22 olefins. Selective catalytic hydrogenation of this double bond could not be carried out due to the more facile reduction of the ring-B olefin. However, the aldol product (**46**) with sodium borohydride in pyridine gave the corresponding side chain saturated diol (**47**). B-Ring deblocking afforded the corresponding hydroxylated provitamin (**49**).

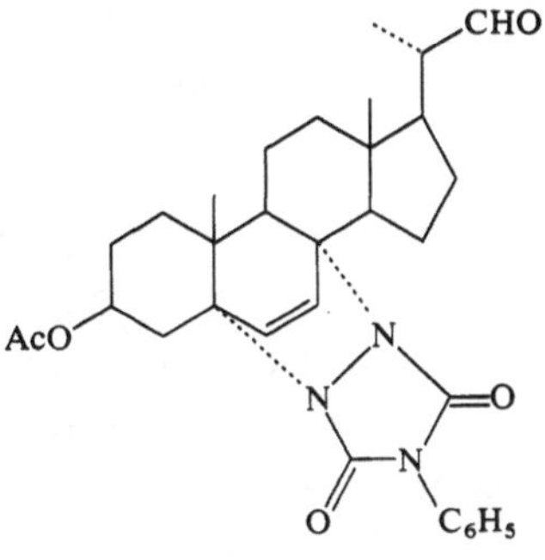

(43)

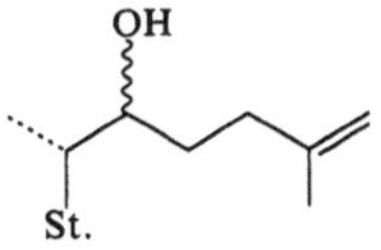

(44)

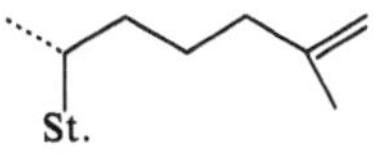

(45)

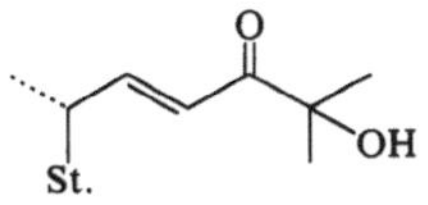

(46)

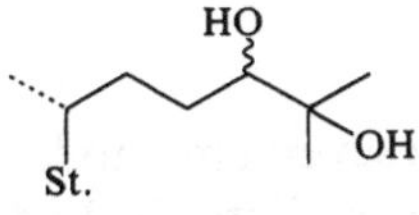

(47)

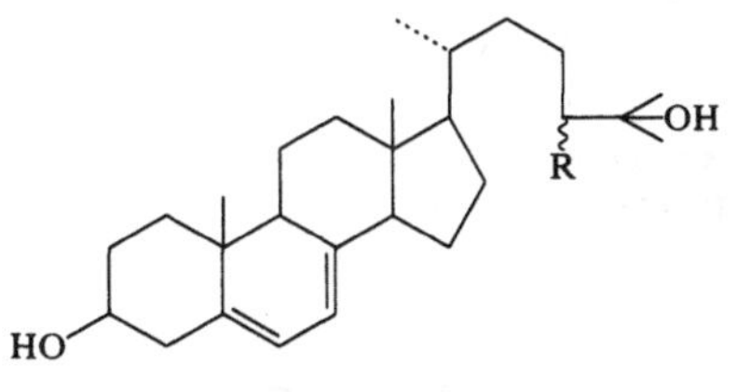

(48) R = H
(49) R = OH

From the commercial point of view, the cheap and readily available plant sterol, stigmasterol (**50**), presently appears to be the most desirable starting material for the preparation of vitamin D metabolites. Since the two double bonds of stigmasterol exhibit nearly equal reactivity toward oxidizing agents it has been necessary to first protect the 5,6-double

bond by conversion of stigmasterol to the 3,5-cyclo-6β-methoxy derivative (51). Ozonization of (51) results in the aldehyde (52) which the
Hoffman-LaRoche group has converted in two steps to the tosylate (53)
(141). This material which is very prone to displacement reaction with
halide ions was treated with the halide-free lithium reagent (56) in
dioxane to give a 90% yield of the cholstyne (57). Hydrogenation
followed by reversal of the protecting groups resulted in 25-hydroxy-
cholesterol in greater than 30% yield from the starting (50).

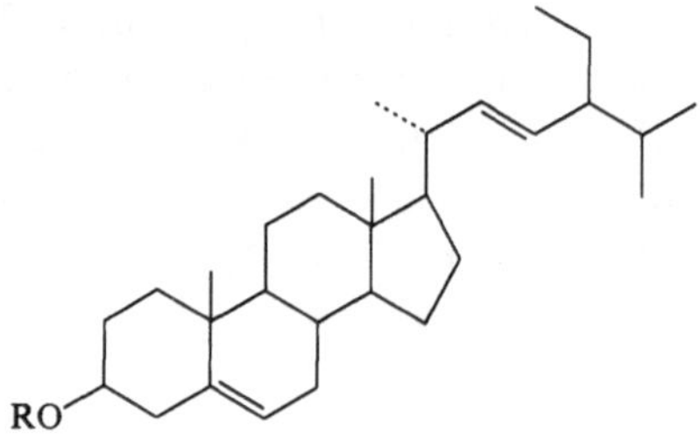

(50) R = H

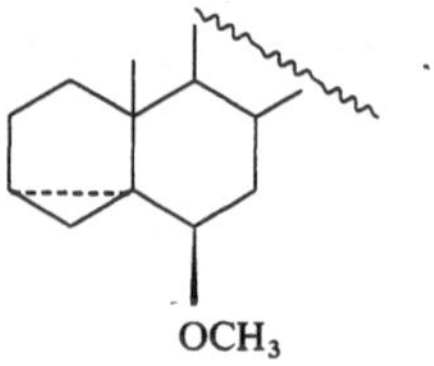

(51)

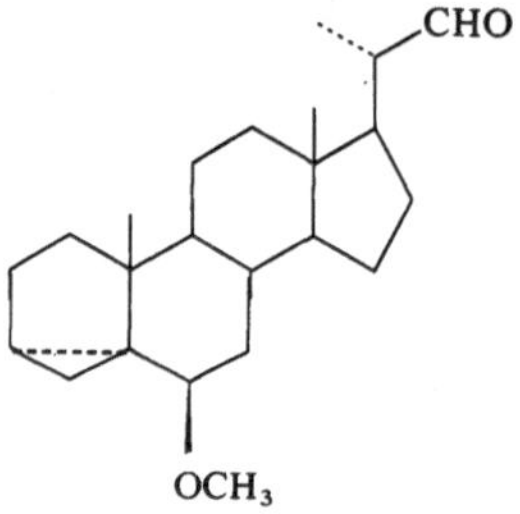

(52)

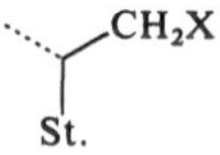

CH₂X

St.

(53) X = O₃SC₆H₄CH₃
(54) X = Br
(55) X = I

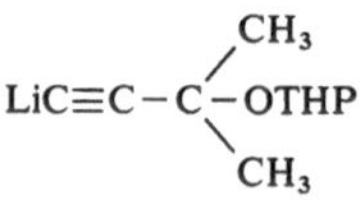

$LiC{\equiv}C-C-OTHP$ with CH₃ groups

(56)

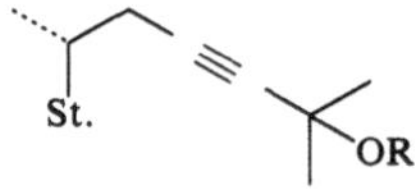

(57) R = THP
(58) R = H

Salmond of the Upjohn Company has reported several methods for the
preparation of 25-hydroxycholesterol from the intermediate aldehyde (52).
In a most efficient process the aldehyde was condensed with the Wittig
reagent (59) obtained in three easy steps from isoprene. The product
diene (61) could be selectively oxidized at the 24,25-double bond to give
epoxide (62) as a mixture epimeric at C-24. Catalytic hydrogenation
concomitantly reduced the 22,23-double bond and the epoxide to give
the 25-hydroxylated cholesterol side chain. Reversal of the i-steroid ether

resulted in the preparation of 25-hydroxycholesterol or its acetate in greater than 50% yield from stigmasterol (165). The epoxide (62) when reacted with methyl magnesium chloride gave a mixture of 24(R)- and 24(S)-methyl-25-hydroxy compounds. The 24(R)-isomer has been transformed into 25-hydroxyvitamin D₂.

The reaction of methylene triphenylphosphorane with isobutylene oxide gave the oxaphosphorane (60). This material in a Wittig reaction with the aldehyde (52) afforded the unsaturated 25-hydroxy side chain (63) as an intermediate in the most direct synthesis of 25-hydroxycholesterol from stigmasterol yet described (162). In another approach the aldehyde (52) was converted to the dichlorovinyl derivative (64) by reaction with the modified Wittig reagent (65). Reaction of (64) with butyl lithium led to an intermediate lithium acetylide which reacted with isobutylene epoxide in the presence of hexamethylphosphoric triamide to give the 25-hydroxycholestyne (66). Hydrogenation and A-ring reversal led to (14) essentially as described above for (58) (166). The halides (54) and (55) have been reacted with 2-lithiodithiane to give the intermediate (67) which, as its conjugate base, condensed with isobutylene oxide to give the 25-hydroxycompound (68). The dithiane residue was removed with lithium aluminum hydride-titanium tetrachloride to give

$$(C_6H_5)_3P=CH-CH=C(CH_3)_2$$

(59)

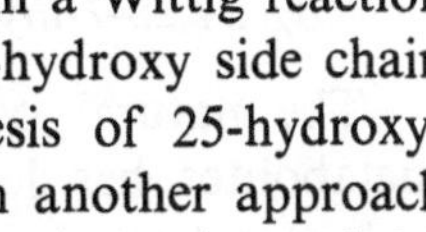

$$(C_6H_5)_3P \overset{CH_2-CH_2}{\underset{O}{<}} C(CH_3)_2$$

(60)

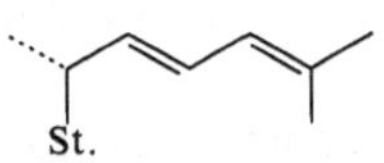

(61)

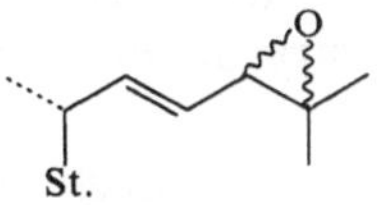

(62)

(63)

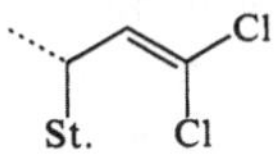

(64)

$$[(CH_3)_2N]_3PCCl_2$$

(65)

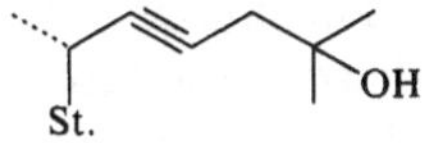

(66)

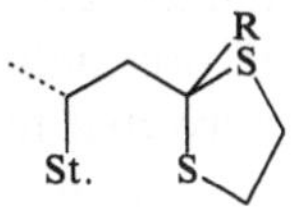

(67) R = H

$$(68)\ R = CH_2C\overset{OH}{\underset{(CH_3)_2}{<}}$$

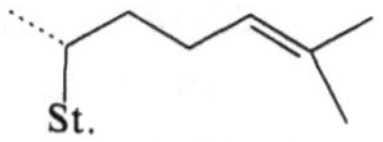

(69)

the desired 25-hydroxycholesterol side chain (*164*). The iodide (**55**) on reaction with bis-π-dimethyl-allyl nickel dibromide affords the C-24,25-olefin (**69**) which could serve as a precursor for the 25-hydroxy derivatives as described above for desmosterol (*38*).

### b) 24,25-Dihydroxycholesterol and 25,26-Dihydroxycholesterol

The natural occurrence of 24,25- and 25,26-dihydroxyvitamin D metabolites has led to studies of their synthesis and configurational assignment *via* the initial preparation of the correspondingly substituted cholesterol derivatives. Initial approaches to these compounds were carried out on oxidized forms of Δ24- or Δ25-steroids (*98, 142, 154, 155, 169, 170*) or by reduction of the ketol (**46**) (*53*). In these cases, almost equal amounts of the diasteroisomeric epoxides or diols were obtained because the newly formed asymmetric position was too distant from a chiral center (*i. e.*, at C-17 or C-20) for stereoselective induction.

Groups from Japan (*169*) and Hoffman-LaRoche (*142*) independently separated the diastereoisomeric C-24,25-epoxides obtained from the desmosterol derivatives (**41**) and (**69**), respectively. Acidic methanolysis of each epoxide from (**41**) gave the corresponding 24-hydroxy-25-methoxy steroid with retention of C-24 stereochemistry. The absolute stereochemistry at C-24 was then established by measuring the relative rate of reaction of the alcohol with the enantiomers · of (±) α-phenylbutyric anhydride (Horeau's method) (*169*). Alternatively, the diol obtained from each epoxide of (**69**) gave strong, split Cotton effects in the presence of the lanthanide complexing agent, Eu(fod)$_3$, which permitted the assignment of stereochemistry. Finally, X-ray analysis of a 24-hydroxy intermediate with the 24(R)-configuration has been performed thus proving conclusively the stereochemistry of this position in the vitamin D$_3$ metabolites (*142*).

The acetylenic side chain of (**58**) has been hydrogenated in the presence of Lindler catalyst to give the Z-allylic alcohol (**70**). This olefin was stereoselectively epoxidized at −78° with *t*-butyl hydroperoxide and a catalytic amount of vanadyl acetoacetate to give an 85 : 15 mixture of products (**71**) and (**72**). The major isomer (**71**) was separated and converted to 24(R),25-dihydroxycholesterol (**73**) by the subsequent steps of lithium aluminum hydride reduction and A-ring deprotection (*143*). These steps provide the means for selectively preparing 24,25-dihydroxy-vitamin D compounds with the natural configuration at the 24-position.

No method of stereoselectively preparing either isomer of the 25ε,26-dihydroxy side chain has been found. Only recently has separation of the isomers of 5-cholesten-3β,25,26-diacetate (**74**) been accomplished (*156*).

These isomers were converted separately into the corresponding vitamins D₃ but no correlation was made with natural material.

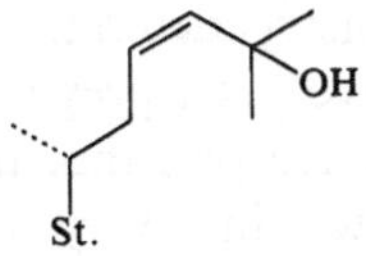

(70)

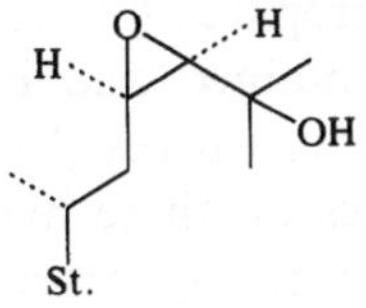

(71)

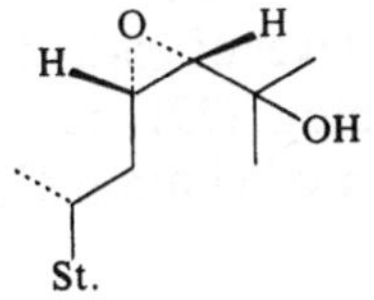

(72)

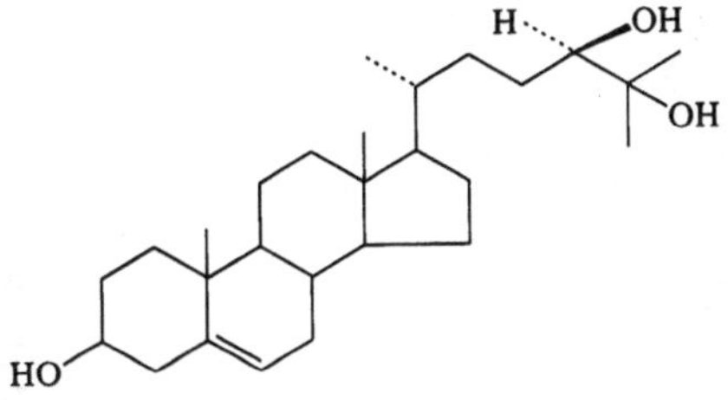

(73)

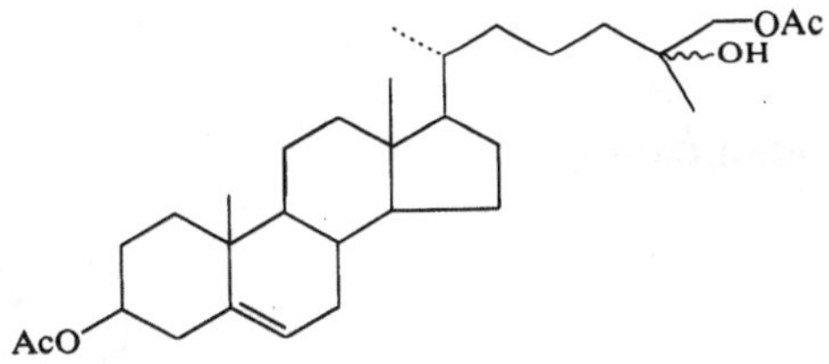

(74)

## c) Side Chain Analogs of Vitamin D

In an effort to elucidate the structural requirements for biological activity of vitamin D, several groups have modified the size and/or substitution pattern of the side chain.

Pregnenolone has been converted to the hydroxylated penta- and hexanor vitamins (75) and (76) (75). Similarly, standard conversions of various 3β-hydroxychol-5-enic acids have afforded the nor-, dinor-, trinor-, and homo-25-hydroxyvitamins (79) through (82), respectively (85). The 27-nor- and 26,27-dinor-25-hydroxyvitamins (83) and (84) have also been prepared (20, 75). Each isomer of 24-hydroxyvitamin $D_3$ (85) has been obtained and its stereochemistry determined (95, 119). Dihydroxylated derivatives (77) and (78) have been prepared (20). It is clear from studies on these hydroxylated derivatives that the structural features necessary for agonistic activity fall within a very narrow range. Only (82) and (85) [24(R)-isomer only] approach vitamin $D_3$ in their ability to enhance calcium absorption. The mono and dinor compounds (83) and (84) show activity at levels that are 10 to 100 times higher than an effective level of 25-hydroxyvitamin $D_3$ (75).

The nor analog (79) has been found to antagonize the calcium mobilizing action of vitamin $D_3$, but not that of 25-hydroxyvitamin $D_3$,

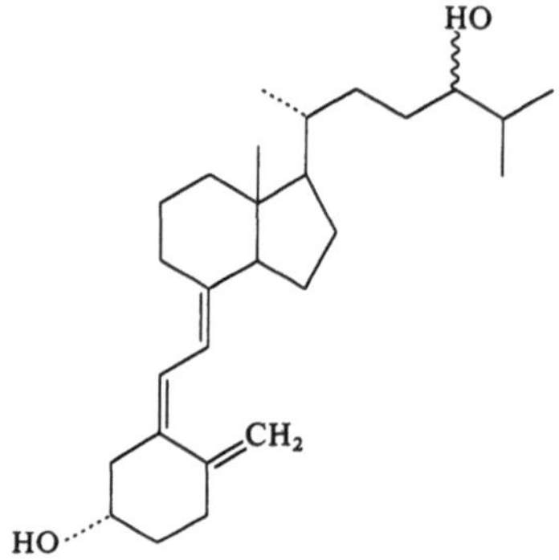

(85)

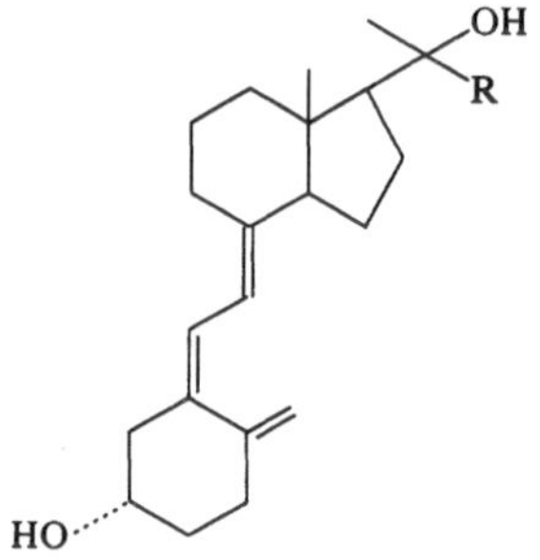

(75) R = H
(76) R = CH₃
(77) R = (CH₂)₃C(CH₃)₂OH
(78) R = (CH₂)₃CHOHCH₃

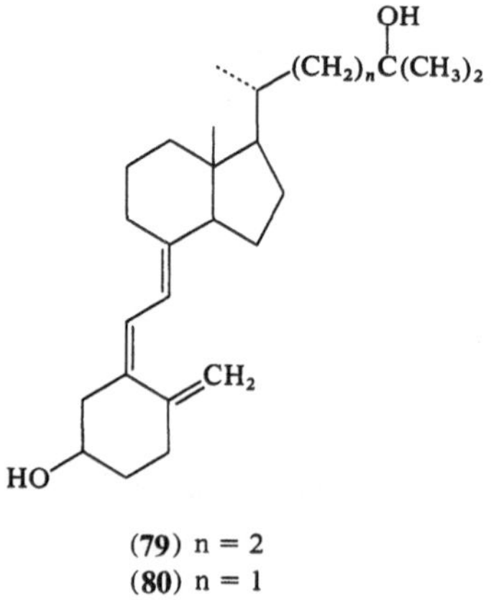

(79) n = 2
(80) n = 1
(81) n = 0
(82) n = 4

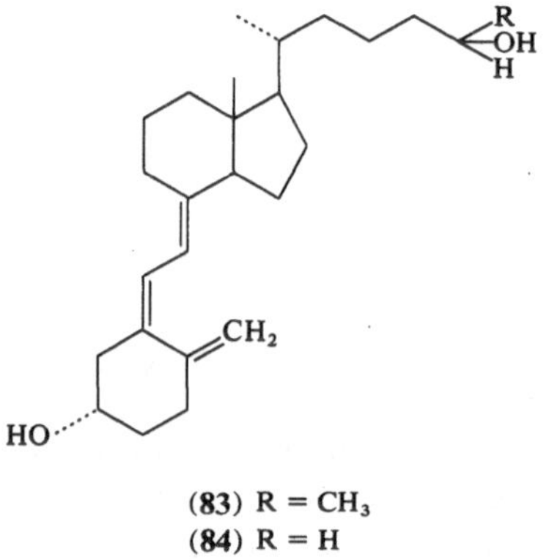

(83) R = CH₃
(84) R = H

seemingly indicating that interference with the liver 25-hydroxylating enzyme was taking place (*86*). In this vein the 24-dehydro-, 25-dehydro-, 25-fluoro-, and 25-aza-analogs of vitamin $D_3$ were prepared to determine if direct inhibition of the enzyme could be effected by modification of the substrate site (*138, 200*).

## 3. Ring Modified Sterols

### a) 1α-Hydroxycholesterols

The standard method of introducing a functionality at the 1α-position of steroids is to prepare a $\Delta^1$-3-ketone and then to carry out an appropriate Michael-type addition reaction which gives the adduct of proper 1α-stereochemistry. One complication of this method when working toward the preparation of $\Delta^5$-sterols is that the usual conditions of the conjugate addition to the unsaturated ketone are sufficient to equilibrate the $\Delta^5$-double bond to the more stable $\Delta^4$-3-ketone. Conversely, conditions for deconjugation of the resulting $\Delta^4$-3-ketones are unsuitable for retaining substituents at C-1 which can be eliminated by β-elimination. The methods outlined below are adaptations which have been devised to circumvent these difficulties.

Early syntheses utilized C-6 oxygenated sterols as intermediates where the elimination of a C-6 substituent served as a method of generating a $\Delta^5$-double bond after incorporation of the required 1α-functionality (*59, 78, 117, 122, 125, 149, 158, 171*). The C-6 position of $\Delta^5$-sterols can be oxidized to a ketone by a sequence of reactions involving successively nitration, reduction and hydrolysis (*5*); or by hydroboration and a two-stage oxidation (*125*). Hydride reduction of the C-6 ketone then gives the axial 6β-hydroxyl group which can be selectively dehydrated at the appropriate time to give the desired $\Delta^5$ olefin (Scheme I). In this way, the C-6 substituted 3-ketones (**86**) have been prepared. Bromination and dehydrohalogenation gave the corresponding unsaturated ketones. Epoxidation with hydrogen peroxide in alkaline medium cleanly converted these compounds into their 1α,2α-epoxyderivatives (**87**). Simultaneous epoxide opening and ketone reduction with lithium aluminum hydride was not stereoselective at C-3 as a large percentage of the undesired 3α-isomer was formed. More selective formation of the 3β-hydroxyl group was attained by sodium borohydride reduction either prior to epoxide opening or after its conversion to a 1α-hydroxy-3-ketone (**88**). The use of appropriate protecting groups at the 1α- and 3β-positions and/or leaving groups at C-6 then permitted the preparation of the corresponding 1α-hydroxycholesterol (89) and its analogs. If a C-25 hydroxyl group is to be carried through this sequence of reactions, it apparently needs a protecting group to survive dehydrative steps (*125, 171*) or, if lost, it can be reintroduced by oxymercuration-demercuration of the diacetate of the resulting 1α-hydroxydehydrocholesterol (*158*).

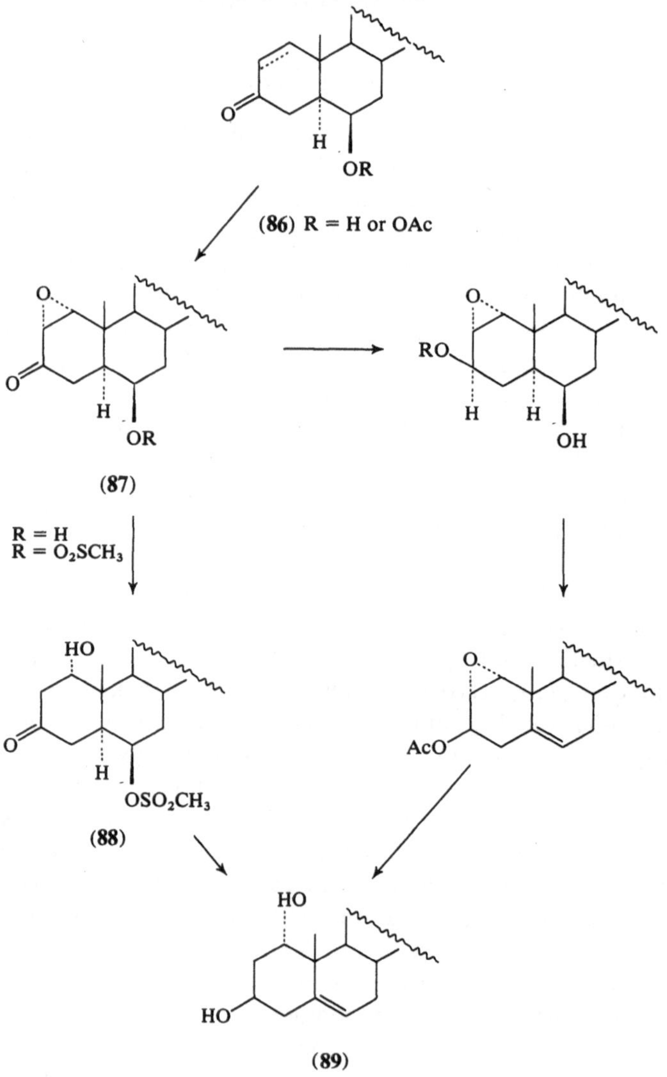

Scheme 1

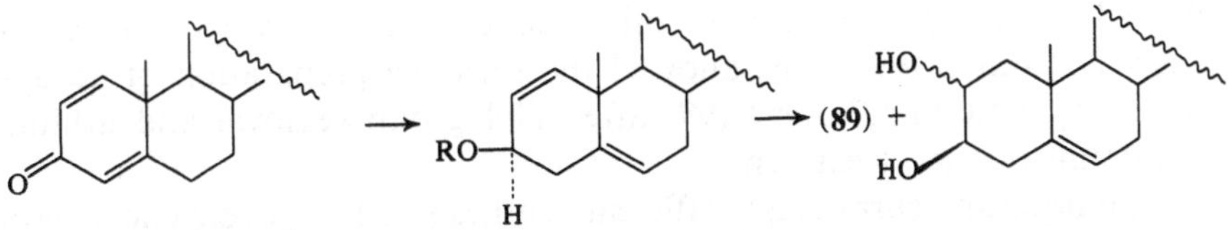

(90)

(91) R = H
(92) R = Ac

(93) 2α-hydroxy
(94) 2β-hydroxy

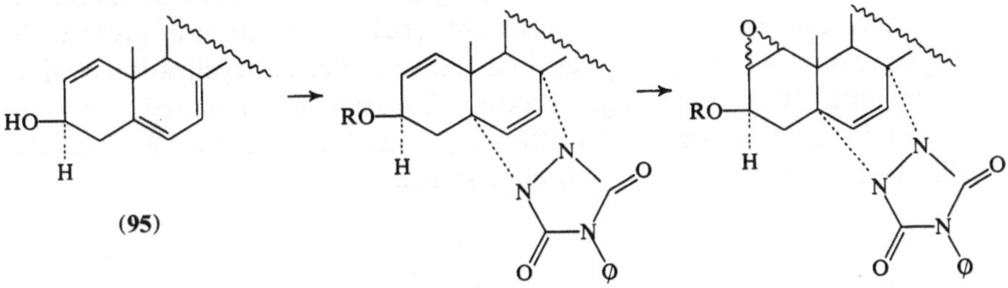

(89) +

(95)

(96) R = H
(97) R = tBu(CH₃)₂Si

(98)

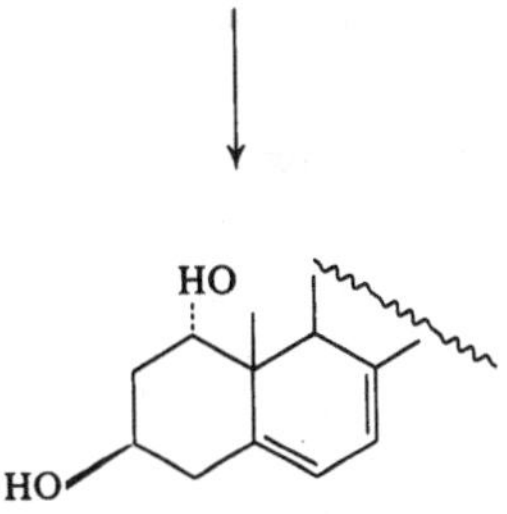

(99)

A shorter but less selective route to 1α-hydroxycholesterol has been described by KANEKO (90). Cholesta-1,4-dien-3-one (90) was deconjugated to its $\Delta^{1,5}$-isomer and then reduced with borohydride to give 3β-hydroxy-cholesta-1,5-diene (91). Selective hydroboration followed by alkaline peroxide treatment afforded a 30% yield of a separable 1 : 1 mixture of 1α- and 2α-hydroxycholesterol (89) and (93). If the acetate (92) was hydrated by an oxymercuration-demercuration procedure, a mixture of the 2β-hydroxycholesterol (94) (14%) and 1α-hydroxycholesterol (89) (26%) was isolated after saponification (118). When starting with the readily available cholesta-1,4,6-trien-3-one, the deconjugation and reduction afforded 3β-hydroxycholesta-1,5,7-triene (95) (89). The 5,7-diene system was then protected as its 4-phenyl-1,2,4-triazoline-3,5-dione (PTD) adduct (96). Treatment of (96) with m-chloroperbenzoic acid gave a 5 : 3 mixture of the 1β,2β- and 1α,2α-epoxides (98). More recently, WHALLEY and his co-workers have found that if the dimethyl-tbutylsilyl ether (97) is used in this reaction, the 1α,2α-epoxide is exclusively formed (49). This allowed the selective preparation of 1α,3β-dihydroxycholesta-5,7-diene (99) after silyl group removal and lithium aluminum hydride treatment.

A unique and surprisingly efficient synthesis of 1α-hydroxycholesterol has been described by Mihailovic and co-workers (114). The 5α-hydroxycholestanyl acetate (100), obtainable in high yield from cholesterol, was converted by irradiation with visible light in the presence of mercuric oxide and iodine into the 5,10-secosterol (101). Irradiation of (101) with a high pressure mercury lamp resulted in intramolecular cyclization to give the oxetane (102) as the major product. Treatment of this material with hydroiodic acid in acetic acid afforded 1α-hydroxycholesterol 3-acetate (103) in 20% overall yield from cholesterol.

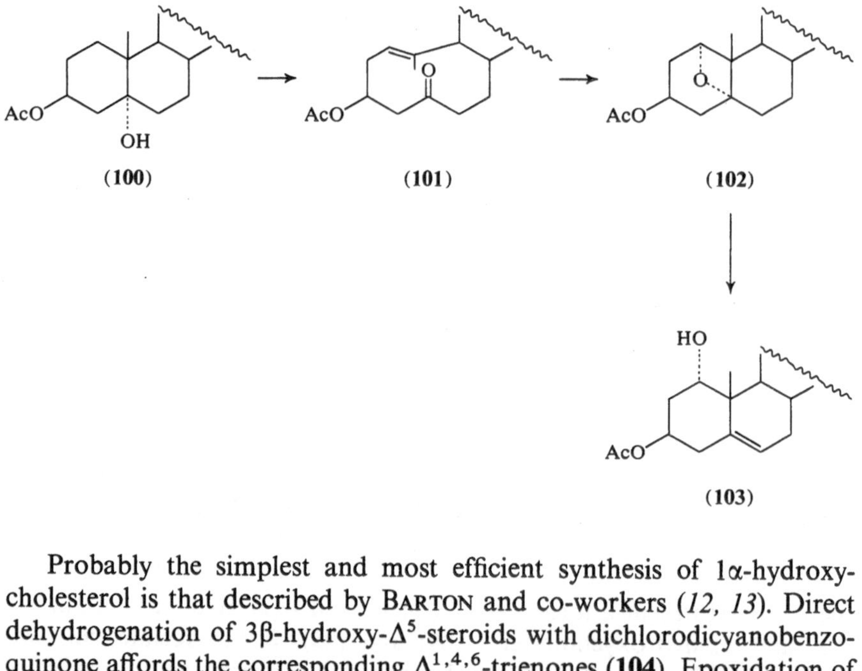

(100)          (101)          (102)

(103)

Probably the simplest and most efficient synthesis of 1α-hydroxycholesterol is that described by Barton and co-workers (12, 13). Direct dehydrogenation of 3β-hydroxy-Δ⁵-steroids with dichlorodicyanobenzoquinone affords the corresponding $\Delta^{1,4,6}$-trienones (104). Epoxidation of (104) with basic hydrogen peroxide gives the 1α,2α-epoxide (105) in high yield. The epoxide (105) derived from egosterol (99), cholesterol or its 25-hydroxy derivative upon treatment with a large excess each of lithium and ammonium chloride in ammonia-tetrahydrofuran gave the desired 1α,3β-dihydroxy-Δ⁵-sterol (89) in a single step in high yield. Apparently what is happening in this reaction is that the epoxide is reductively opened, then reduction of the 4,6-dienone occurs to give an intermediate 3-keto-Δ⁵-system which is further reduced to the 3β-hydroxy-$\Delta^5$-steroid before double bond equilibration can occur. Some care must be employed, however, as MAZUR, in carrying out this reaction under somewhat modified conditions, found that 1$\alpha$,3$\beta$-dihydroxy-5$\alpha$-cholest-6-ene (106) was the major product (45%) accompanied by lesser amounts of (89) and cholesterol (57). It was suggested that the intermediate 1$\alpha$-hydroxy group promotes hydrogen addition at C-5 in the reduction of the diene system under these conditions thereby favoring formation of the $\Delta^6$-unsaturated isomer. This material has also served as a useful intermediate in the preparation of 1$\alpha$-hydroxyvitamin D.

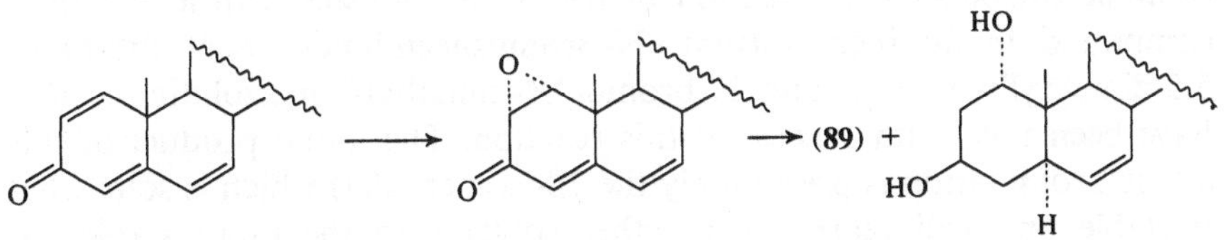

(104)          (105)          → (89) +          (106)

### b) Other A-Ring Modifications

The C-3 position has been quite extensively altered. Early work led to the preparation of 3$\alpha$-methylprovitamin D$_3$, 7-dehydrothiocholesterol, methyl 7-dehydrocholesteryl-3-carboxylate and 3-hydroxymethylcholesta-5,7-diene (179). More recently, 3-epivitamin D$_3$ (1) and the 3$\beta$-fluoro analog of vitamin D$_3$ (199) have been prepared. LAM and OKAMURA and their co-workers have prepared 1$\alpha$-hydroxy-3-deoxyvitamin D$_3$ derivatives by several routes and have found that the C-3 hydroxy group is not required for biological activity when a 1$\alpha$-hydroxyl substituent is present (99, 135, 137). Similarly, replacement of the C-3 hydroxyl group in these compounds by a methoxyl (97), 3$\alpha$-methyl (134) or 3$\alpha$-hydroxy (136) group does not destroy their vitamin action. The preparation of 4$\alpha$-hydroxyvitamin D$_3$ from 4$\alpha$-hydroxycholesterol diacetate has been reported but attempts to prepare the isomeric 4$\beta$-hydroxy compound failed when the diene system of the provitamin could not be generated (147).

The preparation of 4,4-dimethyl vitamin D$_3$ has been described (131). The 1$\alpha$-hydroxy-4,4-dimethylprovitamins D$_2$ and D$_3$ have recently been prepared (4, 30). The presence of the geminal methyl groups in these latter compounds caused the hydride reduction of the intermediate epoxyketone to give mainly the 3$\alpha$-hydroxy epimer. The alternative 1$\alpha$-hydroxy-3-ketone when reduced with lithium aluminum hydride gave a more favorable (1 : 1) ratio of 3$\beta$- to 3$\alpha$-hydroxyl group formation.

## c) Reactions in the B-Ring of Sterols

The 5,7-diene system of the provitamin is a relatively unstable moiety and, because of this, it must be generated at a late stage in the preparation of vitamin Ds from sterols or be suitably protected to survive chemical manipulations elsewhere in the molecule. Except for ergosterol the 5,7-diene system (109) has been generated from precursors having one double bond in the B-ring. The most common precursor is a $\Delta^5$ sterol ester which by allylic activation is converted into a C-7 substituted derivative. The C-7 substituent utilized most often is bromine introduced by reaction of the $\Delta^5$ sterol ester with a N-bromo compound in an inert solvent. N-Bromosuccinimide, N,N'-dibromo-5,5-dimethylhydantoin and N-bromo-4,4-dimethyl-2-oxazolidinone (88) have been used with success in this reaction. The initial product of this allylic bromination is presumably the 7β-isomer (107) which is somewhat unstable and will rearrange in ether solvents to the more stable 7α-bromide (108) (160). The bromide so obtained usually is not purified extensively but is subjected to dehydrohalogenation after simple separation of reaction by-products. The choice of dehydrohalogenation reagents has been limited to basically two: trimethylphosphite and s-collidine. The trimethylphosphite procedure works best in refluxing xylene and the product diene isolation is facilitated when the sterol is esterified as a benzoate (79). With s-collidine a refluxing hydrocarbon solvent is used. With either of these mothods a significant amount of the 4,6-diene (110) formation occurs, complicating the isolation of pure provitamin and at the same time reducing the efficiency of the process. Some workers have found that mild acid treatment of the mixture will eliminate the ester of the 4,6-diene while leaving the ester of the 5,7-diene intact thus simplifying the isolation of the desired material. Salmond

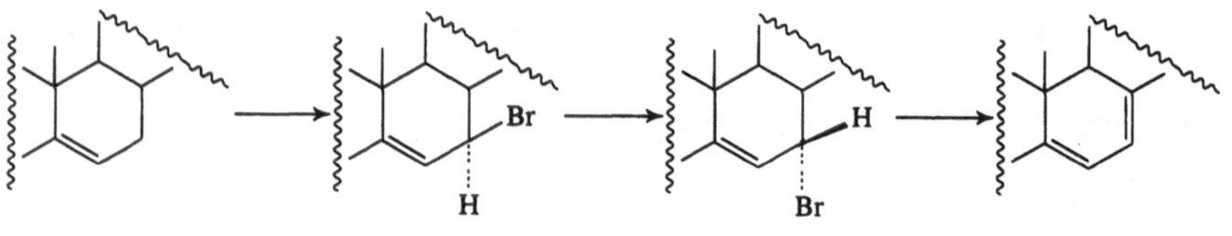

(107)          (108)          (109)

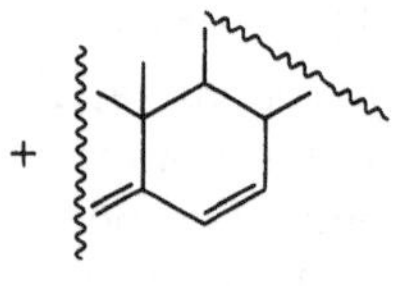

(110)

and co-workers in an apparent attempt to make the diene synthesis more selective reacted the 7α-bromo-compound with sodium phenyl selenolate to give the 7β-selenide (111) (161). Oxidation of this selenide with hydrogen peroxide afforded a 1 : 1 mixture of the diene (109) and the 5β-hydroxy-Δ⁶-compound (112), a product of an intermediate 2,3-sigmatropic shift. The product mixture is explained on steric considerations of the decomposition of selenoxides of different chirality on selenium.

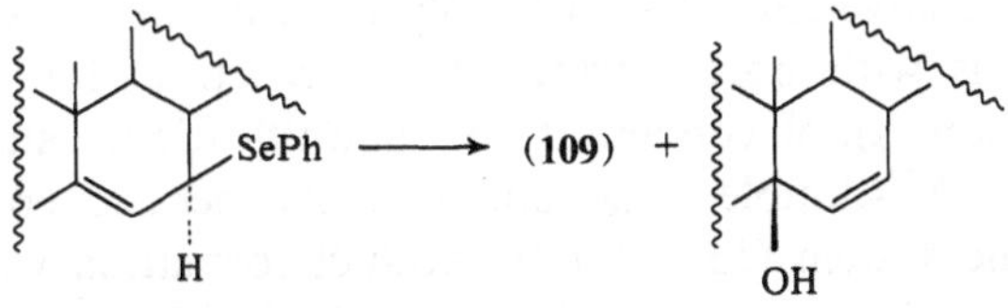

SePh ⟶ (109) +

H                          OH

(111)                      (112)

An alternate route to the 5,7-diene system is by the hydride induced decomposition of Δ⁵,⁷-toluenesulfonylhydrazones (113) (31, 41). The corresponding 7-oxo-steroids (114) can be prepared by allylic oxidation of Δ⁵-sterol esters with chromium trioxide-amine complexes in methylene chloride at room temperature (163). The attractive feature of this method is that the product formed in the toluenesulfonyl hydrazone decomposition is virtually free of the 4,6-diene isomer.

The cholest-6-en-1α,3β-diol (106) obtained in MAZUR's liquid ammonia reduction of 1α,2α-epoxycholesta-4,6-dien-3-one has been brominated to give the corresponding 6,7-dibromide which was dehydrohalogenated in hexamethylphosphoramide containing triethyl methyl ammonium dimethylphosphate to give a mixture (1.3 : 1) of the 5,7- and 4,6-diene compounds (57). As mentioned above treatment of steroidal 1,4,6-trien-3-ones with potassium butoxide in dimethylsulfoxide followed by quenching in mildly acidic water results in the formation of the unstable 1,5,7-trien-3-one system (89). Immediate borohydride reduction serves to "freeze" the 5,7-diene in place and to give the corresponding 3β-ol derivative (95).

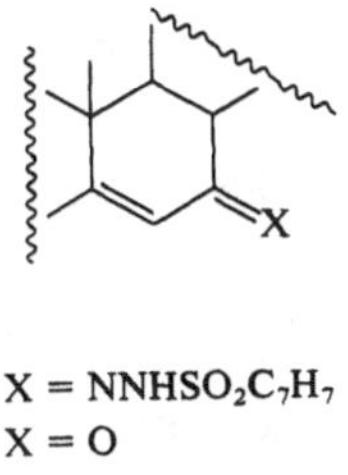

X

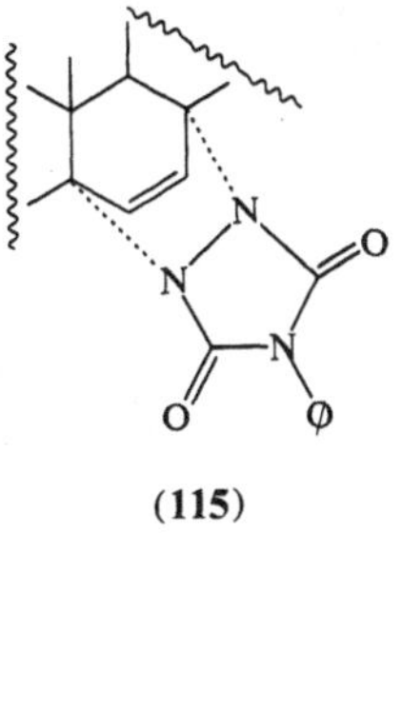

(113) X = NNHSO₂C₇H₇
(114) X = O                          (115)

In the latter case and in similar ones originating from ergosterol and cholesta-1,4,7-trien-3-one, it was found necessary to block the diene system before further transformations could be carried out on other portions of the molecule. Barton and co-workers in studying the Diels-Alder adduct, (115), of 4-phenyl-1,2,4-triazoline-3,5-dione (PTD) with ergosterol, found that the blocking group was easily removed by reduction with lithium aluminum hydride to regenerate the parent 5,7-diene system (11, 15). This adduct protects the B-ring system under conditions of selective side chain ozonolysis, reaction of side chain and A-ring carbonyls with borohydride, Wittig reaction, low temperature Grignard reaction, aldol condensation, saponification, catalytic hydrogenation of the $\Delta^{22}$-double bond and selective peracid epoxidation of the 3β-oxy-1-ene system (53, 54, 61). Adduct formation with PTD has also been reversed under conditions of alkyl lithium reactions and treatment with the strong organic base diazabicyclononane. PTD adds to vitamin $D_3$ but its removal is difficult and thus offers no advantage as a protecting function for its triene system (2). Selective two-step hydration of the $\Delta^5$-double bond and tricarbonyliron complexes have also been used as effective protecting groups for the 5,7-diene system in ergosterol (11, 14).

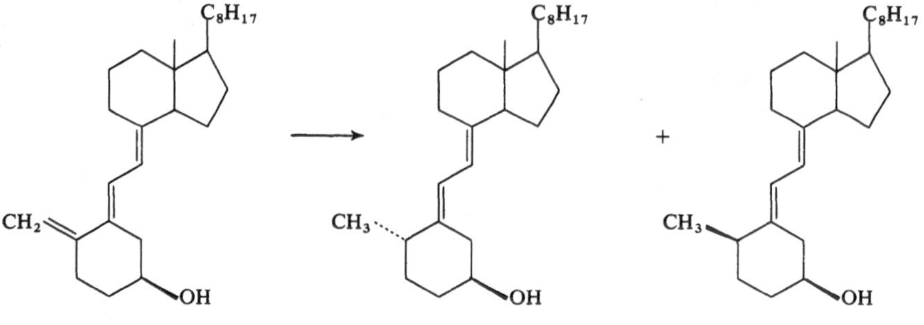

(116)                  (117)                  (118)

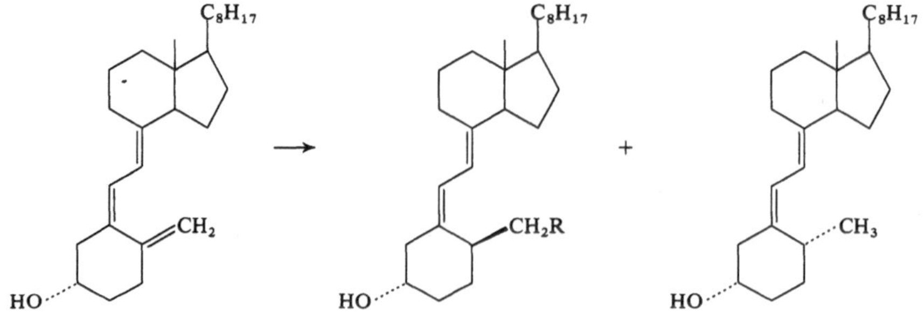

(2b)             (119) R = H (121) R = OH             (120)

The preparative photochemical rearrangement of 5,7-dienes to give the 9,10-secotriene system of the previtamin can be carried out in a variety of oxygen-free, ultraviolet transparent solvents at low temperature (near 0° C) using a medium pressure mercury vapor lamp as the light source. It is generally necessary to keep the exposure time to a minimum so as to prevent extensive irreversible conversion of the previtamin (171) to other photoproducts *(vide infra)*. The initial photoequilibrium mixture generally contains the previtamin (171), its 5,6(*E*)-isomer (tachysterol) (173), the starting 5,7-diene (provitamin) (1) and its 9β,10α-isomer (lumisterol) (172). EYLEY and WILLIAMS have found that the preparative yield of previtamin can be increased by a second irradiation of the photolysis mixture in the presence of fluorenone as a photosensitizer (*52*). The fluorenone allows a more favorable ratio of tachysterol: previtamin to be formed. As tachysterol readily forms Diels-Alder adducts it can be reacted with maleic anhydride to permit a more convenient separation from the previtamin.

Thermal equilibration of the previtamin (171) with the vitamin (2) occurs on brief warming of a solution in an oxygen-free atmosphere. This equilibration generally gives the vitamin as the major isomer but seldom is complete thus requiring chromatographic separation if pure vitamin is desired.

## 4. Analogs Prepared by Modification of Vitamin D and Related Compounds

In a review of vitamin D chemistry in a previous volume of this series, INHOFFEN has described the early work on the direct chemical modification of the secosteroids (*82*). Of the compounds which have been obtained directly from the vitamin, the most interesting from the standpoint of biological activity has been dihydrotachysterol, DHT, (117), which has been found to be clinically effective in the treatment of certain hypocalcemic disorders. DHT is obtained by catalytic or dissolving metal reduction of vitamin D; the process involving saturation of the C-10,19 bond as well as $Z$ to $E$ isomerization of the C-5,6 olefin. OKAMURA and co-workers have recently described studies of compounds obtained by selective reduction of vitamin D (2b) and C-5(*E*) vitamin D (116) with 9-borobicyclo[3:3:1]nonane followed by acetic acid treatment (*123, 133*). Under these conditions, isomerization about the C-5,6 bond was not observed and about an equal quantity of compounds isomeric about the C-10 carbon were obtained. NMR studies of each of the reduced products then allowed the definitive assignment of C-10 as $S$ in $DHT_3$ and as $10(R)$ in its isomer (118) (DHV$_3$-IV). Similarly, in the reduction products of vitamin D$_3$, C-10 was assigned as $S$ in (119) DHV$_3$-II and $R$ in (120)

DHV$_3$-III. Cleavage of the intermediate boranes in these reactions by acetic acid was not very efficient making this a poor preparative method for these compounds. If the boranes were cleaved by alkaline hydrogen peroxide, however, high yields of the corresponding mixture of 19-hydroxy derivatives were obtained. One of these, (121), has shown interesting anti-vitamin D$_3$ activity.

Oxidation of 1α-hydroxy- or 1α,25-dihydroxyvitamin D$_3$ with activated manganese dioxide results in the formation of the corresponding 1-oxo-previtamin (122) (139, 172), the reversion to the previtamin structure presumably occurring because of the additional stabilization of the linear triene system by the carbonyl function. Sodium borohydride reduction of (122) leads selectively to the 1β-hydroxy previtamin (123) which on thermolysis affords the 1β-hydroxyvitamin (124). Neither [(124), R=H] nor [(124), R=OH] has shown appreciable vitamin D-like biological activity.

Solvolysis of vitamin D$_3$ tosylate in a buffered aqueous medium affords the A-ring bicyclic C-6(S) alcohol (125) in 60% yield (173, 175). Aqueous acid converts (125) into a mixture of vitamin D$_3$ and trans-

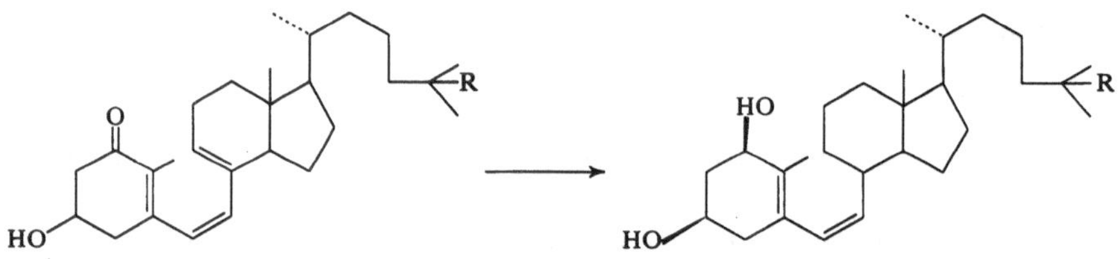

(122)          (123) R = H or OH

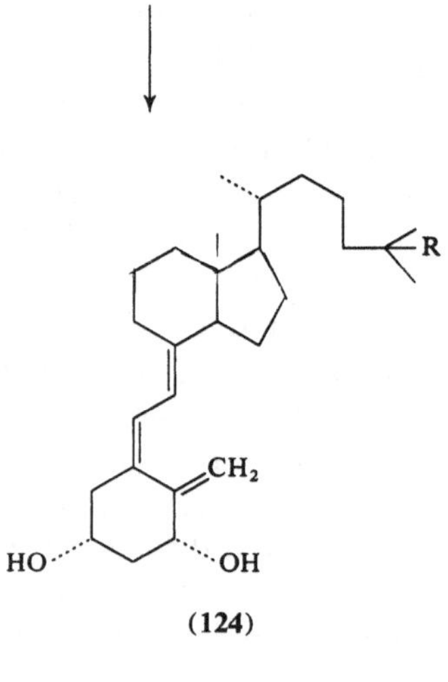

(124)

vitamin D₃ (116). Oxidation of (125) with activated manganese dioxide affords the corresponding ketone (126). A mixture of isomeric C-6 methyl alcohols is formed on reaction of (126) with methyl lithium (*174*). Either isomer solvolyzes in acidic aqueous medium to give a mixture of 6-methylvitamin D₃ (128) and 6-methyl-5,6-transvitamin D₃ (129). The presence of the 6-methyl group destabilizes the vitamin structure to such an extent that at thermal equilibrium exclusive formation of the pre-vitamin form results.

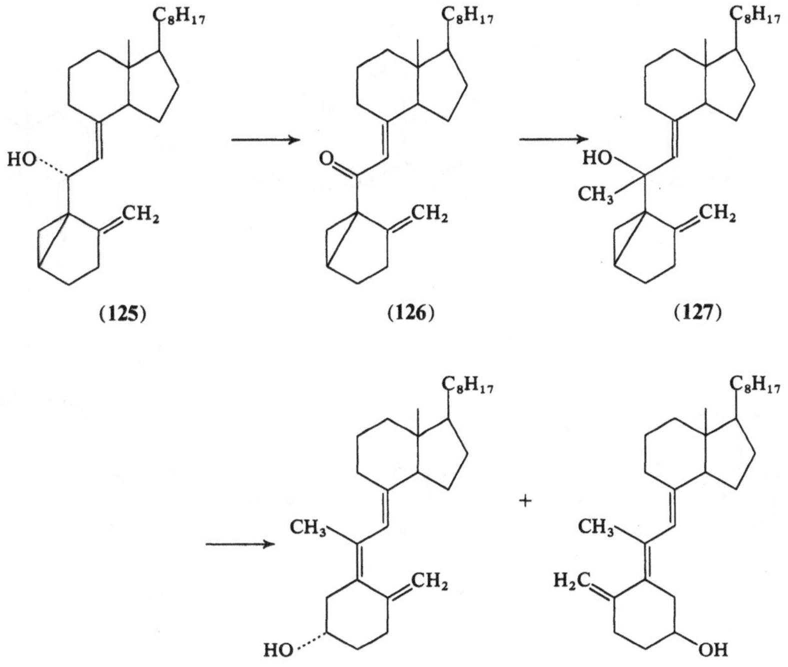

(125)    (126)    (127)

(128)    +    (129)

## 5. Total Synthesis

Although the preparation of vitamin D-like compounds from the rigid framework of naturally occurring sterol precursors offers the advantage of high stereochemical control of new substituent intro-duction, it suffers from the disadvantage that all transformations must be done sequentially and are dependent on structural features already present. Total synthesis, however, can be done in a convergent fashion allowing greater flexibility and variety as to the positioning and separate manipulation of desired functionalities. The total syntheses which have been reported to date are, as yet, not competitive with those utilizing natural precursors but they offer considerable promise that such approaches will provide reasonably efficient routes to more difficultly obtained metabolites and analogs.

The strategy for total synthesis of the 9,10-secosteroid skeleton of vitamin D would seem logically to involve the separate construction of fragments corresponding to the A-ring, a C-D-bicyclic ring system and a side chain which could be linked at various stages of the synthesis. This is indeed the approach taken by the two groups who have reported

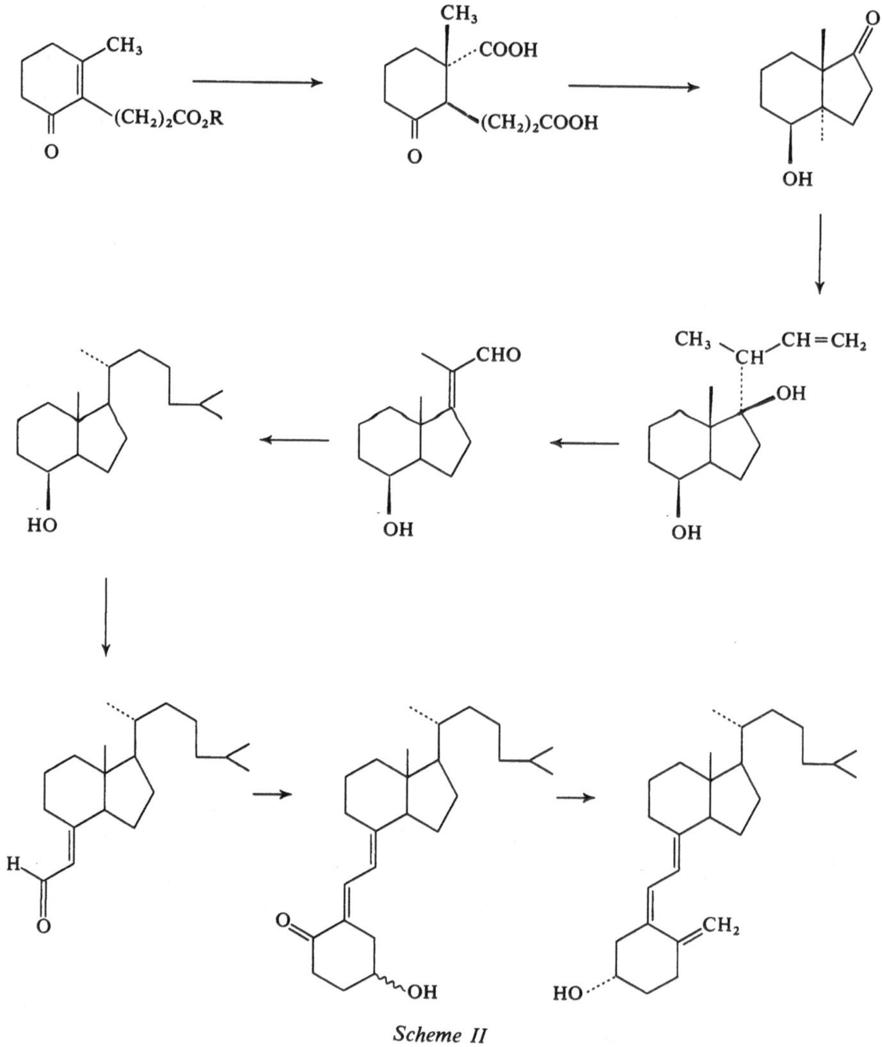

*Scheme II*

success in this area. INHOFFEN and his group at Braunschweig reported the first synthesis of cholecalciferol in 1958. This synthesis, which is shown in abbreviated form in Scheme II, was described in detail in an earlier volume of this series (82) and will not be discussed further here. LYTHGOE and his group at Leeds, active over a number of years in the synthesis of vitamin D compounds, have reported the synthesis of tachysterol in 1963, precalciferol in 1970, 1α-hydroxyvitamin $D_3$ in 1973 and new syntheses of vitamin $D_2$ and $D_3$ in 1977.

The maintenance of stereochemical integrity of the individual parts is of importance in a convergent synthesis. The formation of diastereo-isomeric pairs automatically reduces the yield of the desired final product and often results in tedious separation problems. It is useful then to work with optically active material as early as possible in the synthesis and to use processes and intermediates which are amenable to stereochemical control. The key features of the LYTHGOE group syntheses described below are the early resolution of A-ring and C-ring precursor carboxylic acids and the use of bicyclic intermediates and stereospecific Claisen-type rearrangements to control the chirality of newly introduced functionalities. The syntheses are unique in vitamin D chemistry in that they do not require photocatalyzed rearrangements to create the 9,10-secotriene system of the final product.

### a) A-Ring Precursors

Lithium-liquid ammonia reduction of 5-methoxy-2-methyl-benzoic acid followed by hydrolysis of the intermediate enol ether afforded the ketoacid (130) (Scheme III) which was resolved as its quinine salt. Boro-hydride reduction of (130) afforded the bicyclic lactone (131). Treatment of (131) with sodium methoxide gave an α,β-unsaturated ester (132) which was reduced to the diol (133) (46). The allylic hydroxyl group of (133) when converted to the corresponding triphenyl phosphonium salt gave the A-ring precursor in the synthesis of tachysterol (42). Selective oxidation of (133) gave the corresponding aldehyde (134). When the lactone (131) was epoxidized, a single epoxide was formed selectively which on methoxide treatment gave the α,β-unsaturated ester (135) (65). This was converted into the aldehyde (136). The aldehydes were converted to the acetylenic compounds (137)—(138) by dehydrohalogenation of intermediate chlorovinyl derivatives obtained from (134) and (136) by a Wittig reaction with chloromethylenetriphenylphosphorane (43, 65). The trimethylsilyl protected acetylides of (137) and (138) as their lithium derivatives were reacted with a chlorinated C-D fragment (153) to give the chlorohydrin (139) (44, 65). Treatment with chromium (II) reagent afforded an intermediate dienyne which on hydrogenation with Lindlar

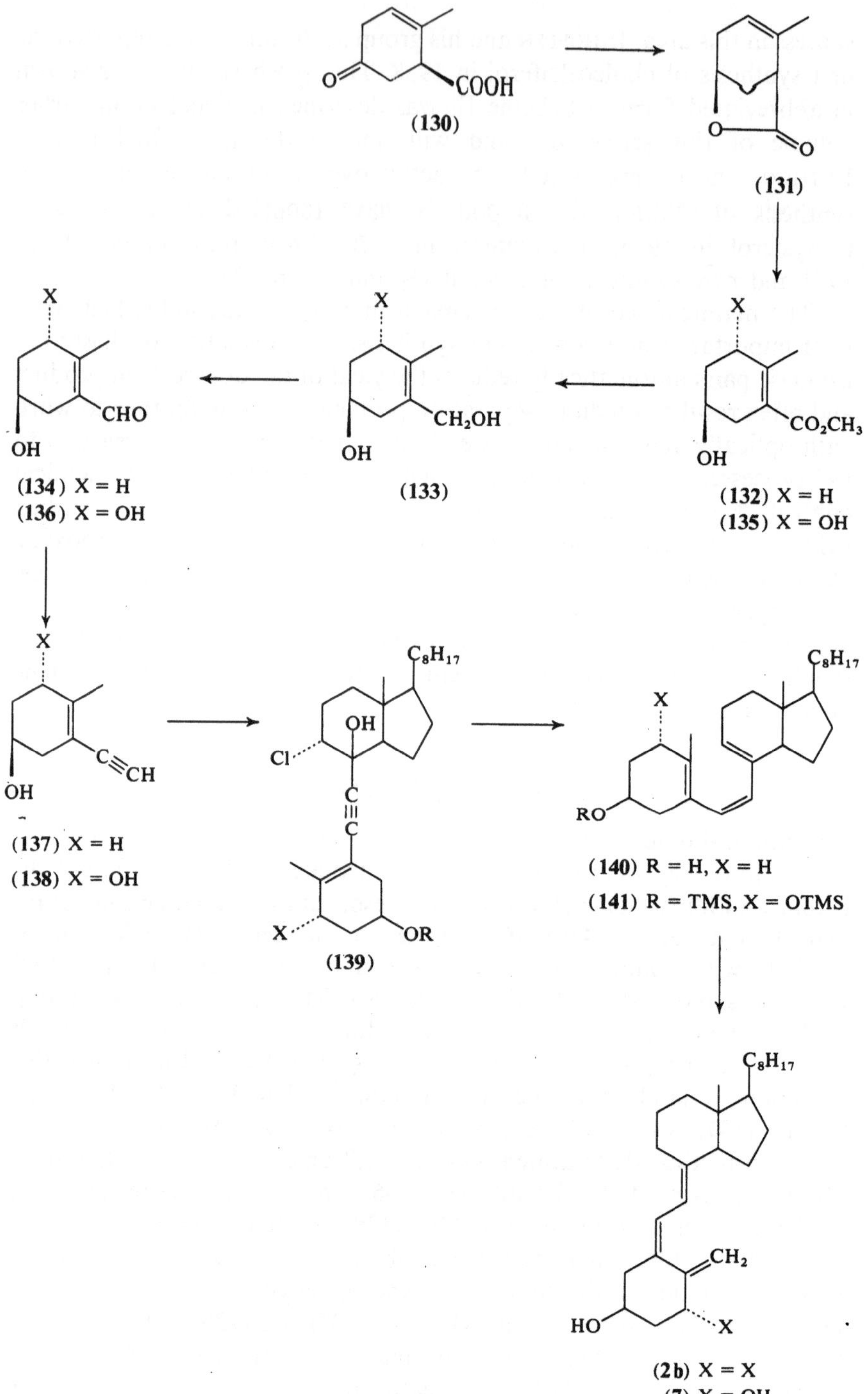
COOH
(130)
(131)
X
CHO
OH
(134) X = H
(136) X = OH
CH₂OH
(133)
CO₂CH₃
(132) X = H
(135) X = OH
C≡CH
(137) X = H
(138) X = OH
C₈H₁₇
Cl
OH
OR
(139)
RO
(140) R = H, X = H
(141) R = TMS, X = OTMS
CH₂
HO
(2b) X = X
(7) X = OH

*Scheme III*

catalyst gave the precalciferols (140) and (141). The thermal equilibration, and protecting goup removal in the case of (141), afforded vitamin D₃ (2 b) and 1α-hydroxyvitamin D₃ (7).

More recently an A-ring precursor which provides the vitamin framework directly on coupling with a C-D fragment has been described (Scheme IV) (*107, 108*). The diol (142) which can be obtained optically pure from the Diels-Alder adduct of butadiene with chiral fumarate esters was converted in seven steps to the lactone (143). The primary hydroxyl group was converted to an iodide and this was eliminated using diazabicycloundecane. The resulting enelactone gave predominantly a single thioether on reaction of the lithium enolate with diphenyldisulfide. Reduction and benzoylation gave the intermediate (144). Thermolyses of

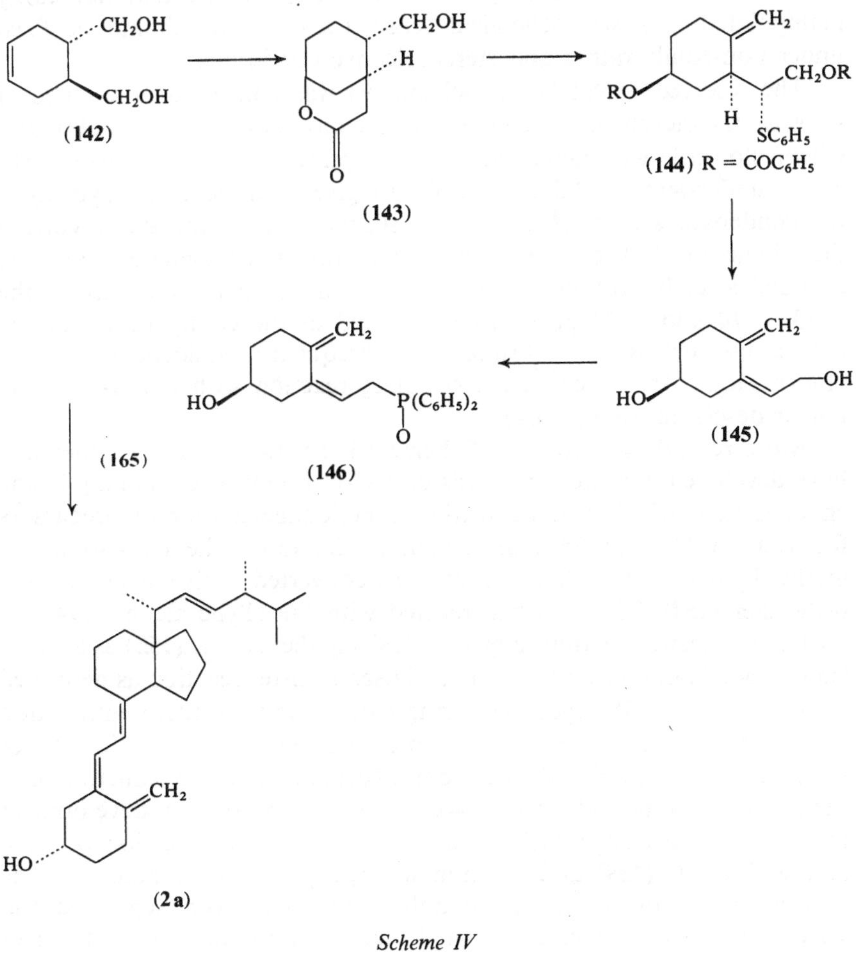

*Scheme IV*

the sulfoxide and hydrolysis afforded (145). The primary hydroxyl group was converted *via* the chloride into the phosphine oxide (146) which reacted with the C-D fragment ketone under conditions of the Wittig-Horner reaction to give vitamin $D_2$. The unique feature of this synthesis is the synthesis and preservation of the Z-geometry of C-5,6 double bond while the final step of the Wittig-Horner reaction allows the selective formation of E-geometry for the C-7,8 olefin.

### b) C-D Ring and Side Chain Fragments

The stereoselective preparation of the four contiguous assymetric centers at C-14, C-13, C-17 and C-20 of the vitamin D structure presents a formidable synthetic challenge. As the *trans* ring junction is less stable than the *cis* and the C-20 position is freely rotating and not easily manipulated in a stereochemical sense, these centers should be fixed under non-equilibrating and stereoselective conditions.

The resolved diacid (147) (Scheme V) was converted in a series of steps *via* the bicyclic lactone (148) into the diol monobenzoate (149) (*19*). Citronellonitrile was converted into the optically active orthoester (150) which underwent a Claisen transfer to give after benzoate hydrolysis the condensed system (151) (*18*). A second Claisen transfer involving the allylic alcohol of (151) and 1-dimethylamino-1-methoxyethylene afforded after hydrolysis the dibasic acid in which the chirality of the C-13, C-14 and C-20 positions were fixed in the configuration of the natural product. Ring D closure and subsequent modifications gave the chloroketone (153) used in the coupling reaction with the A-ring precursor described above (*104*).

More recently as shown in Scheme VI, LYTHGOE and his colleagues have developed a related synthesis of a C-D fragment containing a side chain moiety which can be further functionalized to give precursors for vitamin $D_2$ and 25-hydroxyvitamin $D_3$ (*34*). The (+)-isomer of methyl hydrogen β-methylglutarate was converted in five steps into the orthoester (154). This material reacted with the allylic alcohol (149) in a Claisen transfer reaction to give exclusively the lactone (155) as a single isomer. Methanolysis and a second Claisen transfer reaction as described for (151) gave (156). Hydrolysis, ring closure and reductive elimination of the ring oxygen gave the C-D ring fragment (157). Epoxidation of the benzyl ether of (157) *via* the bromohydrin and hydride reduction gave the hydroxy compound (158) ($R = CH_2\emptyset$) (*109*). Side chain development then followed a design earlier studied by Sucrow. Conversion of (158) to the aldehyde (159) and reaction with propynyl magnesium bromide gave a mixture of acetylenic alcohols (160). These were separated and selectively reduced to give the vinyl alcohols (161) and (162). Each of these compounds on reaction with ethyl orthopropionate in a Claisen reaction gave the same side chain (163) in which the methyl group at the ultimate C-24 is of the same orientation as in ergosterol. The ester mixture was reduced to give a diol (164). Selective tosylation of the C-26 hydroxyl group followed by hydride reduction and oxidation of the C-8

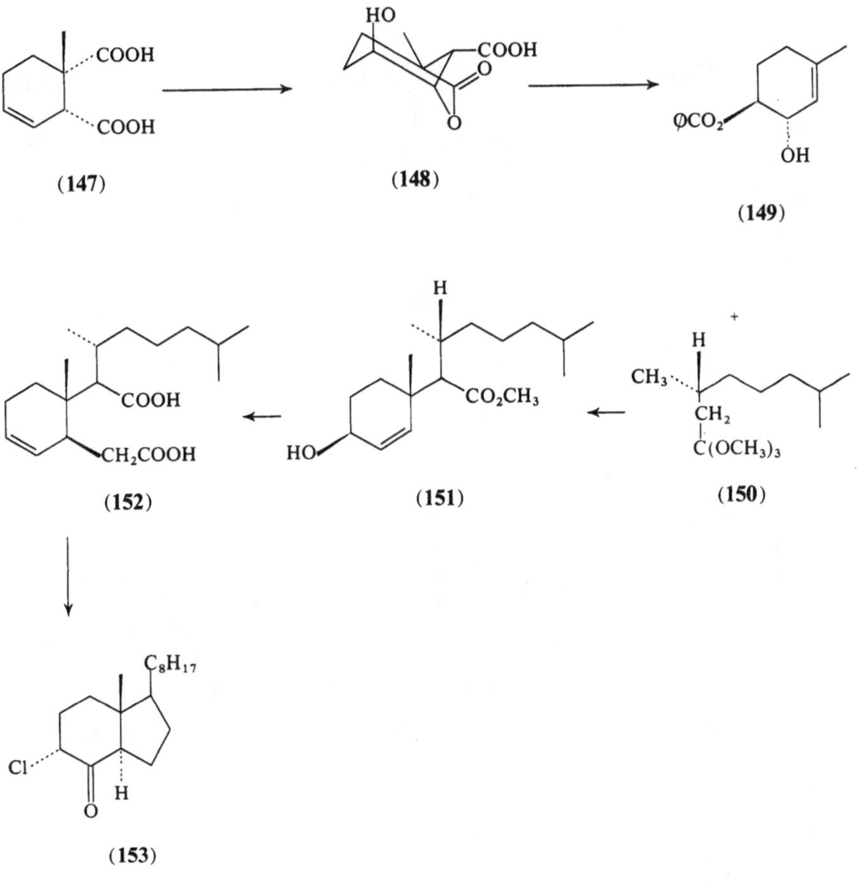

(147)          (148)          (149)

(152)          (151)          (150)

(153)

*Scheme V*

hydroxyl group gave the ketone (165) used in final step of the preparation of Vitamin D$_2$. Alternately, when the tosylate (158) (R=O$_3$SC$_7$H$_7$) was coupled with 3-methylbut-3-enylmagnesium chloride in the presence of dilithium tetrachlorocuprate, the olefin (166) was formed. Oxymercuration followed by demercuration with sodium borohydride afforded the 25-hydroxylated side chain intermediate (167).

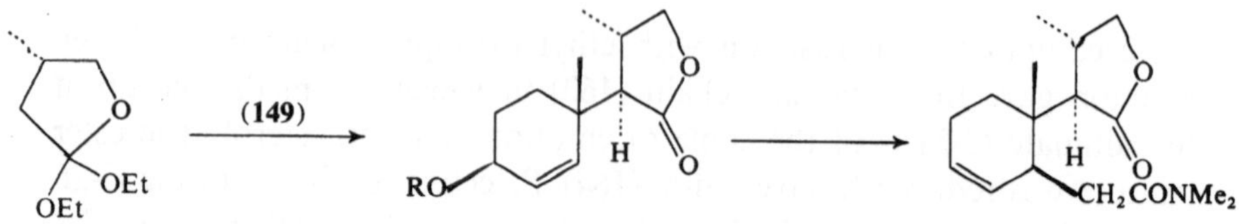

(154)

(149)

(155) R = C₆H₅CO

(156)

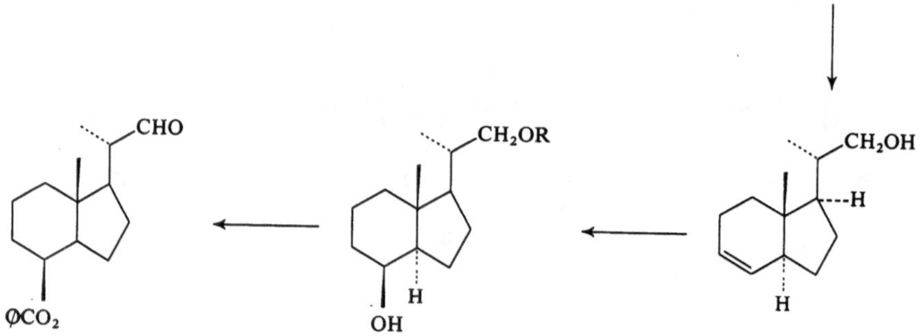

(159)

(158)

(157)

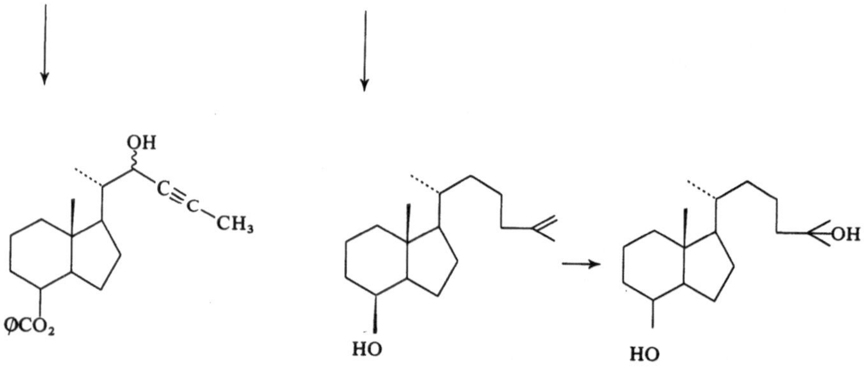

(160)

(166)

(167)

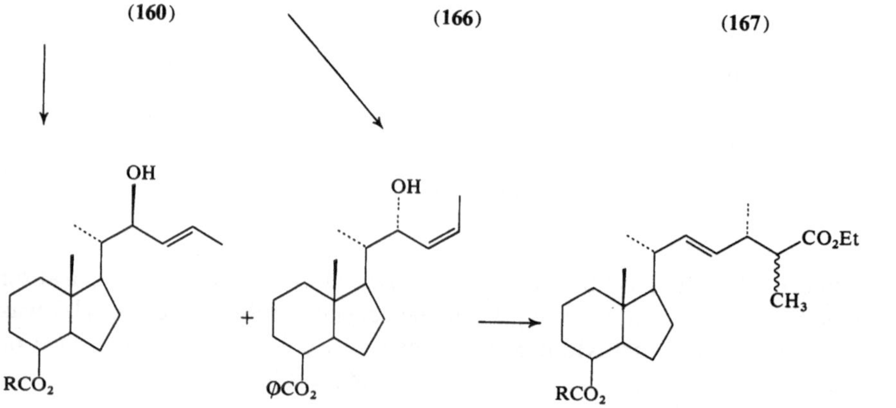

(161)

(162)

(163)

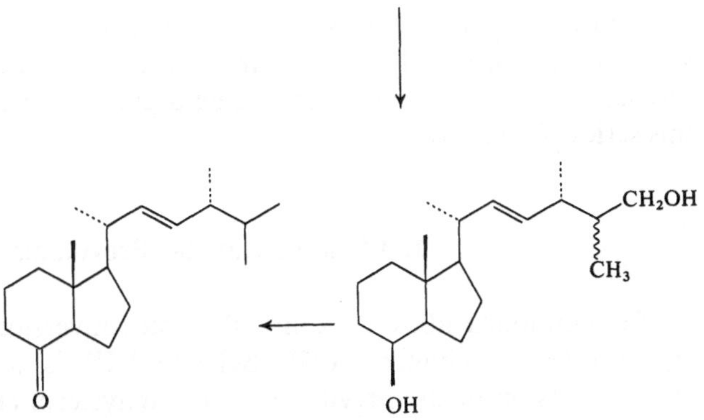

(165)                                        (164)

*Scheme VI*

## c) Other Approaches

TROST and co-workers have reported the coupling of 7-methyloct-Z-2-en-1-yl bromide with the dianion of methyl acetoacetate to give (168) (R = H₂) (*186*). The diazoderivative (168) (R = N₂) was thermolyzed in the presence of copper powder to give the cyclopropane (169). Ring cleavage of the cyclopropyl ring using lithium dimethyl cuprate resulted in stereospecific formation of the cholesterol side chain where the relative orientation of the ultimate C-17 and C-20 positions in (170) is the same as that of the natural product.

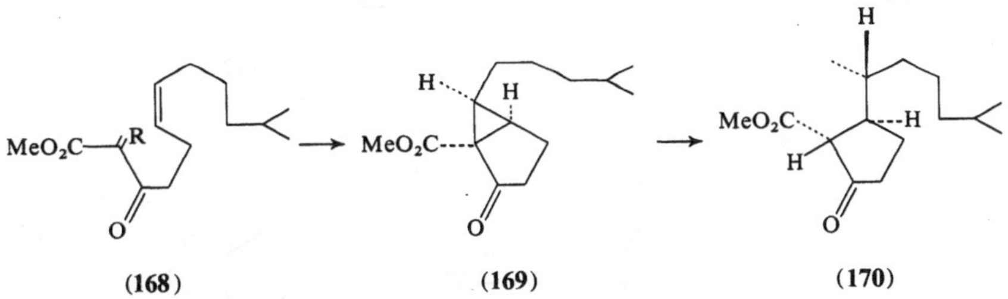

(168)                    (169)                    (170)

## III. Photolytic and Thermal Conversions of the Vitamin D Isomers

The conversion of ergosterol (provitamin D₂) (1a) or 7-dehydrocholesterol (provitamin D₃) (1b) into the corresponding vitamins by sequential photolysis and thermolysis is an important commercial process and has received considerable study. Also, the nature of the steroid molecule, its conformational restrictions and rigid backbone, makes it an unique and interesting system in which to test the results of orbital symmetry as they apply to the rearrangements found. The topic has been discussed in several reviews including a chapter in an earlier volume of this series (70, 92, 167).

## 1. Photolysis of the Provitamin

As mentioned above, the initial photoconversion of the provitamin (**1**) into the previtamin (**171**) (Scheme VII) is followed by a facile interconversion of the previtamin with tachysterol (**173**), and to a lesser extent with lumisterol (**172**) and the starting provitamin. These secondary interconversions are difficult to avoid as the ultraviolet absorption spectrum of the previtamin completely overlaps that of the starting (**1**). Optimization of previtamin formation can be attained by short irradiation times at low temperature (55), by operating at a wavelength close to 295 nm (3) and by a second irradiation in the presence of a triplet sensitizer which will catalyze a more favorable ratio of previtamin to tachysterol to be formed (52). If the photolysis is carried out on a

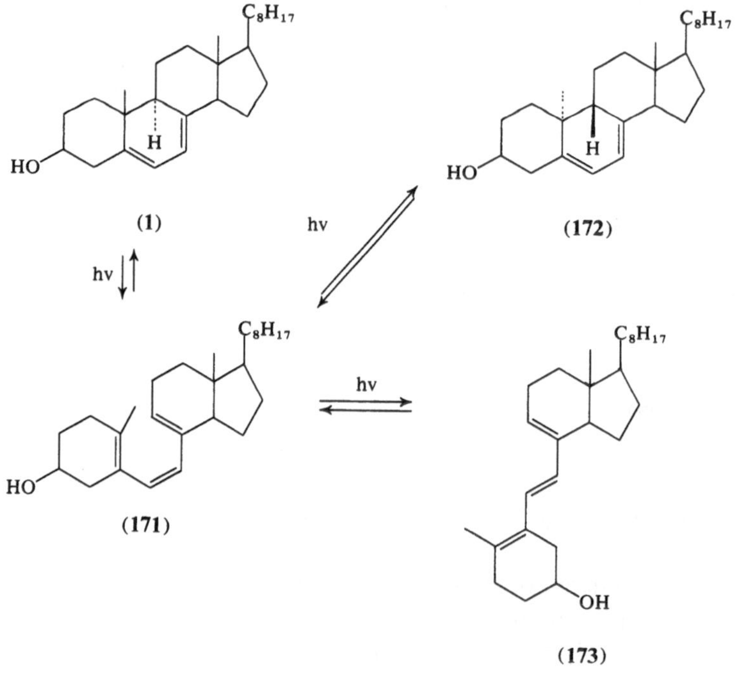

Scheme VII

relatively concentrated solution of the provitamin, significant coupling of the steroidal diene occurs to give a mixture of bischolestadienols linked at the 7,7'-position (22). Of the other 5,7-diene systems, lumisterol (172), which has the *trans* 9β,10α-configuration, is in photoequilibrium with the previtamin and thus gives the same irradiation products as the provitamin. The *cis* 9,10-isomers, pyrocalciferol (174) and isopyrocalciferol (175) *(vide infra)*, however, cannot undergo conrotary ring opening and are photoisomerized (Scheme VIII) to the corresponding bicyclo[2,2,0] hexene derivatives (176) and (177) (40). These strained systems revert to the starting dienes upon heating.

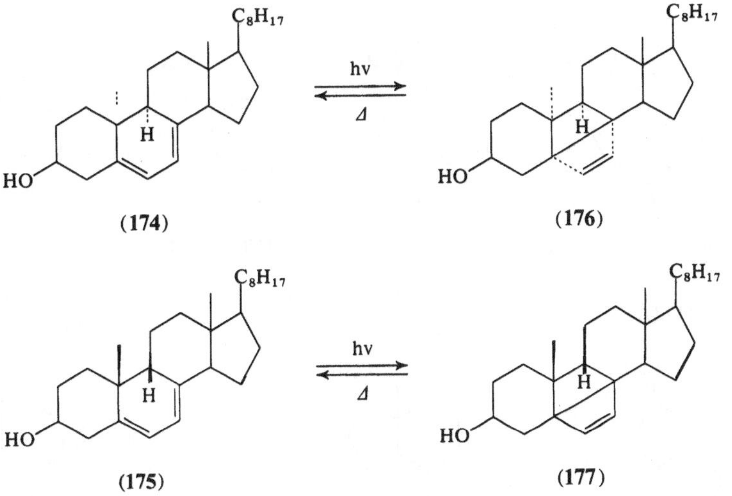

Scheme VIII

## 2. Photolysis of the Previtamin

It has been found that prolonged irradiation of the initial "quasi" stationary state results in new products being formed in an irreversible process. The structure of these products, derived primarily from the previtamin and called toxisterols, have recently been investigated in two laboratories. The nature of the product mixture obtained depends on the temperature and the type of solvent used in the photolysis as well as on the wavelength of the light used.

In BARTON's laboratory, ergosterol was irradiated in a quartz apparatus at room temperature with a high pressure mercury arc lamp until the starting sterol absorption reached a minimum (24 hours in ethanol, 72 hours in hexane) (*10*). Under these conditions some conversion of the initially formed precalciferol to ergocalciferol took place as the vitamin and suprasterol *(vide infra)* were formed in addition to other compounds (**178—184**). The spiro toxisterols (**178**) were most likely formed from the previtamin by a C-4 to C-9 hydrogen shift followed by ring closure between C-4 and C-8. Internal or external addition of hydroxyl functions at the C-9 and C-10 positions afforded the ethers (**182**) (*9*) and (**183**). The hydrocarbons (**179**), (**180**) and (**184**) and ethers (**181**) presumably arose by photochemical fragmentation of various olefins generated in the mixture followed by secondary reactions or dimerization in the case of (**184**).

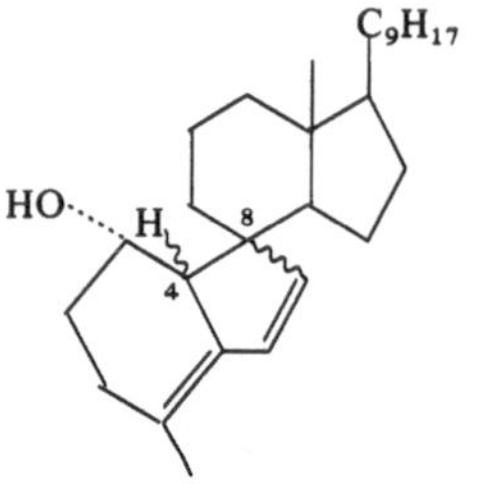

(**178**)

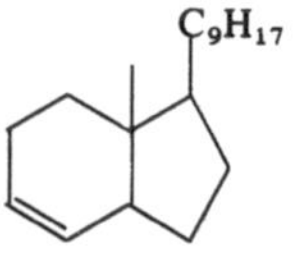

(**179**)

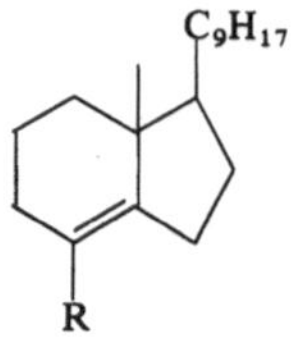

(**180**) R = H, CH$_3$

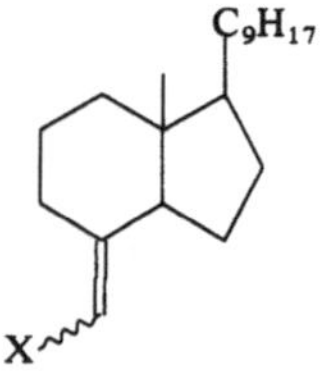

(**181**)

X = H, OC$_2$H$_5$

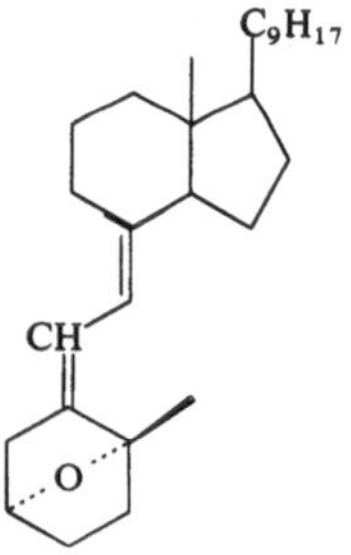

(**182**)
5(*E*) and 5(*Z*)

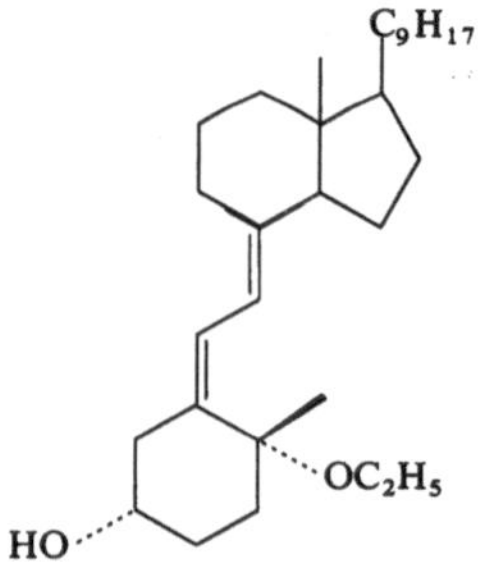

(**183**)

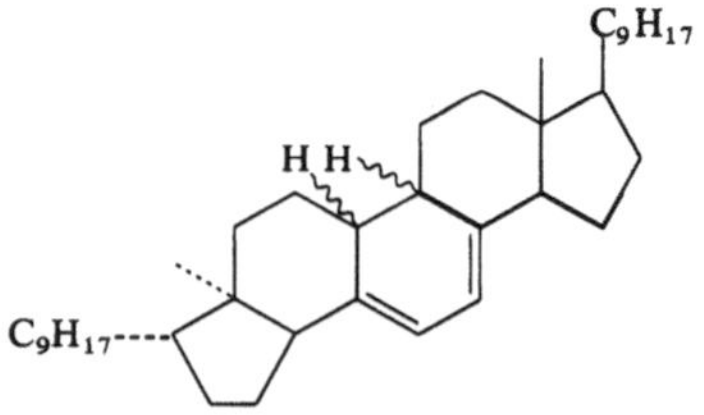

(**184**)

In HAVINGA's laboratory, the toxisterol mixture was generated by prolonged irradiation of 7-dehydrocholesterol at 5° C in diethyl ether or alcoholic solvents (*21, 84*). The irradiations were carried out at wavelengths longer than, 300 nm to prevent destruction of those toxi-sterols showing absorption bands near 250 nm. Under these conditions, thirteen products were isolated and identified. The major toxisterol formed in alcohol or ether solvent was the bicyclo[3.1.0]hexenyl steroid of structure (**186**) (two isomers) formed by an internal cycloaddition of the triene system of previtamin D. The deconjugated triene (**187**) results from a photocatalyzed internal [1,5] hydrogen shift from C-19 to C-7 of the previtamin. The cyclobutenyl compound (**188**), found only in aprotic irradiation solvents, apparently arises from the photolysis of tachysterol. It can be smoothly converted to previtamin D in a thermal isomerization. As in the photolysis of ergosterol described above, three isomers of the spirosystem (**185**) were found. In alcohol solvent three alcohol addition products (**189**) were isolated. These latter compounds proved to be somewhat labile to the isolation conditions and lost alcohol to afford the corresponding 1(10),5,7- and 5(10),6,8(14)-triene systems. Some photochemical reduction occurs on prolonged irradiation in alcohol to give the dihydrocompound (**190**).

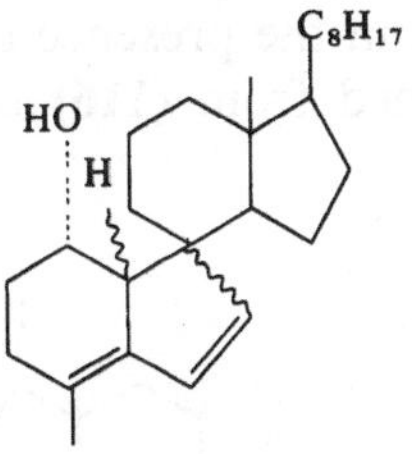

(**185**)

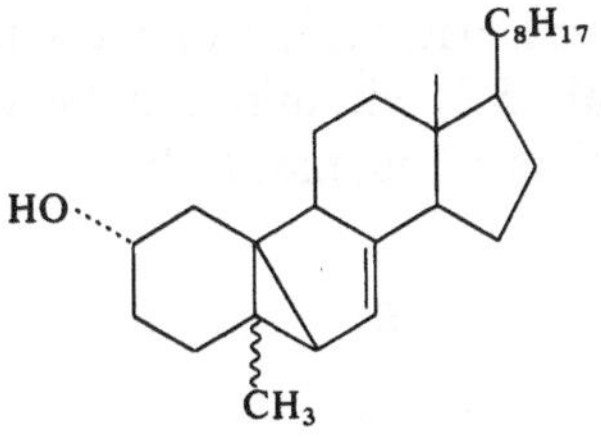

(**186**)

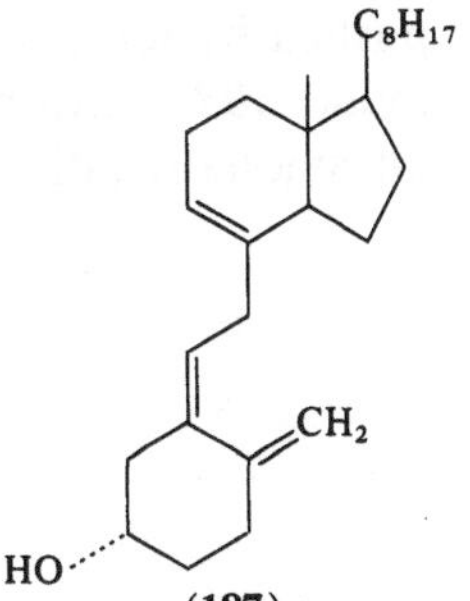

(**187**)

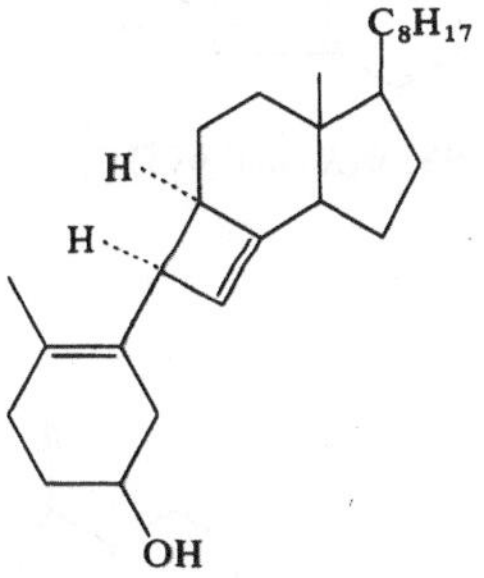

(**188**)

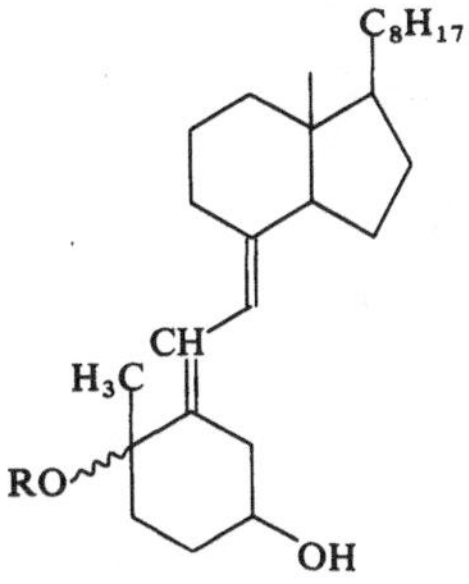

(**189**) 5(*E*), 10(*S*)
5(*E*), 10(*R*)
5(*Z*), 10(*R*)

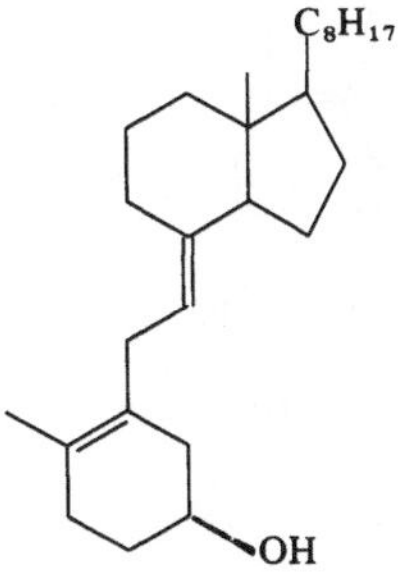

(**190**)

## 3. Photolysis of Vitamin D

The triene system of vitamin D is very sensitive to ultraviolet irradiation. Because of this sensitivity, it is necessary to carry out the photolytic conversion of the provitamin to previtamin at low temperature and then to thermally convert the previtamin to the vitamin in a second step, thus avoiding exposure of the ultimate product to intense irradiation. The overirradiation of vitamin D leads to compounds which are devoid of ultraviolet absorption maxima above 230 nm. The first of these which were isolated and characterized were called suprasterols. Two isomers were obtained which correspond to the bicyclo[3.1.0]hexenyl structures (191) and (192) (39). Later studies resulted in the isolation and identification of the cyclobutenes (193) and the allenes (194) (8). These structures are what would be anticipated from photoisomerization of a hexatriene containing the cZt conformation of vitamin D. The existence of pairs of isomers can be explained by their formation from ring A conformers in which the C-10,19 methylene group is oriented either up or down relative to the plane of the C-D ring system.

The selective formation of *trans* (5E) vitamin D (116) is best accomplished by long wavelength irradiation of the vitamin in the presence of iodine (82). The natural (5Z) vitamin can be obtained from (116) by ultraviolet irradiation of its 5E isomer (82).

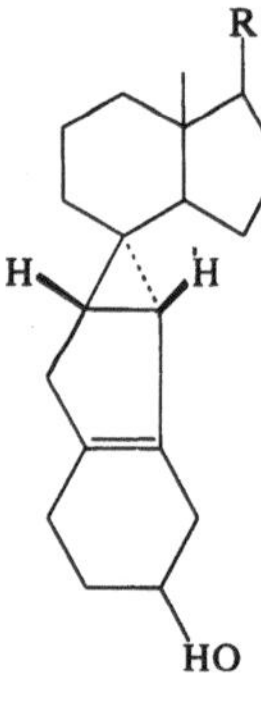

(191)

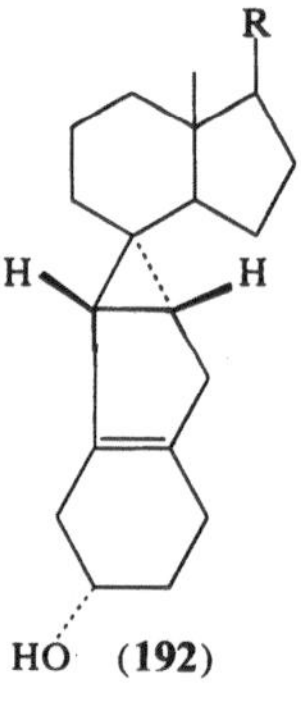

(192)

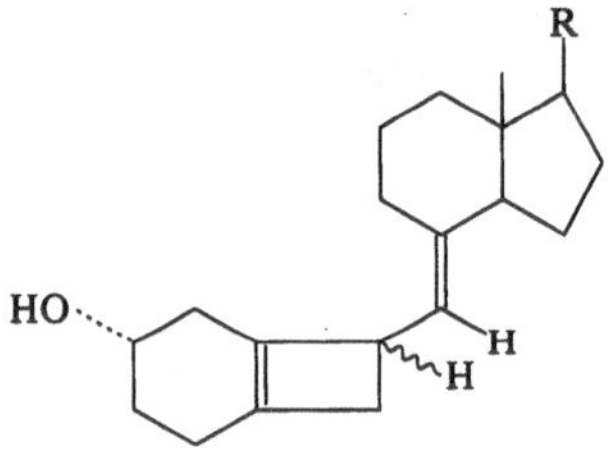

(193) 6(R) and 6(S)

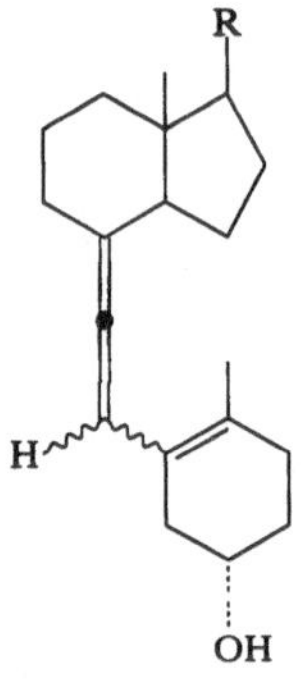

(194)
7(R) and 7(S)

## 4. Thermolysis of Previtamin D

The facile (1—7) hydrogen shift between C-19 and C-9 found on heating the previtamin at moderate (20—100° C) temperature affords an equilibrium mixture in which a 4 : 1 excess of vitamin exists. The position of the equilibrium in this reaction appears surprising as one might expect that the formation of three exocyclic double bonds to six-membered rings would be an energetically uphill process. This, indeed, appears to be the case if the C-ring were not *trans* linked to the 5-membered D-ring. The strain induced by the *trans* linked D-ring significantly destabilizes the endocyclic C-8 double bond relative to the exocyclic C-7 double bond and thus allows the rearrangement to proceed. Other perturbations such as a C-6 methyl group as in (128) or a carbonyl group at C-1 as in (122) lead to equilibrium mixtures which favor the triene configuration of the previtamin.

Heating of the previtamin or vitamin at relatively high temperature leads to an irreversible thermal ring closure in which the 5,7-diene system of a 9—10 *syn* steroid is formed (150). These compounds, pyro- and isopyrocalciferols (174) and (175), are found to be the major components eluted in the GLC analysis (> 200°) of vitamin D or its previtamin.

### References

1. ABERHART, D. J., J. Y. R. CHU, and A. C. T. HSU: Synthesis of 3-Epicholecalciferol. J. Organ. Chem. (U.S.A.) 41, 1067—1069 (1976).
2. ABERHART, D. J., and A. C. T. HSU: Studies on the Adduct of 4-Phenyl-1,2,4-triazoline-3,5-dione with Vitamin $D_3$. J. Organ. Chem. (U.S.A.) 41, 2098 (1976).
3. ABILLON, E., and R. MERMET-BOUVIER: Effect of Wavelength on Production of Previtamin $D_2$. J. Pharm. Sci. 62, 1688 (1973).
4. AHMAD, R., D. HANDS, S. L. LEUNG, J. M. MIDGLEY, H. SAFWAT, and W. B. WHALLEY: Unsaturated Steroids. Part 8. Synthesis of Ergosta-5,7-diene-1α,3β-diol, the 4,4-Dimethyl Analogue, and 4,4-Dimethylergosta-5,7-dien-3β-ol. J. Chem. Soc. Perkin I 1978, 74.
5. ANAGNOSTOPOULOS, C. E., and L. F. FIESER: Nitration of Unsaturated Steroids. J. Amer. Chem. Soc. 76, 532 (1954).
6. ASKEW, S. A., H. M. BOURDILLON, H. M. BRUCE, R. K. CALLOW, S. T. PHILPOT, and T. A. WEBSTER: Crystalline vitamin D. Proc. Roy. Soc. (B) 109, 488 (1932).
7. ATKINS, D.: A Possible Role of 24,25-Dihydroxycholecalciferol in Bone Resorption. Proc. Soc. Endocr. 144, 28 P (1976).
8. BAKKER, S. A., J. LUGTENBURG, and E. HAVINGA: Studies on Vitamin D and Related Compounds XXII. New Reactions and Products in Vitamin $D_3$ Photochemistry. Rec. Trav. Chim. Pays-Bas 91, 1459 (1972).
9. BARRETT, A. G. M., D. H. R. BARTON, and R. A. RUSSELL: Structure of the Toxisterols: X-Ray Crystal Structure of Toxisterol$_2$-D Epoxide. J. Chem. Soc. Chem. Comm. 1976, 659.

10. Barrett, A. G. M., D. H. R. Barton, R. A. Russell, and D. A. Widdowson: Photochemical Transformations. Part 34. Structures of the Toxisterols. J. Chem. Soc. Perkin I **1977**, 631.

11. Barton, D. H. R., A. Gunatilaka, T. Nakanishi, H. Patin, D. A. Widdowson, and B. R. Worth: Synthetic Uses of Steroidal Ring B Diene Protection: 22,23-Dihydro-ergosterol. J. Chem. Soc. Perkin I **1976**, 821.

12. Barton, D. H. R., R. H. Hesse, M. M. Pechet, and E. Rizzardo: A Convenient Synthesis of 1α-Hydroxy-Vitamin $D_3$. J. Amer. Chem. Soc. **95**, 2748 (1973).

13. — —    — Convenient Synthesis of Crystalline 1α,25-Dihydroxyvitamin $D_3$. J. Chem. Soc. Chem. Comm. **1974**, 203.

14. Barton, D. H. R., and H. Patin: Chemistry of the Tricarbonyliron Complexes of Calciferol and Ergosterol. J. Chem. Soc. Perkin I **1976**, 829.

15. Barton, D. H. R., T. Shioiri, and D. A. Widdowson: Biosynthesis of Terpenes and Steroids. Part V. The Synthesis of Ergosta-5,7,22,24(28)-tetraen-3β-ol, a Biosynthetic Precursor of Ergosterol. J. Chem. Soc. (C) **1971**, 1968.

16. Baum, J., F. Holden, M. Roginsky, M. I. Cohen, and L. Finberg: 25-Hydroxy-cholecalciferol in the Management of Rickets Associated with Extra-hepatic Bilary Atresia. J. Pediatrics **88**, 1041 (1976).

17. Blunt, J. W., H. F. DeLuca, and H. K. Schnoes: 25-Hydroxycholesterol: A Biologically Active Metabolite of Vitamin $D_3$. Biochemistry **7**, 3317 (1968).

18. Bolton, I. J., R. G. Harrison, and B. Lythgoe: Calciferol and its Relatives. Part XIV. Total Synthesis of Des-AB-cholestane-8β,9α-diol(1β-[(1R)-1,5-Dimethylhexyl]-7αβ-methyl-*trans*-perhydroindane 4β,5α-diol). J. Chem. Soc. (C) **1971**, 2950.

19. Bolton, I. J., R. G. Harrison, B. Lythgoe, and R. S. Manwaring: Calciferol and its Relatives. Part XIII. Derivatives of Enantiomeric-4-methylhex-3-ene-1, *trans*-2-diols. J. Chem. Soc. (C) **1971**, 2944.

20. Bontekoe, J. S., A. Wignall, M. P. Rappoldt, and J. R. Roborgh: Hydroxylated Vitamin D Analogues. Internat. J. Vit. Res. **40**, 589 (1970).

21. Boomsma, F., H. J. C. Jacobs, E. Havinga, and A. van der Gen: The "Overirradiation Products" of Previtamin D and Tachysterol: Toxisterols. Rec. Trav. Chim. Pays-Bas **96**, 104 (1977).

22. — — — — Vitamin D and Compounds. XXIII. Structure of Bis(cholestadienol) isomers. Rec. Trav. Chim. Pays-Bas **92**, 1361 (1973).

23. Bordier, P., M. M. Pechet, R. Hesse, P. Marie, and H. Rosenson: Response of Adult Patients with Osteomalacia to Treatment with Crystalline 1α-Hydroxyvitamin $D_3$. New Engl. J. Med. **291**, 866 (1974).

24. Boris, A., J. F. Hurley, and T. Trimal: Relative Activities of Some Metabolites and Analogs of Cholecalciferol in Stimulation of Tibia Ash Weight in chicks otherwise deprived of vitamin D. J. Nutr. **107**, 194 (1977).

25. Boyle, I. T., R. W. Gray, and H. F. DeLuca: Regulation by calcium of in vivo Synthesis of 1,25-Dihydroxycholecalciferol and 21,25-Dihydroxycholecalciferol. Proc. Nat. Acad. Sci. (U.S.A.) **68**, 2131 (1971).

26. Boyle, I. T., L. Miravet, R. W. Gray, M. F. Holick, and H. F. DeLuca: The Response of Intestinal Calcium Transport to 24-Hydroxy and 1α,25-Dihydroxy-vitamin D in Nephrectomized Rats. Endocrin. **90**, 605 (1972).

27. Boyle, I. T., J. L. Omdahl, R. W. Gray, and H. F. DeLuca: The Biological Activity and Metabolism of 24,25-Dihydroxyvitamin $D_3$. J. Biol. Chem. **248**, 4174 (1973).

28. Brickman, A. S., J. W. Coburn, and A. W. Norman: Action of 1,25-dihydroxy-cholecalciferol, A Potent Kidney Produced Metabolite of Vitamin $D_3$ in Uremic Man. New Engl. J. Med. **287**, 891 (1972).

29. Brockman, H., and A. Busse: Crystalline Vitamin D from Tuna Fish Liver Oil. Naturwiss. **26**, 122 (1938).

30. BRYNJOLFFSSEN, J., J. M. MIDGLEY, and W. B. WHALLEY: Unsaturated Steroids. Part 4. Some Steroidal Hydroxy-4,4-dimethyl-5,7-dienes and 4,4-Dimethyl-5,7,14(15)-trienes. J. Chem. Soc. Perkin I **1977**, 812.

31. CAGLIOTI, L., P. GRASSELLI, and G. MAINA: Modification of the Bamford-Stevens Reaction: A New Route to 7-Dehydrocholesterol. Chim. Ind. (Milan) **45**, 559 (1963).

32. CHALMERS, P. M., N. W. DAVIES, J. O. HUNTER, K. F. SZAZ, B. PELC, and E. KODICEK: 1α-Hydroxycholatecalciferol as a Substitute for the Kidney Hormone 1,25-Dihydroxycholecalciferol in Chronic Renal Failure. Lancet **2**, 696 (1973).

33. CHAN, J. C. N., S. B. OLDEN, M. F. FOLEY, and H. F. DELUCA: 1α-Hydroxyvitamin $D_3$ in Chronic Renal Failure. J. Amer. Med. Assoc. **234**, 47 (1975).

34. CHAPELO, C. B., P. HALLETT, B. LYTHGOE, I. WATERHOUSE, and P. W. WRIGHT: Calciferol and its Relatives. Part 19. Synthetic Applications of Cyclic Orthoesters: Stereospecific Synthesis of a Bicyclic Alcohol Related to the Vitamins D. J. Chem. Soc. Perkin I **1977**, 1211.

35. COHEN, Z., E. KEINAN, Y. MAZUR, and A. ULMAN: Hydroxylation with Ozone on Silica Gel. The Synthesis of 1,25-Dihydroxyvitamin $D_3$. J. Organ. Chem. (U.S.A.) **41**, 2651 (1976).

36. CORK, D. J., M. R. HAUSSLER, M. J. PITT, E. RIZZARDO, R. H. HESSE, and M. M. PECHET: 1α-Hydroxyvitamin $D_3$: A Synthetic Sterol which is Highly Active in Preventing Rickets. Endocrin. **94**, 1337 (1974).

37. CORRADINO, R. A.: Embryonic chicken testing in organ culture: Response to Vitamin $D_3$ and its Metabolites. Science **179**, 402 (1973).

38. DASGUPTA, S. K., D. R. CRUMP, and M. GUT: New Preparation of Desmosterol. J. Organ. Chem. (U.S.A.) **39**, 1658 (1975).

39. DAUBEN, W. G., and P. BAUMAN: Photochemical Transformations. IX. Total Structure of Suprasterol II. Tetrahedron Lett. **1961**, 565.

40. DAUBEN, W. G., and G. J. FONKEN: The Structure of Photoisopyrocalciferol and Photopyrocalciferol. J. Amer. Chem. Soc. **81**, 4060 (1959).

41. DAUBEN, W. G., and D. S. FULLERTON: Steroids with Abnormal Internal Configuration. A Stereospecific Synthesis of 8α-Methyl Steroids. J. Organ. Chem. (U.S.A.) **36**, 3277 (1971).

42. DAVIDSON, R. S., S. M. WADDINGTON-FEATHER, D. H. WILLIAMS, and B. LYTHGOE: Calciferol and its Relatives. Part IX. The Synthesis of Tachysterol₃. J. Chem. Soc. (C) **1967**, 2534; Tetrahedron Lett. **1963**, 1413.

43. DAWSON, M. T., J. DIXON, P. S. LITTLEWOOD, and B. LYTHGOE: Calciferol and its Relatives. Part XII. (S)-3-Ethynyl-4-methylcyclohex-3-en-1-ol. J. Chem. Soc. (C) **1971**, 2352.

44. DAWSON, T. M., J. DIXON, P. S. LITTLEWOOD, B. LYTHGOE, and A. K. SAKSENA: Calciferol and its Relatives. Part XVI. Total Synthesis of Precalciferol₃. J. Chem. Soc. (C) **1971**, 2960.

45. DIEM, K.: Documenta Geigy. Scientific Tables. New York: Geigy Pharmaceuticals, Ardsley.

46. DIXON, J., B. LYTHGOE, I. A. SIDDIQUI, and J. TIDESWELL: Calciferol and its Relatives. Part XI. Improved Routes to (S)-3-Hydroxymethyl-4-methylcyclohex-3-en-1-ol. J. Chem. Soc. (C) **1971**, 1301.

47. DOXIADIS, S. A., and P. D. LAPATSANIS: 1α-Hydroxyvitamin D in Neonatal Hypocalcaemia. Lancet **1**, 426 (1977).

48. EBEL, J. G., A. N. TAYLOR, and R. H. WASSERMANN: Vitamin D Induced Calcium Binding Protein of Intestinal Mucosa. Amer. J. Clin. Nutr. **22**, 431 (1969).

49. EMKE, A., D. HANDS, J. M. MIDGLEY, W. B. WHALLEY, and R. AHMAD: Unsaturated steroids. Part 6. A Route to Cholesta-5,7-diene-1α,3β-diol; Preparation of Steroidal 4,6,8(14)-Trienes. J. Chem. Soc. Perkin I **1977**, 820.

50. Emtage, J. S., D. E. M. Lawson, and E. Kodicek: The Response of the Small Intestine to Vitamin D. Biochem. J. **144**, 339 (1974).
51. Evans, I. M. A., M. Boulton-Jones, F. H. Doyle, G. F. Jocwin, M. Lockwood, E. W. Matthews, and I. MacIntyre: The Clinical Use of 1,25-Dihydroxycholecalciferol. Proceedings of XI European Symposium on Calcified Tissues. Elsinore, Denmark, **1976**, 236.
52. Eyley, S. C., and D. H. Williams: Photolytic Production of Vitamin D. The Preparative Value of a Photosensitiser. J. Chem. Soc. Chem. Comm. **1975**, 858.
53. — — Synthesis of 24ε,25-Dihydroxyprovitamin D₃. J. Chem. Soc. Perkin I **1976**, 727.
54. — — Synthesis of 25-Hydroxyprovitamin D₃ and 25ε,26-dihydroxyprovitamin D₃. J. Chem. Soc. Perkin I **1976**, 731.
55. Fischer, M.: Industrial Applications of Photochemical Syntheses. Angew. Chem. Int. Ed. **17**, 16 (1978).
56. Fraser, D. R., and E. Kodicek: Unique Biosynthesis by Kidney of a Biologically Active Vitamin D Metabolite. Nature **228**, 764 (1970).
57. Freeman, D., A. Acher, and Y. Mazur: Synthesis of 1α-Hydroxyprovitamin D₃. Tetrahedron Lett. **1975**, 261.
58. Freund, T., and F. Bronner: Stimulation *in vitro* by 1,25-Dihydroxyvitamin D₃ of Intestinal Cell Calcium Uptake and Calcium Binding Protein. Science **190**, 1300 (1975).
59. Fürst, A., L. Labler, W. Meier, and K.-H. Pfoertner. Synthesis of 1α-Hydroxycholecalciferol. Helv. Chim. Acta **56**, 1708 (1973).
60. Georghiou, P. E.: The Chemistry of "Vitamin" D: The Hormonal Calciferols. Chem. Soc. Rev. **6**, 83 (1977).
61. Georghiou, P. E., and G. Just. Cycloadducts of Ergosterol with Azo-type Dienophiles, and their Chemical Reactivities. J. Chem. Soc. Perkin I **1973**, 888.
62. Gray, R. W., H. P. Weber, J. H. Dominquez, and J. Lemann. The Metabolism of Vitamin D₃ and 25-Hydroxyvitamin D₃ in Normal and Anephric Humans. J. Clin. Endocr. Metab. **39**, 1045 (1974).
63. Harmeyer, J., and H. F. DeLuca: Calcium Binding Protein and Calcium Absorption after Vitamin D Administration. Arch. Biochem. Biophys. **133**, 247 (1969).
64. Harrison, I. T., and B. Lythgoe: Calciferol and its Relatives. Part III. Partial Synthesis of Calciferol and of Epicalciferol. J. Chem. Soc. (London) **1958**, 837.
65. Harrison, R. G., B. Lythgoe, and P. W. Wright: Calciferol and its Relatives. Part XVIII. Total Synthesis of 1α-Hydroxyvitamin D₃. J. Chem. Soc. Perkin I **1974**, 2654.
66. Haussler, M. R.: Vitamin D₃ Metabolism, Mode of Action, and Assay of Circulating Hormonal Form. In: Vitamin D and Problems Related to Uremic Bone Disease (Norman, A. W., ed.), p. 25. Berlin: Walter de Gruyter and Co. 1975.
67. Haussler, M. R., D. W. Boyle, E. T. Littledike, and H. Rasmussen: A Rapidly Acting Metabolite of Vitamin D₃. Proc. Nat. Acad. Sci. (U.S.A.) **68**, 177 (1971).
68. Haussler, M. R., J. F. Myrtle, and A. W. Norman: The Association of a Metabolite of Vitamin D₃ with Intestinal Mucosa Chromatin *in vivo*. J. Biol. Chem. **243**, 4055 (1968).
69. Haussler, M. R., J. E. Verwekh, R. H. Hesse, E. Rizzardo, and M. M. Pechet: Biological Activity of 1α-Hydroxycholecalciferol, A Synthetic Analogue of the Hormonal Form of Vitamin D₃. Proc. Nat. Acad. Sci. (U.S.A.) **70**, 2248 (1973).
70. Havinga, E.: Vitamin D, Example and Challenge (Review). Experientia **29**, 1181 (1973).
71. Henderson, R. G., J. G. E. Levinghaf, D. O. Oliver, R. Smith, R. J. Wolton, D. J. Small, C. Preston, G. T. Warner, and A. W. Norman: Effects of 1,25-

Dihydroxycholecalciferol on Calcium Absorption, Muscle Weakness and Bone Disease in Chronic Renal Failure. Lancet I, 379 (1974).

72. HESS, A. F., and M. B. GUTMAN: The Cure of Infantile Rickets with Sunlight. J. Amer. Med. Assoc. **78,** 29 (1922).

73. HIELBRON, I. M., R. N. JONES, K. M. SAMANT, and F. S. SPRING: Studies in the Sterol Group, Part XXIV. The Constitution of Calciferol. J. Chem. Soc. (London), **1936,** 905.

74. HILL, L. F., M. DAVIES, C. M. TAYLOR, and S. W. STANBURY: Treatment of Hyperparathyroidism with 1,25-Dihydroxycholecalciferol. Clin. Endocrinol. **5,** 167S (1976).

75. HOLICK, M. F., M. GARABEDIAN, H. K. SCHNOES, and H. F. DELUCA: Relationship of 25-Hydroxyvitamin $D_3$ Side Chain Structure to Biological Activity. J. Biol. Chem. **250,** 226 (1975).

76. HOLICK, M. F., A. KLEINER-BOSALLIER, H. K. SCHNOES, P. M. KASTEN, I. T. BOYLE, and H. F. DELUCA: 1,24,25-Trihydroxyvitamin $D_3$. J. Biol. Chem. **248,** 6691 (1973).

77. HOLICK, M. F., H. K. SCHNOES, H. F. DELUCA, T. SUDA, and R. J. COUSINS: Isolation and Identification of 1,25-Dihydroxycholecalciferol: A Metabolite of Vitamin D Active in the Intestine. Biochemistry **10,** 2799 (1971).

78. HOLICK, M. F., E. J. SEMMLER, H. K. SCHNOES, and H. F. DELUCA: 1α-Hydroxy Derivative of Vitamin $D_3$: A Highly Potent Analogue of 1α,25-Dihydroxyvitamin $D_3$. Science **180,** 190 (1973).

79. HUNZIKER, F., and F. X. MÜLLNER: Dehydrohalogenierung mit Trialkylphosphiten: Neue Methode zur Herstellung von 7-Dehydrocholesterin. Helv. Chim. Acta **41,** 70 (1958).

80. IKAN, R., A. MARKUS, and Z. GOLDSCHMIDT: Synthesis of Steroidal Cyclopropanes. J. Chem. Soc. Perkin I **1972,** 2423.

81. IKEKAWA, N., M. MOTISAKI, N. KOIZUMI, Y. KATO, and T. TAKESHITA: Synthesis of active forms of vitamin D. VIII. Synthesis of [24R]- and [24S]-1α,24,25-trihydroxyvitamin $D_3$. Chem. Pharm. Bull. **23,** 695 (1975).

82. INHOFFEN, H. H., and K. IRMSCHER: Fortschritte der Chemie der Vitamine D und ihrer Abkömmlinge (Review). Fortschr. Chem. organ. Naturstoffe **17,** 70 (1959).

83. INHOFFEN, H. H., K. IRMSCHER, H. HIRSCHFIELD, U. STACHE, and A. KREUTZER: Partial Synthese der Vitamine $D_2$ und $D_3$. Chem. Ber. **91,** 2309 (1958).

84. JACOBS, H. J. C., F. BOOMSMA, E. HAVINGA, and A. VAN DER GEN: The Photochemistry of Previtamin D and Tachysterol. Rec. Trav. Chim. Pays-Bas **96,** 113 (1977).

85. JOHNSON, R. L., S. C. CAREY, A. W. NORMAN, and W. H. OKAMURA: Studies on Vitamin D and Its Analogues. 10. Side Chain Analogues of 25-Hydroxyvitamin $D_3$. J. Med. Chem. **20,** 5 (1977).

86. JOHNSON, R. L., W. H. OKAMURA, and A. W. NORMAN: Mode of Action of Calciferol. X. 24-Nor-23-hydroxyvitamin $D_3$, An Analog of 25-Hydroxyvitamin $D_3$ Having Antivitamin Activity. Biochem. Biophys. Res. Comm. **67,** 797 (1975).

87. JUTTMANN, J. R., J. D. BATHS, and J. C. BIRKENHAGER: Treatment of Anticonvulsant Osteomalacia with 1α-Hydroxycholecalciferol. Brit. Ned. J. **1977,** 551.

88. KAMINSKI, J. J., and N. BODOR: 3-Bromo-4,4-dimethyloxazolidinone, Preparation and Investigation of a New Brominating Agent. Tetrahedron **32,** 1097 (1976).

89. KANEKO, C., A. SUGIMOTO, Y. EGUCHI, S. YAMADA, and M. ISHIKAWA: A New Synthetic Method of 1α-Hydroxy-7-dehydrocholesterol. Tetrahedron **30,** 2701 (1974).

90. KANEKO, C., S. YAMADA, A. SUGIMOTO, Y. EGUCHI, M. ISHIKAWA, T. SUDA, M. SUZUKI, S. KAKUTA, and S. SASAKI: Synthesis and Biological Activity of 1α-Hydroxyvitamin $D_3$. Steroids **23,** 75 (1974).

91. Kaneko, C., S. Yamada, A. Sugimoto, M. Ishikawa, S. Sasaki, and T. Suda: 1α-Hydroxylation of Cholesterol and the Related 3-Hydroxysteroids. Tetrahedron Lett. **1973**, 2339.

92. Kaupp, G.: Photochemical Rearrangements and Framentations of Alkenes and Polyenes. Angew. Chem. Int. Ed. **17**, 150 (1978).

93. Kodicek, E.: The Biosynthesis of $C^{14}$-Labelled Ergocalciferol. Biochem. J. **60**, XXV (1955).

94. Kodicek, E., D. E. M. Lawson, and P. W. Wilson: Biological Activity of a Polar Metabolite of Vitamin $D_3$. Nature **228**, 763 (1970).

95. Koizumi, N., M. Morisaki, and N. Ikekawa: Absolute Configurations of 24-Hydroxycholesterol and Related Compounds. Tetrahedron Lett. **1975**, 2203.

96. Kooh, S., W. D. Fraser, R. Toon, and H. F. DeLuca: Response of Protracted Neonatal Hypocalcaemia to 1α,25-Dihydroxyvitamin $D_3$. Lancet **II**, 1105 (1976).

97. Lam, H.-Y., B. L. Onisko, H. K. Schnoes, and H. F. DeLuca: Synthesis and Biological Activity of 3-Deoxy-1α-hydroxyvitamin $D_3$. Biochem. Biophys. Res. Comm. **59**, 845 (1974).

98. Lam, H.-Y., H. K. Schnoes, and H. F. DeLuca: Synthesis and Biological Activity of 25ε,26-Dihydroxycholecalciferol. Steroids **25**, 247 (1975).

99. — — — Synthesis of 1α-Hydroxyergocalciferol. Steroids **30**, 671 (1977).

100. Lawson, D. E. M.: Abstracts. Third Workshop on Vitamin D, California, 64 (1977).

101. Lawson, D. E. M., D. R. Fraser, E. Kodicek, H. R. Morris, and D. H. Williams: Identification of 1,25-Dihydroxycholecalciferol, A New Kidney Hormone Controlling Calcium Metabolism. Nature **230**, 228 (1971).

102. Lawson, D. E. M., P. W. Wilson, and E. Kodicek: New Vitamin D Metabolite Localized in Intestinal Cell Nuclei. Nature **222**, 171 (1969).

103. — — — Metabolism of Vitamin D: A new Cholecalciferol Metabolite Involving Loss of Hydrogen at C-1 in Chick Intestinal Nuclei. Biochem. J. **115**, 269 (1969).

104. Littlewood, P. S., B. Lythgoe, and A. K. Saksena: Calciferol and its Relatives. Part XV. The preparation of Des-AB-cholest-8-ene-8-carbaldehyde {1β-[(1 R)-1,5-Dimethylhexyl]-3aα,6,7,7aβ-tetrahydro-7a-methylindane-4-carbaldehyde} and 9α-Chloro-des-AB-cholestan-8-one {5α-Chloro-1β-[(1 R)-1,5-dimethylhexyl]-7aβ-methyl-*trans*-perhydroindan-4-one}. J. Chem. Soc. (C) **1971**, 2956.

105. Long, R. G., K. Skinner, M. R. Wells, and S. Sherlock: Serum 25-Hydroxyvitamin D and Untreated Parenchymal and Cholestatic Liver Disease. Lancet **2**, 650 (1976).

106. Lund, B., L. Hjorth, I. Kjaer, I. Reimann, T. Friis, R. B. Anderson, and O. H. Sorenson: Treatment of Osteoporosis of Aging with 1α-Hydroxycholatecalciferol. Lancet **2**, 1168 (1975).

107. Lythgoe, B., R. Manwaring, J. R. Milner, T. A. Moran, M. E. N. Nambudiry, and J. Tideswell: Calciferol and its Relatives. Part 21. A Synthesis of (S)-(Z)-2-(5-Hydroxy-2-methylenecyclohexylidene)ethanol. J. Chem. Soc. Perkin I **1978**, 387.

108. Lythgoe, B., M. E. Nambudiry, and J. Tideswell: Direct Total Synthesis of Vitamins $D_2$ and $D_3$. Tetrahedron Lett. **1977**, 3685.

109. Lythgoe, B., D. A. Roberts, and I. Waterhouse: Calciferol and its Relatives. Part 20. A Synthesis of Windhaus and Grundmann's $C_{19}$ Ketone. J. Chem. Soc. Perkin I **1977**, 2608.

110. McCollum, E. V., N. Simmonds, J. E. Becker, and P. G. Shipley: Studies on Experimental Rickets XXI. J. Biol. Chem. **53**, 293 (1922).

111. McMorris, T. C., and S. R. Schow: A Convenient Synthesis of 25-Oxo-27-nor-cholesteryl Acetate. J. Organ. Chem. (U.S.A.) **41**, 3759 (1976).

112. McMorris, T. C., S. R. Schow, and G. R. Weike: Evidence for a C-29 Hydroxyl Group in Oogoniol. Tetrahedron Lett. **1978**, 335.

113. Mellanby, E.: An Experimental Investigation of Rickets. Lancet **196**, 407 (1919).

*114.* MIHAILOVIĆ, M. LJ., LJ. LORENC, and V. PAVLOVIĆ: A Convenient Synthesis of 1α- and 1β-Hydroxycholesterol. Tetrahedron Lett. **1977**, 441.

*115.* MIRAVET, L., J. R. DEL, M. CARRE, M. L. QUEILLE, and P. BORDIER: The Biological Activity of Synthetic 25,26-Dihydroxycholecalciferol and 24,25-Dihydroxycholecalciferol in Vitamin D-Deficient Rats. Calcif. Tiss. Res. **21**, 145 (1976).

*116.* MORII, H., J. LUND, P. NEVILLE, and H. F. DELUCA: Biological Activity of a Vitamin D Metabolite. Arch. Biochem. Biophys. **120**, 508 (1967).

*117.* MORISAKI, M., K. BANNAI, and N. IKEKAWA: Synthesis of Active Forms of Vitamin D. II. Synthesis of 1α-Hydroxycholesterol. Chem. Pharm. Bull. **21**, 1853 (1973).

*118.* — — — Studies on Steroids. XXXVIII. A New Preparation of 1α-Hydroxycholesterol. Chem. Pharm. Bull. **24**, 1948 (1976).

*119.* MORISAKI, M., N. KOIZUMI, and N. IKEKAWA: Synthesis of Active Forms of Vitamin D. Part IX. Synthesis of 1α,24-Dihydroxycholecalciferol. J. Chem. Soc. Perkin I **1975**, 1422.

*120.* MORISAKI, M., J. RUBIO-LIGHTBOURN, and N. IKEKAWA: Synthesis of Active Forms of Vitamin D. I. A Facile Synthesis of 25-Hydroxycholesterol. Chem. Pharm. Bull. **21**, 457 (1973).

*121.* MORISAKI, M., J. RUBIO-LIGHTBOURN, N. IKEKAWA, and T. TAKESHITA: Synthesis of Active Forms of Vitamin D. V. A Practical Route to 1α,25-Dihydroxycholesterol. Chem. Pharm. Bull. **21**, 2568 (1973).

*122.* MORISAKI, M., A. SAIKA, K. BANNAI, M. SAWAMURA, J. RUBIO-LIGHTBOURN, and N. IKEKAWA. Synthesis of Active Forms of Vitamin D. X. Synthesis of 1α-Hydroxyvitamin $D_3$. Chem. Pharm. Bull. **23**, 3272 (1975).

*123.* MOURIÑO, A., and W. H. OKAMURA: Studies on Vitamin D and Its Analogues. 14. On the 10,19-Dihydrovitamins Related to Vitamin $D_2$ Including Dihydrotachysterol. J. Organ. Chem. (U.S.A.) **43**, 1653 (1978).

*124.* MYRTLE, J. F., and A. W. NORMAN: Vitamin D: A Cholecalciferol Metabolite Highly Active in Promoting Intestinal Calcium Transport. Science **171**, 79 (1971).

*125.* NARWID, T. A., J. F. BLOUNT, J. A. IACOBELLI, and M. R. USKOKOVIC: Vitamin $D_3$ Metabolites. III. Synthesis and X-ray Analysis of 1α,25-Dihydroxycholesterol. Helv. Chim. Acta **57**, 781 (1974).

*126.* NARWID, T. A., K. E. COONEY, and M. R. USKOKOVIC: Vitamin $D_3$ Metabolites. II. Further Syntheses of 25-Hydroxycholesterol. Helv. Chim. Acta **57**, 771 (1974).

*127.* NORMAN, A. W.: The Mode of Action of Vitamin D. Biol. Rev. Cambridge Phil. Soc. **43**, 97 (1968).

*128.* NORMAN, A. W., and H. F. DELUCA: The Preparation of $H^3$-Vitamins $D_2$ and $D_3$ and Their Localization in the Rat. Biochemistry **2**, 1160 (1963).

*129.* NORMAN, A. W., J. LUND, and H. F. DELUCA: Biologically Active Forms of Vitamin $D_3$ in the Kidney and Intestine. Arch. Biochem. Biophys. **108**, 12 (1964).

*130.* NORMAN, A. W., J. F. MYRTLE, R. J. MIDGETT, H. G. NOWICKI, V. WILLIAMS, and G. POPJAK: 1,25-Dihydroxycholecalciferol: Identification of the Proposed Active Form of Vitamin $D_3$ in the Intestine. Science **173**, 51 (1971).

*131.* OHKI, E.: Steroidal Studies. XV. Synthesis of 4,4-Dimethylcholecalciferol and Its Conversion into 10-Isosteroid. Chem. Pharm. Bull. **8**, 46 (1960).

*132.* OCHI, K., I. MATSUNAGA, M. SHINDO, and C. KANEKO: Synthesis of Desmosterol and Epidesmosterol from Hyodeoxycholic Acid. Steroids **30**, 2204 (1977).

*133.* OKAMURA, W. H., M. L. HAMMOND, A. REGO, A. W. NORMAN, and R. M. WING: Studies of 5,6-trans-Vitamin $D_3$ and the Stereoisomers of 10,19-Dihydrovitamin $D_3$ Including Dihydrotachysterol. J. Organ. Chem. (U.S.A.) **42**, 2284 (1977).

*134.* OKAMURA, W. H., M. N. MITRA, M. R. PIRIO, A. MOURINO, S. C. CAREY, and A. W. NORMAN: Studies on Vitamin D and Its Analogs. 13. 3-Deoxy-3α-methyl-1α-hydroxyvitamin $D_3$, 3-Deoxy-3α-methyl-1α,25-dihydroxyvitamin $D_3$ and 1α-Hydroxy-3-

epivitamin $D_3$. Analogues with Conformationally Biased A Rings. J. Organ. Chem. (U.S.A.) **43**, 574 (1978).

135. Okamura, W. H., M. N. Mitra, R. M. Wing, and A. W. Norman: Chemical Syntheses and Biological Activity of 3-Deoxy-1α-hydroxyvitamin $D_3$ an Analog of 1α,25-Dihydroxyvitamin $D_3$, the Active Form of Vitamin $D_3$. Biochem. Biophys. Res. Comm. **60**, 179 (1974).

136. Okamura, W. H., and M. R. Pirio: Vitamin D and Its Analogs. IX. 1α-Hydroxy-3-epivitamin $D_3$, Its Synthesis and Conformational Analysis. Tetrahedron Lett. **1975**, 17.

137. Onisko, B. L., H.-Y. Lam, L. E. Reeve, H. K. Schnoes, and H. F. DeLuca: Synthesis and Bioassay of 3-Deoxy-1α-hydroxyvitamin $D_3$, an Active Analog of 1α,25-Dihydroxyvitamin $D_3$. Bioorg. Chem. **6**, 203 (1977).

138. Onisko, B. L., H. K. Schnoes, and H. F. DeLuca: Synthesis of Potential Vitamin D Antagonists. Tetrahedron Lett. **1977**, 1107.

139. Paaren, H. E., H. K. Schnoes, and H. F. DeLuca: Synthesis of 1β-Hydroxyvitamin $D_3$ and 1β,25-Dihydroxyvitamin $D_3$. a) J. Chem. Soc. Chem. Comm. **1977**, 890; b) Proc. Nat. Acad. Sci. (U.S.A.) **75**, 2080 (1978).

140. Palm, T. A.: The Geographical Distribution and Etiology of Rickets. Practitioner **45**, 270 (1890).

141. Partridge, J. J., S. Faber, and M. R. Uskokovic: Vitamin $D_3$ Metabolites. I. Synthesis of 25-Hydroxycholesterol. Helv. Chim. Acta **57**, 764 (1974).

142. Partridge, J. J., S.-J. Shiney, E. G. Baggiolini, B. Hennessy, and M. R. Uskokovic: A Stereoselective Synthesis of 1α,24$R$, 25-Trihydroxycholecalciferol, A Metabolite of Vitamin $D_3$. In: Vitamin D: Biochemical, Chemical and Clinical Aspects Related to Calcium Metabolism (Norman, A. W., et al., ed.), p. 47. Berlin: Walter de Gruyter and Co. 1977.

143. Partridge, J. J., V. Toome, and M. R. Uskokovic: A Stereoselective Synthesis of the 24($R$), 25-Dihydroxycholesterol Side Chain. J. Amer. Chem. Soc. **98**, 3739 (1976).

144. Peacock, M., J. C. Gallagher, and B. E. C. Nordin: Action of 1α-Hydroxyvitamin $D_3$ on Calcium Absorption and Bone Resorption in Man. Lancet **I**, 385 (1974).

145. Pechet, M. M., and R. H. Hesse: Metabolic and Clinical Effects of Pure Crystalline 1α-Hydroxyvitamin $D_3$ and 1α,25-Dihydroxyvitamin $D_3$. Amer. J. Med. **57**, 13 (1974).

146. — — The Biological Activities of Pure Crystalline 1α-Hydroxyvitamin $D_3$ and 1α,25-Dihydroxyvitamin $D_3$. Mol. Cell Endo. **1**, 305 (1974).

147. Pelc, B.: 4β-Hydroxycholecalciferol and an Attempted Synthesis of its 4β-Hydroxy-epimer. J. Chem. Soc. Perkin I **1974**, 1436.

148. — The Selenium Dioxide Oxidation of Cholecalciferol. Steroids **30**, 193 (1977).

149. Pelc, B., and E. Kodicek: 1α-Hydroxycholesterol. J. Chem. Soc. (C) **1970**, 1624.

150. Pelc, B., and D. H. Marshall: Thermal Transformation of Cholecalciferol Between 100—170°. Steroids **31**, 23 (1978).

151. Ponchon, G., and H. F. DeLuca: The Role of the Liver in the Metabolism of Vitamin D. J. Clin. Invest. **48**, 2373 (1969).

152. Proscal, D. A., H. L. Henry, G. J. Friedlander, and A. W. Norman: Studies on the Mode of Action of Calciferol. Arch. Biochem. Biophys. **179**, 229 (1977).

153. Raisz, L. G., and H. F. DeLuca: Effects of Thyrocalcitonin and Phosphate Ion on the Parathyroid Hormone Stimulated Resorption of Bone. Endocrin. **1972**, 479.

154. Redel, J., N. Bazely, Y. Calando, and F. Delbarre: The Synthesis of 24,25-Dihydroxycholecalciferol, A Metabolite of Vitamin $D_3$. J. Steroid Biochem. **6**, 117 (1975).

155. Redel, J., P. A. Bell, N. Bazely, Y. Calando, F. Delbarre, and E. Kodicek: The Synthesis and Biological Activity of 25,26-Dihydroxycholecalciferol, A Polar Metabolite of Vitamin $D_3$. Steroids **24**, 463 (1974).

156. REDEL, J., L. MIRAVET, N. BAZELY, Y. CALANDO, M. CARRE, and F. DELBARRE: Synthesis and Biological Activity of Diastereoisomers of 25 R- and 25 S-25,26-Dihydroxycholecalciferol. Compt. Rend. Acad. Sc. (Paris) D **285**, 443 (1977).

157. ROTMAN, A., and Y. MAZUR: C-25 Hydroxylation of Cholesterol Derivatives. J. Chem. Soc. Chem. Comm. **1974**, 15.

158. RUBIO-LIGHTBOURN, J., M. MORISAKI, and N. IKEKAWA: Synthesis of Active Forms of Vitamin D. III. Synthesis of 1α,25-Dihydroxycholesterol. Chem. Pharm. Bull. **21**, 1854 (1973).

159. RUSSELL, R. G. G., R. SMITH, R. J. WALTON, C. PRESTON, R. BASSON, R. E. HENDERSON, and A. W. NORMAN: 1,25-Dihydroxycholecalciferol and 1α-Hydroxycholecalciferol in Hyperparathyroidism. Lancet **II**, 14 (1974).

160. SALMOND, W. G.: Approaches to the Synthesis of Vitamin D Metabolites and Analogs. In: Vitamin D: Biochemical, Chemical and Clinical Aspects Related to Calcium Metabolism (NORMAN, A. W., K. SCHAEFER, J. W. COBURN, H. F. DELUCA, D. FRASER, H. G. GRIGOLEIT, and D. V. HERRATH, eds.), p. 61. Berlin: W. de Gruyter. 1977.

161. SALMOND, W. G., M. A. BARTA, A. M. CAIN, and M. C. SOBALA: Alternative Modes of Decomposition of Allylic Selenoxides Diastereoisomeric at Selenium. Preparation of Δ5,7- and 5β-Hydroxy-Δ6-steroids. Tetrahedron Lett. **1977**, 1683.

162. SALMOND, W. G., M. A. BARTA, and J. L. HAVENS: A Stereoselective Wittig Reagent and Its Application to the Synthesis of 25-Hydroxylated Vitamin D Metabolites. J. Organ. Chem. (U. S. A.) **43**, 790 (1978).

163. — — — Allylic Oxidation with 3,5-Dimethylpyrazole: Chromium Trioxide Complex. Steroidal $\Delta^{5,7}$-Ketones. J. Organ. Chem. (U. S. A.) **43**, 2057 (1978).

164. SALMOND, W. G., and K. D. MAISTO: A Synthesis of 25-Hydroxycholesterol. Tetrahedron Lett. **1977**, 987.

165. SALMOND, W. G., and M. C. SOBALA: An Efficient Synthesis of 25-Hydroxycholesterol from Stigmasterol. Tetrahedron Lett. **1977**, 1695.

166. SALMOND, W. G., M. C. SOBALA, and K. D. MAISTO: A Synthesis of 25-Hydroxycholesterol. Tetrahedron Lett. **1977**, 1237.

167. SANDERS, G. M., J. POT, and E. HAVINGA: Some Recent Results in the Chemistry and Stereochemistry of Vitamin D and its Isomers (Review). Fortschr. Chem. organ. Naturstoffe **27**, 131 (1969).

168. SCHNOES, H. K., and H. F. DELUCA: Vitamin D: Chemistry and Biochemistry of a New Hormonal System (Review). In: Bioorganic Chemistry, Vol. II (TAMELEN, E. E. VAN, ed.), chap. 12, p. 299. New York: Academic Press. 1978.

169. SEKI, M., N. KOIZUMI, M. MORISAKI, and N. IKEKAWA: Synthesis of Active Forms of Vitamin D. VI. Synthesis of (24 R)- and (24 S)-24,25-Dihydroxyvitamin $D_3$. Tetrahedron Lett. **1975**, 15.

170. SEKI, M., J. RUBIO-LIGHTBOURN, M. MORISAKI, and N. IKEKAWA: Synthesis of Active Forms of Vitamin D. IV. Synthesis of 24,25- and 25,26-Dihydroxycholesterol. Chem. Pharm. Bull. **21**, 2783 (1973).

171. SEMMLER, E. J., M. F. HOLICK, H. K. SCHNOES, and H. F. DELUCA: The Synthesis of 1α,25-Dihydroxycholecalciferol — A Metabolically Active Form of Vitamin $D_3$. Tetrahedron Lett. **1972**, 4147.

172. SHEVES, M., N. FRIEDMAN, and Y. MAZUR: Conformational Equilibria in Vitamin D. Synthesis of 1β-Hydroxyvitamin $D_3$. J. Organ. Chem. (U. S. A.) **42**, 3597 (1977).

173. SHEVES, M., and Y. MAZUR: The Vitamin D-3,5-Cyclovitamin D Rearrangement. J. Amer. Chem. Soc. **97**, 6249 (1975).

174. — — Equilibria in Vitamin $D_3$. Preparation and Properties of 6-Methylvitamin $D_3$ and its Isomers. J. Chem. Soc. Chem. Comm. **1977**, 21.

175. — — Epimerization of Vitamin $D_3$. The Cholecalciferyl Ion. Tetrahedron Lett. **1976**, 1913.

176. Silverburg, Z. S., K. B. Bettcher, G. B. Dossetor, T. R. Overton, M. F. Foley, and H. F. DeLuca: Effect of 1,25-Dihydroxycholecalciferol in Renal Osteodystrophy. Can. Med. Ass. J. **112**, 190 (1975).

177. Spencer, R., M. Charman, P. Wilson, and E. Lawson: Vitamin D-Stimulated Intestinal Calcium Absorption May Not Involve Calcium Binding Protein Directly. Nature **263**, 161 (1976).

178. Stern, P. H., C. L. Trummel, H. K. Schnoes, and H. F. DeLuca: Bone Resorbing Activity of Vitamin D Metabolites and Congeners *in vitro:* Influence of Hydroxyl Substituents in the A-Ring. Endocrin. **97**, 1552 (1976).

179. Strating, J.: Compounds Related to Provitamin D$_3$. IV. 3-Methylcholesterol and the Corresponding Provitamin. Rec. Trav. Chim. Pays-Bas **71**, 822 (1952).

180. Suda, T., H. F. DeLuca, H. K. Schnoes, Y. Tanaka, and M. F. Holick: 25,26-Dihydrocholecalciferol, A Metabolite of Vitamin D$_3$ with Intestinal Calcium Transport Activity. Biochemistry **9**, 4776 (1970).

181. Takeshita, T., S. Ishimoto, and N. Ikekawa: Preparation of Desmosterol from Fucosterol. Chem. Pharm. Bull. **24**, 1928 (1976).

182. Tanaka, Y., L. Castillo, H. F. DeLuca, and N. Ikekawa: The 24-Hydroxylation of 1,25-Dihydroxyvitamin D$_3$. J. Biol. Chem. **252**, 1421 (1977).

183. Tanaka, Y., and H. F. DeLuca: Biological Activity of 1,25-Dihydroxyvitamin D in the Rat. Endocrin. **92**, 417 (1973).

184. Tanaka, Y., H. F. DeLuca, A. Akaiwa, M. Morisaki, and N. Ikekawa: Synthesis of 24S- and 24R-Hydroxy-[24-$^3$H]-vitamin D$_3$ and their Metabolism in Rachitic Rats. Arch. Biochem. Biophys. **177**, 615 (1976).

185. Tanaka, Y., H. F. DeLuca, N. Loizomi, and N. Ikekawa: Importance of the Stereochemical Position of the 24-Hydroxyl to Biological Activity of 24-Hydroxyvitamin D$_3$ and 1α,25-Dihydroxyvitamin D$_3$ on Intestinal Calcium Uptake. Proc. Nat. Acad. Sci. (U.S.A.) **72**, 229 (1975).

186. Trost, B. M., D. F. Taber, and J. B. Alper: An Approach to the Stereocontrolled Creation of an Acyclic Side Chain of Some Natural Products. Tetrahedron Lett. **1976**, 3857.

187. Trost, B. M., and T. R. Verhoeven: New Synthetic Reactions. Catalytic vs. Stoichiometric Allylic Alkylation. Stereocontrolled Approach to Steroid Side Chain. J. Amer. Chem. Soc. **98**, 630 (1976).

188. Trummel, C. L., L. G. Raisz, and R. B. Hallick: 25-Hydroxydihydrotachysterol — Stimulation of Bone Resorption in Tissue Culture. Biochem. Biophys. Res. Comm. **44**, 1096 (1971).

189. Tschesche, R., B. Goossens, G. Piestert, and A. Tospfer: Synthesis of 27-Nor-25-oxocholest-5-en-3β-yl Acetate and 27-Nor-25-oxocholestanol from Pregnenolone. Tetrahedron Lett. **1977**, 735.

190. Wagenfeld, J. B., B. A. Nenchausky, M. Bolt, J. Vander Horst, J. L. Boyer, and I. H. Rosenberg: Comparison of Vitamin D and 25-Hydroxyvitamin D in the Therapy of Primary Biliary Cirrhosis. Lancet **II**, 391 (1976).

191. Wassermann, R. H.: Mechanism of Intestinal Calcium Absorption. Abstracts, Third Workshop on Vitamin D, California, **1976**, 76.

192. Wicha, J., and K. Bal: Synthesis of 25-Hydroxycholesterol from 3β-Hydroxyandrost-5-en-17-one. A Method for Stereospecific Construction of a Sterol Side-Chain. J. Chem. Soc. Chem. Comm. **1975**, 968.

193. — — Synthesis of Preg-17(20)-en-21-oic Acid Derivatives. Syn. Comm. **7**, 215 (1977).

194. Windhaus, A., O. Linsert, A. Luttringhaus, and G. Weidlich: Über das kristallisierte Vitamin D$_2$. Ann. **492**, 226 (1932).

195. Windhaus, A., Fr. Schenk, and F. von Werder: Über das antirachitisch wirksame Bestrahlungsprodukt aus 7-Dehydrocholesterin. Z. Physiol. Chem. **241**, 100 (1936).

196. WINDHAUS, A., and W. THIELE: Über die Konstitution des Vitamin $D_2$. Ann. **521**, 160 (1936).
197. WONG, R. G., J. F. MYRTLE, and H. C. TSAI: Studies on Calciferol Metabolism. V. The Occurrance and Biological Activity of 1,25-Dihydroxyvitamin D in Bone. J. Biol. Chem. **247**, 5728 (1972).
198. WONG, R. G., A. W. NORMAN, C. R. REDDY, and J. W. COBURN: Biological Effects of 1,25-Dihydroxycholecalciferol (A Highly Active Vitamin D Metabolite) in Acutely Uremic Rats. J. Clin. Invest. **51**, 1287 (1972).
199. YAKHIMOVICH, R. I., N. F. FURSAEVA, and G. M. SEGAL: Synthesis of the Fluoroanalog of Vitamin $D_3$. Bioorg. Khim. **1976**, 1526.
200. YANG, S. S., C. P. DORN, and H. JONES: Synthesis of 25-Fluorovitamin D. Tetrahedron Lett. **1977**, 2315.

*(Received August 30, 1978)*

# Stereochemistry of Naturally Occurring Carotenoids

By S. Liaaen-Jensen, Organic Chemistry Laboratories,
Norwegian Institute of Technology, University of Trondheim, Norway

With 2 Figures

## Contents

# I. General Introduction

Carotenoids form a class of isoprenoid polyenes widely distributed in Nature. They are synthesized *de novo* by all photosynthetic organisms and certain bacteria, yeasts and fungi, whereas various animals have the capacity to modify the structure of dietary carotenoids. Carotenoids serve many important functions including protection of photosynthetic organisms against photodynamic destruction. They serve as auxiliary light absorbers for photosynthesis and phototaxis and as pro-Vitamin A in mammals. The role of Vitamin A in the visual process is well established. Whereas carotenoid pigments appear to have no biological activity, several of their presumed metabolites are biologically active. Synthetic carotenoids are used as food additives (*101*).

Particular aspects of carotenoids have been reviewed before in this series: Geometrical isomerism was treated by Zechmeister (*181*) in 1960 and the application of spectroscopic methods in structural elucidation by Weedon (*172*) in 1969.

More recent developments are covered by Isler's comprehensive monograph of 1971 (*101*) which covered all aspects of carotenoids and has been up-dated by the published main lectures from the IUPAC carotenoid symposia in 1972 (*97*), 1975 (*99*), and 1978 (*100*). Entry to the original literature through the middle of 1975 is readily obtained through Straub's (*162*) compilation of naturally occurring carotenoids. Recent reviews of special topics such as carotenoproteins (*123 c*), carotenoids of photosynthetic bacteria (*127, 154*) and marine carotenoids (*128*) are available.

New rules for carotenoid nomenclature are now approved (*98*). In the semirational names a double Greek prefix is used to indicate the $C_9$ end groups, Scheme 1. Thus in the new nomenclature $\alpha$-carotene is $\beta,\varepsilon$-carotene. The numbering of the carbon skeleton is indicated and further principles of the new nomenclature rules are illustrated by the semirational name of neoxanthin (**1**) = (3*S*,5*R*,6*R*,3′*S*,5′*R*,6′*S*)-5′,6′-epoxy-6,7-didehydro-5,6,5′,6′-tetradehydro-$\beta,\beta$-carotene-3,5,3′-triol.

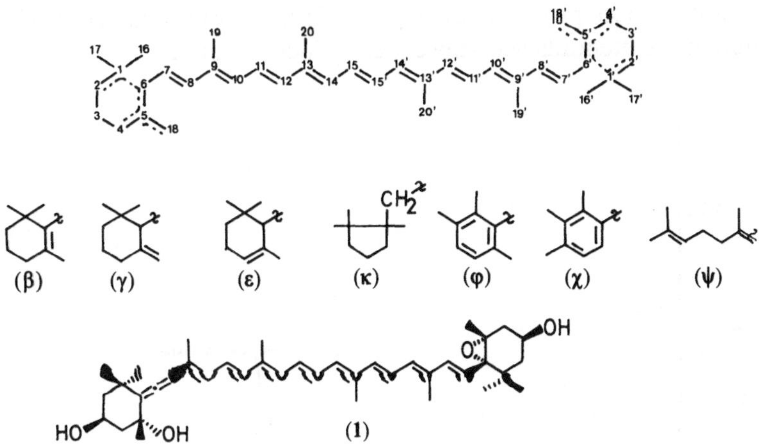

*Scheme 1*

Since 1971 approximately 150 new carotenoids have been described and the number of references has about tripled from the previous 800. Fig. 1 illustrates the trends in carotenoid research. The new methods largely responsible for such progress are indicated to the left and new structural features established for naturally occurring carotenoids are given to the right. The latter comprise diapo ($C_{30}$), homo ($C_{45}$ and $C_{50}$), nor and apo ($< C_{40}$) skeletons and a variety of functional groups, including allenic and triple bonds; epoxy, hydroxy, aldehyde, keto, carboxylic acid, ester and lactone functions, as well as aryl and phenolic end groups. So far no hetero elements other than oxygen have been encountered in naturally occurring carotenoids.

Of the around 450 different carotenoids characterized to date about half are chiral with 1—6 chiral centres in the carotenoid moiety. In the majority of cases the all-*trans* configuration of the polyene chain is preferred. However, some *cis* isomers occur naturally. Certain substituent effects may lead to preferred *cis* configuration.

The stereochemistry of carotenoids is conveniently treated in two categories:

A) Geometrical *(cis-trans)* isomerism around carbon-carbon double bonds and conformation around single bonds.

B) Absolute configuration of any chiral groups: asymmetric carbons and allenes.

Since much activity has been centered around determination of absolute configuration of carotenoids during the last decade, this review will be devoted mainly to this particular aspect of carotenoid research, but will also include recent advances concerning geometrical isomerism. Literature available through July 1978 has been evaluated for this review.

New information provided during the fifth international IUPAC symposium on carotenoids was subsequently incorporated.

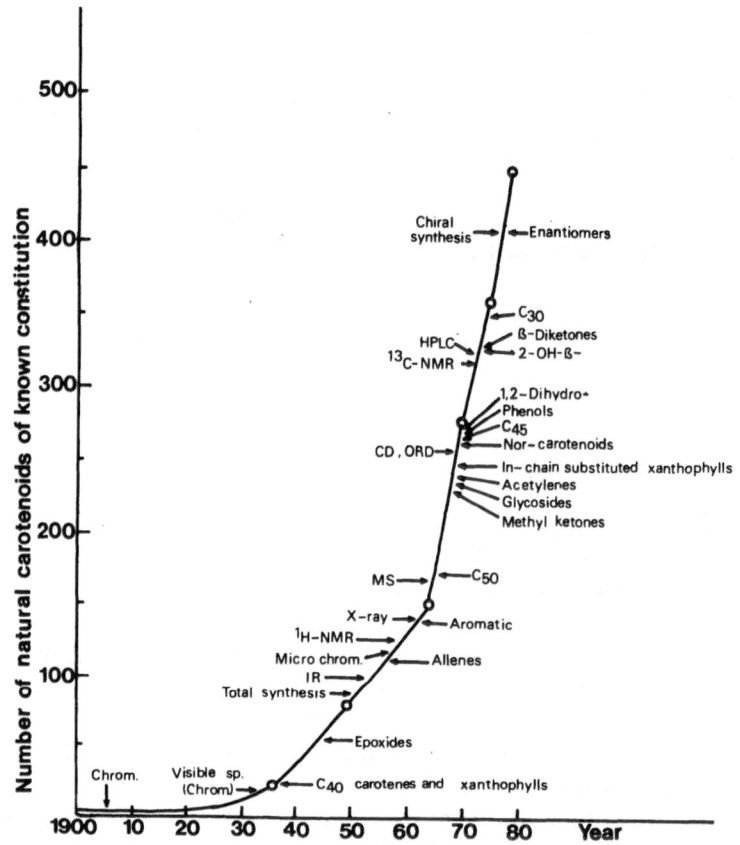

Fig. 1. The number of naturally occurring carotenoids of known constitution as a function of time with important events indicated

## II. Historical

The literature on geometrical isomerism of carotenoids from 1923 to 1970 has been discussed in detail by ZECHMEISTER (*181, 182*) and WEEDON (*174*). Problems related to conformation have also been considered (*174*).

Regarding absolute configuration classical polarimetric studies in the thirties indicated that natural carotenoids occurred as discrete optical isomers (*110*). Pioneering synthetic work in the fifties by KARRER's school (*165, 166*) resulted in preparation of the first optically active synthetic carotenes from $(+)$-$\alpha$-ionone and from $(-)$-$\alpha$-ionone. However, it was not until the late sixties that the chirality of these synthetic carotenes and of naturally occurring $\beta,\varepsilon$-carotene was determined (72). In the meantime oxidative degradation studies of the paprika carotenoids, coupled with spectroscopic studies and syntheses of models by the schools of KARRER and WEEDON, led to the absolute configuration of capsanthin and capsorubin (174). $^1$H NMR spectroscopy, introduced to the carotenoid field around 1960, provides useful information on relative configuration (169). To date only one successful X-ray analysis of a chiral carotenoid has been carried out (167), and only few key degradation products of known crystal structure serve as reference compounds. Correlations during the last decade based on ORD (20) and subsequently CD (126) have been verified by partial syntheses and total syntheses of chiral model carotenoids, and more recently by preparation of pure optical isomers of various natural carotenoids (131). $^{13}$C NMR spectroscopy has lately become a useful tool for establishing geometrical configurations (66, 135).

Whereas WEEDON (174) has covered this topic through 1970, it is with the more recent developments in this field this review will be concerned.

# III. Methods and Application

## 1. X-Ray Crystallography

The stereochemistry of a chiral, intact carotenoid has not yet been solved by X-ray crystallographic analysis (174). Recently the crystal structure of capsanthin (2) di-p-bromobenzoate was reported (167) in support of the previous configurational assignment (see below).

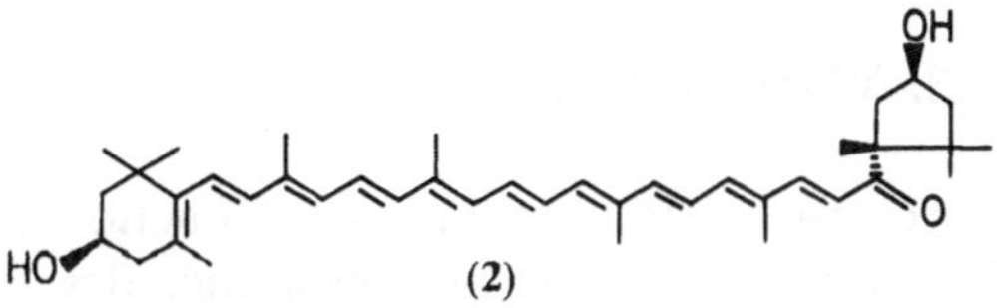

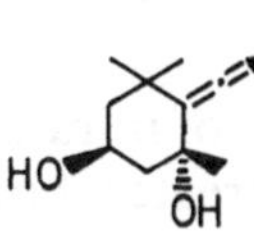

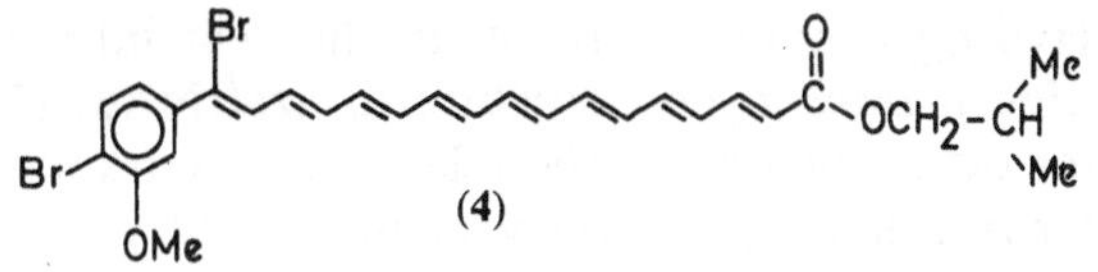

X-ray crystallography of the *p*-bromobenzoate of the allenic ketone (3) (*63*) fòrms the basis for the stereochemical assignment of various carotenoids containing the same end group (*91, 108, 174*) and several carotenoids with 3-hydroxylated β-rings (*64*).

In view of the recent successful structural assignment of the related polyene (4) without centrosymmetry (*9*), future progress in the structural elucidation of chiral carotenoids by X-ray crystallographic analysis is expected. A recent progress report (*135a*) has announced confirmation of the gross geometry of fucoxanthin [(40), R=H], so far without distinguishing between oxygen and carbon atoms of the end groups.

## 2. Electronic Spectra

Details of the absorption spectra in the ultraviolet and visible region serve to indicate the presence and position of *cis* (Z) double bonds in the polyene chain; this subject is covered elsewhere (*174, 181, 182*).

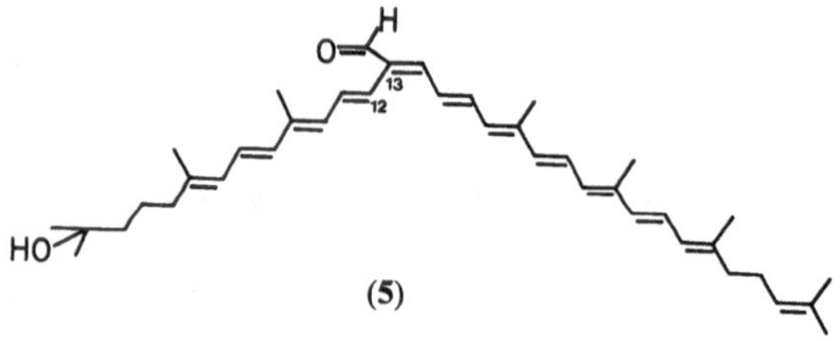

(5)

Recently low-temperature conditions and comparison with calculated spectra were used to prove the 13-*cis*-12-s-*trans* configuration assumed for rhodopin-20-al (5) (*1*). This provided additional information about the out-of-plane twisting of the polyene chain of (5) (*54*).

## 3. Vibrational Spectra

The use of infrared spectroscopy for assignment of functional groups has been treated briefly elsewhere (*2, 18*) and now includes enolized β-diketones (*114*). Hydrogen bonding studies in the near infrared were employed for assigning a *trans*-glycol structure to azafrin (6) (*70, 137*) and more recently for determination of the relative configuration within the triol ring of heteroxanthin (7) and relevant models (*41*).

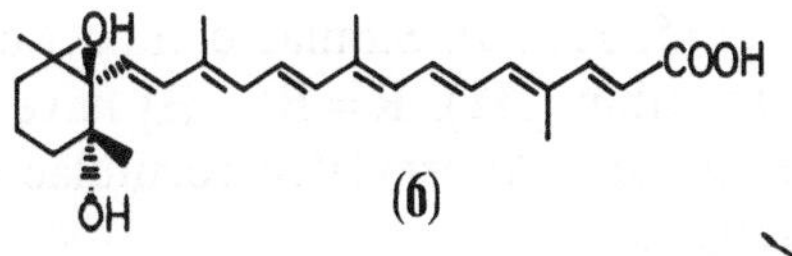

(6)

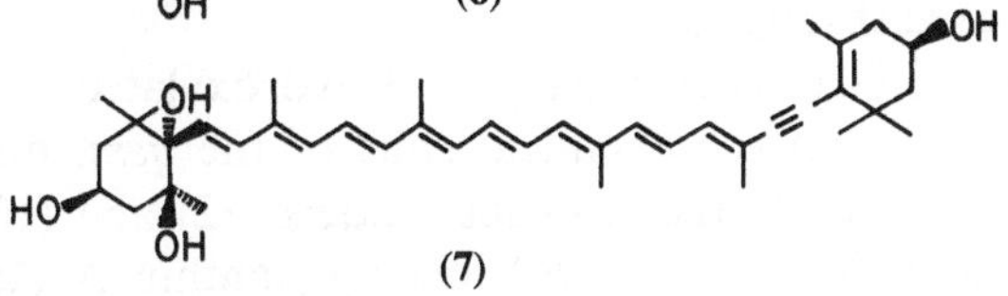

(7)

## 4. Proton Nuclear Magnetic Resonance Spectroscopy

This topic was reviewed in 1971 (*169*), and only more recent applications regarding the relative configuration of carotenoids will be treated here. Fourier transform techniques have greatly reduced the sample requirements and now permit the registration of ${}^1$H NMR spectra of carotenoids on the microgram scale.

Chemical shift considerations for the *gem.* dimethyl groups have been successfully used to assign relative configuration to carotenoids containing 2,6-substituted ε-rings such as the $C_{50}$-carotenoid decaprenoxanthin (**8**) by comparison with synthetic 2,6-*cis* and 2,6-*trans* model $C_{42}$-

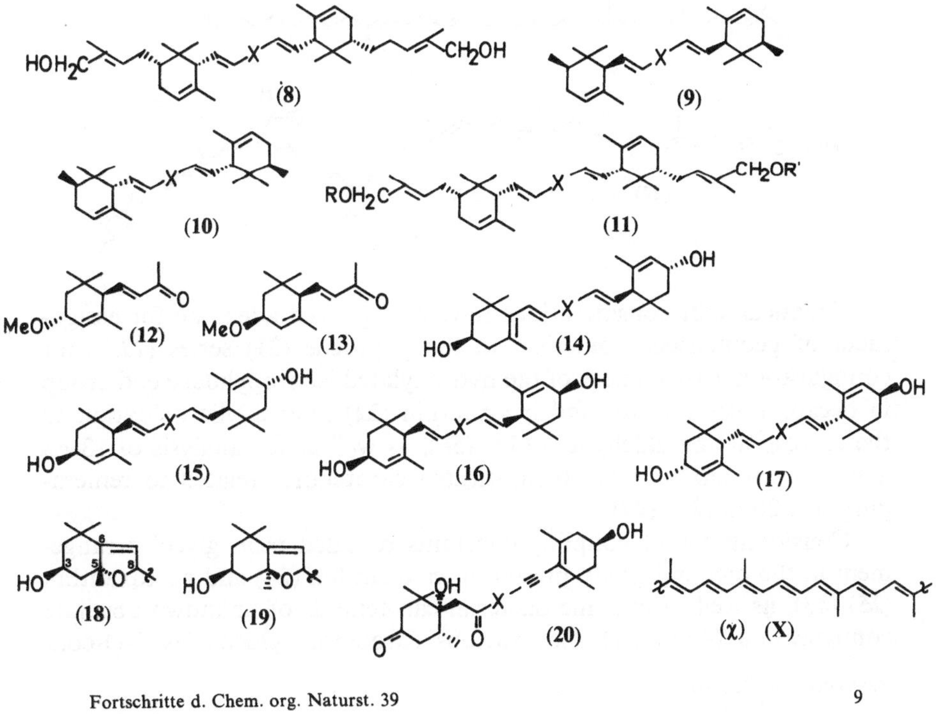

(8)

(9)

(10)

(11)

(12)

(13)

(14)

(15)

(16)

(17)

(18)

(19)

(20)

(χ) (X)

carotenoids (9) and (10) (5, 11, 56). Similar comparisons relevant to the
$C_{50}$-carotenoid sarcinaxanthin [(11), R = R′ = H] have been made in the
2,6-substituted γ-series (6, 90). Abbreviated formulae with X as defined
for (8) will be used throughout.

The relative 3,6-stereochemistry in 3-hydroxylated ε-rings also has
a marked influence on the chemical shift of the *gem.* dimethyl group.
Evidence is available for the relevant models (12) and (13), lutein (14),
subsequently degraded to (12) (40), chiriquixanthin A (15) and B (16)
(24) and tunaxanthin (17) (152).

For 3-hydroxylated 5,8-furanoxides such as (18) and (19) the relative
configuration for C-3,5 may also be deduced by considering the chemical
shift of $CH_3$-16,17,18 (78). Since the stereochemistry at C-5 is retained
during the rearrangement of 5,6-epoxides to 5,8-furanoxides (46), the
relative configuration of carotenoids containing 3-hydroxy-5,6-epoxy end
groups [(18) and (19)] may conveniently be determined. In a very
recent contribution involving aromatic solvent induced shifts (ASIS)
and double resonance experiments some of the original assignments
have been corrected (46a). Shielding effects for isomytiloxanthin (20)
and for appropriate synthetic models were evaluated so as to permit
assignment of stereochemistry to the substituted cyclohexanone ring of
(20) (114, 175).

(21)

(22)

(23)

Chemical shift considerations have also proved adequate for assignment of geometrical isomerism in the phytoene (21) series (115), for
configurational assignment of the hydroxylated isopropylidene end group
in decaprenoxanthin (8) and lycoxanthin (22) after allylic oxidation to
the corresponding aldehydes (133, 156), as well as for analysis of 13-*cis*
and 13-*trans* mixtures of 20-substituted carotenoids related to reniera-
purpurin-20-al (23) (49).

Considerations of coupling constants revealed *trans* glycol arrangement in the diol end group present in caloxanthin (24) and nostoxanthin
(25) (43), as well as in some bacterial carotenoids of unknown absolute
configuration (120, 121). For various carotenoid glucosides β-D-con-

figuration has been indicated from the *trans* coupling of H-1 and H-2 in the glucose moiety (*16, 90, 155*). Assignment of configuration to the sugar moiety of carotenoid rhamnosides is more difficult, but is facilitated by ¹H NMR assignments for methyl triacetyl β- and α-L-rhamnoside (*87*) which are based on chemical shifts and coupling constants. A bacterial carotenoid has been identified as (3 *R*, 3′ *R*)-zeaxanthin (**26**) di-α-L-rhamnoside (*88*).

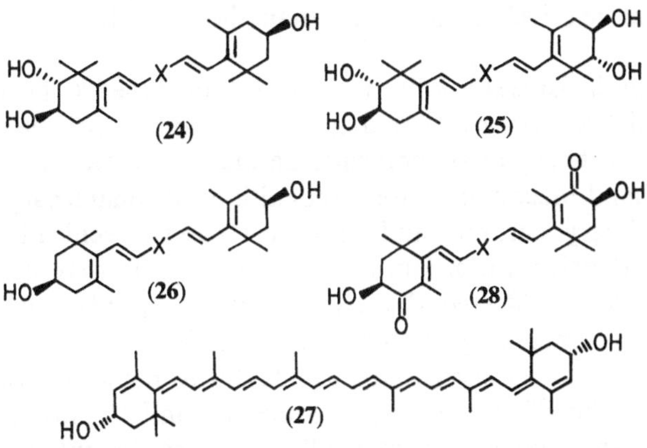

Decoupling techniques have been used when appropriate (*121, 161*), and the first application of ASIS has recently been reported (*46 a*).

Shift reagents have been applied to locate functional groups (*13, 90, 117, 118*), but the use of chiral shift reagents for checking optical purity *etc.* has not been fully explored. However, it has been reported that (3*S*,3′*S*)-astaxanthin (**28**) exhibited larger induced shifts with tris[(3-heptafluorobutyryl)-*d*-camphorato] europium (III) than its 3*R*,3′*R* enantiomer (*12*).

Nuclear Overhauser effects (*142*) of appropriate model compounds were studied for assignment of the all-*trans* configuration of the polyene chain of eschscholtzxanthin (**27**) (*8*); and for identification of the H-7 and H-8 signals of the furanoxides flavoxanthin and crysanthemaxanthin (*46 a*) derived from lutein epoxide (**46**).

## 5. ¹³C Nuclear Magnetic Resonance Spectroscopy

### a) Authentic Carotenoids

Application of ¹³C NMR spectroscopy (*177*) to the carotenoid field was first reported in 1970 by ROBERTS and co-workers (*103*) using proton-noise decoupling and off-resonance partial proton decoupling.

Partly tentative assignments were made for retinyl acetate, 15-*cis*- and all-*trans*-β,β-carotene and 15,15′-didehydro-β,β-carotene. [13]C assignments for phytoene (**20**) are still disputed (*83a, 135*). BREMSER and PAUST (*29*) correctly assigned the [13]C resonances of β,β-carotene as a limiting case of a series of apocarotenals with increasing chain length. The influence of the carbonyl function in apocarotenals on the polarization of the double bonds of the polyene chain was studied. Alternating charge densities along the polyene chain cause the signals of the α,γ *etc.* carbon atoms to resonate at relatively higher field than those of the β,δ *etc.* carbon atoms.

In a comprehensive study ENGLERT (*68*) studied the [13]C NMR spectra of 83 Vitamin A derivatives and 23 β-apo- and C[40]-carotenoids and provided complete signal assignments. Besides CW offset [1]H decoupling and broad-band decoupling for recognition of quaternary, tertiary, secondary/primary carbons and in a few cases specifically deutered derivatives, extensive use of lanthanide shift reagents were used. In LIS experiments which reflected the distance of the specific carbon under study from the complexing site (up to 8 double bonds away), Yb (dpm)[3] offered advantages over europium reagents due to reduced *Fermi* contact contributions. The effect of *cis-trans* isomerism on [13]C shifts of Vitamin A analogues was studied systematically. In a trisubstituted *cis* double bond of the polyene chain the methyl substituent resonated at *ca.* 8 p.p.m. downfield value relative to the corresponding *trans* derivative. Characteristic shielding of adjacent olefinic protons was also observed in the *cis* isomers. Thus [13]C NMR resonance is the preferred method for identification of *cis* double bonds in carotenoids.

In a symposium paper Moss (*135*) published data for a series of methyl β-apocarotenoates and some fifteen C[40]-carotenoids. Long range [13]C-[31]P coupling was noted in allylic phosphonium salts.

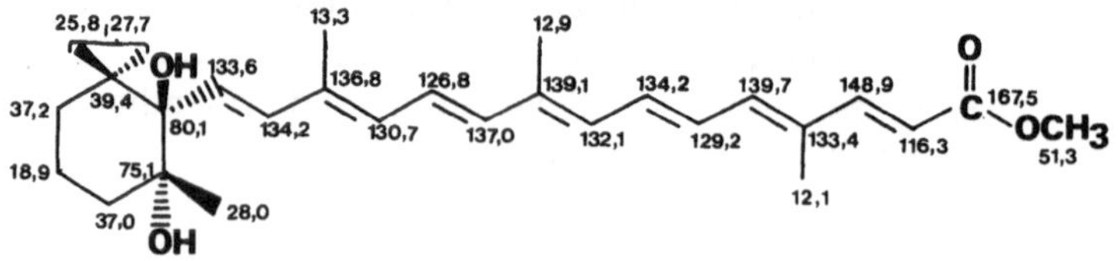

*Scheme 2*

[13]C NMR assignments for (3*S*,3′*S*)-astaxanthin (**28**) have been confirmed by Yb (dpm)[3]-induced shifts (*69*). Data for the 15-*cis* isomer agreed with those previously obtained for 15-*cis*-β,β-carotene (*68, 135*). Recently the [13]C NMR signals of azafrin (**6**) methyl ester and of

$(5R,6R)$-5,6-dihydro-β,β-carotene-5,6-diol (**29**) have been completely assigned (*71*) and those of the C-8 epimeric furanoid rearrangement products of lutein epoxide (**46**), that is flavoxanthin and crysanthema-xanthin, have been partly assigned (*46 a*). As an example chemical shifts in $d_5$-pyridine of azafrin (**6**) methyl ester are cited in Scheme 2 (*71*) which demonstrates the large shift differences between $sp^3$ and $sp^2$ carbon atoms.

## b) Use in Structural Elucidation

$^{13}$C NMR was first used in structural elucidation of carotenoids for peridinin (**30**) (*108, 160, 161*). Support for the presence of the allo-xanthin (**31**) end group in isomytiloxanthin (**20**) was obtained by direct comparison of $^{13}$C NMR spectra (*114, 175*). From the characteristic shifts of the 14-methyl and C-8 signals violeoxanthin was shown to be the 9-*cis* isomer of violaxanthin (**32**) (*136*).

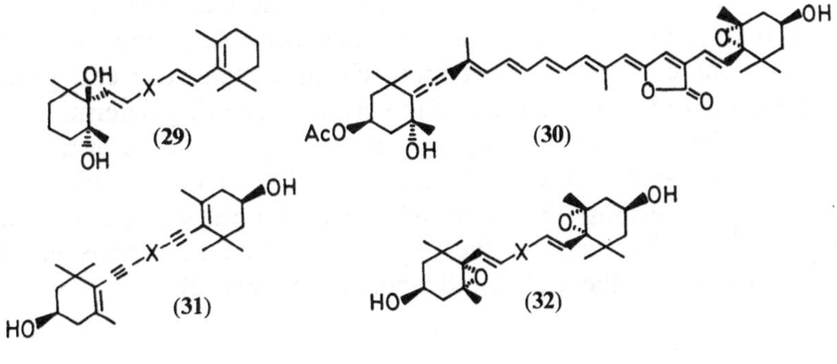

The structure of the C-5 monomethyl ether of azafrin (**6**) methyl ester was determined by partial analysis of the $^{13}$C NMR spectrum (*170*), and a new isomer of phytoene (**21**) has been assigned the $Z,E,Z$ or $Z,E,E$ configuration from $^{13}$C NMR evidence (*18*). Further application of $^{13}$C NMR spectroscopy for the structural elucidation of carotenoids will be facilitated by the assigned spectra of authentic carotenoids gradually available.

## 6. Optical Rotatory Dispersion and Circular Dichroism

Except for measurement of optical rotation at single wavelengths chiroptical properties of carotenoids have only been studied during the last decade.

In a fundamental paper on ORD spectra of carotenoids BARTLETT *et al.* (*20*) classified chiral carotenoids into homodichiral (with two identical chiral end groups), heterodichiral (with two different chiral end groups) and monochiral (with one chiral and one achiral end group) carotenoids. They advanced an additivity hypothesis, stating that the ORD contribution from each end group was additive.

In subsequent work CD spectra rather than ORD curves have been commonly employed (*157*). Comparative CD studies by EUGSTER's group (*35, 37, 44*) on monochiral carotenoids served to establish the chirality of various carotenoids containing unsubstituted ε-rings, using (6′R)-β,ε-carotene as chiroptical standard. Several papers from our laboratory (*2, 4—8, 10, 11, 13—15, 21, 25, 27, 28, 42, 43, 84, 88—91, 105, 108, 118, 151—153, 164*) have applied CD spectroscopy in studies on absolute configuration. The validity of the additivity hypothesis has been demonstrated also for CD spectra (*28, 118*), provided homodichiral or heterodichiral carotenoids of identical chromophores are compared (*13*). For monochiral carotenoids variation in the length of the polyene chain results in predictable shifts of the maxima of the CD curves.

A second important empirical rule has been developed (*7, 118*), stating that for carotenoids containing chiral cyclohexene end groups the preferred chiral conformation of the cyclohexene ring determines the sign of the Cotton effect caused by that end group. Conformational analysis has revealed that key substituents at C-2 or C-3 in substituted β- and ε-rings determine the preferred conformation, whereas an additional hydroxy substituent in the allylic 4-position in β-rings has no marked influence on the CD (see Scheme 3 and Fig. 2).

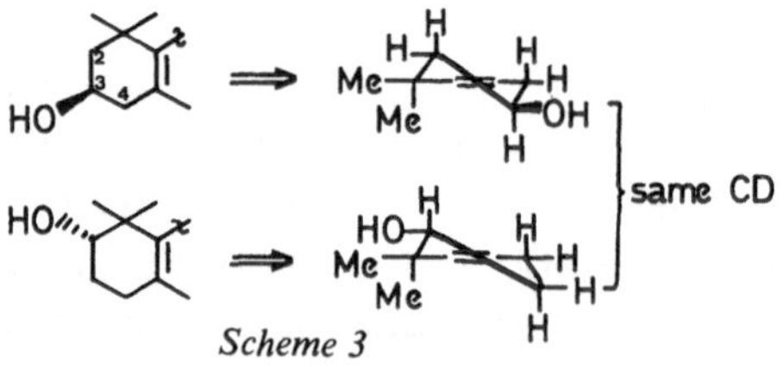

*Scheme 3*

By this approach the absolute configuration of carotenoids such as (2R,2′R)-β,β-carotene-2,2′-diol [(**33**), R = OH], see Fig. 2, and of (3S,3′S)-astaxanthin (**28**) (after LiAlH₄ reduction) has been assigned (*7, 118*) and subsequently proved by independent methods (*38, 116*). In accordance with the conformational rule esterification or glycosidation of a hydroxy group (*21, 88*) of a β-ring or replacement of a 2-hydroxy group in β,ε- or γ-rings with methyl or isopentenyl substituents has little effect on the

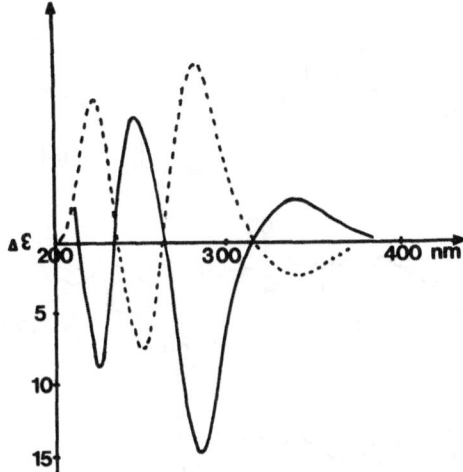

Fig. 2. CD spectra in EPA solution of ———— (3R, 3′R)-zeaxanthin (**26**) and
––– (2R, 2′R)-β, β-carotene-2,2′-diol (**33**, R = OH)

CD (*5, 6, 11, 69, 70, 118*), and a chiral centre at C-3 in an ε-ring has no
particular influence on the Cotton effect caused by that end group (*24, 25,
40, 152*).

Triple bonds in the 7(7′) position greatly reduce the Cotton effect of
a chiral cyclohexene ring. This is exemplified by the CD spectra of
alloxanthin (**31**) (*20, 21*) and (3S,3′S)-7,8,7′,8′-tetradehydroastaxanthin
(**34**) (*21*).

Chiral cyclohexyl end groups such as are found in heteroxanthin (**7**)
(*42*) and chiral aliphatic end groups such as are found in plectaniaxanthin
[(**35**), R = R′ = H] (*28, 153*), aleuriaxanthin (**36**) (*14*) and bacterioruberin
(**37**) (*107*) exhibit only weak Cotton effects compared with the cyclohexene
derivatives discussed above. The same is true for cyclopentane derivatives
such as mytiloxanthin (**38**) (*57*) and trikentriorhodin (**39**) (*3, 39, 57*) *vs.*
cyclopentene derivatives (*139*). Weak Cotton effects are also caused by the
allenic end group in peridinin (**30**) (*108*) and 19′-hexanoyloxyfucoxanthin
[(**40**), R = OCO(CH₂)₄CH₃] (*91*), where oxidative degradation has been
required for stereochemical assignment.

Substituent effects in chiral aliphatic end groups on CD spectra are
not yet fully explored, but appear to be more complex than in cyclohexene derivatives (*28, 107, 153*).

The effect of *cis* double bonds in the polyene chain should be pointed
out. In bicyclic carotenoids such as diatoxanthin (**41**) mono-*cis* configuration at Δ9,9′, 13,13′ and 15 reverses the sign of the Cotton effect
(*20, 36, 69, 126, 136*). Consequently CD spectra are sensitive to impurities
of geometrical isomers. The CD band in the so-called *cis*-peak region

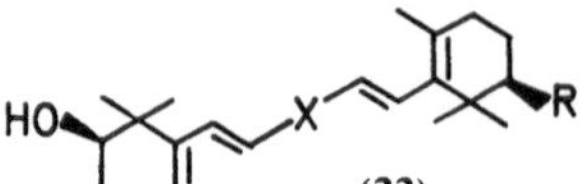

(33)

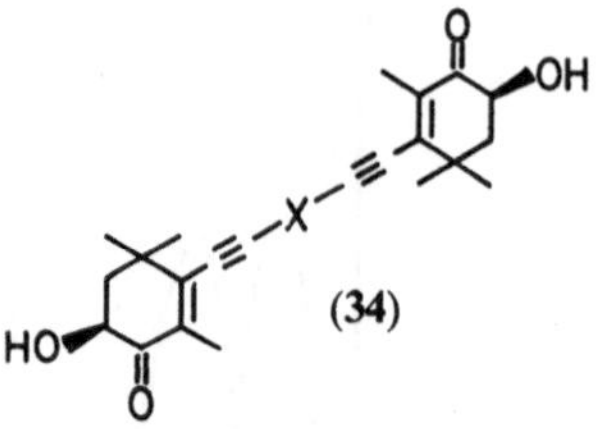

(34)

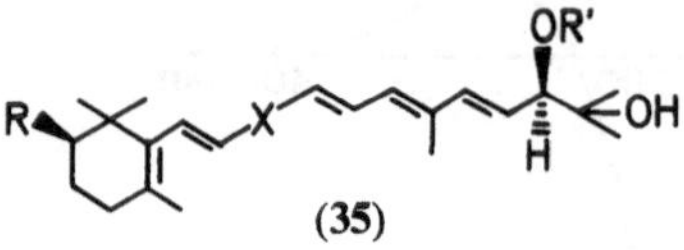

(35)

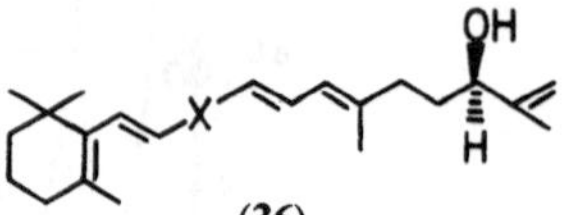

(36)

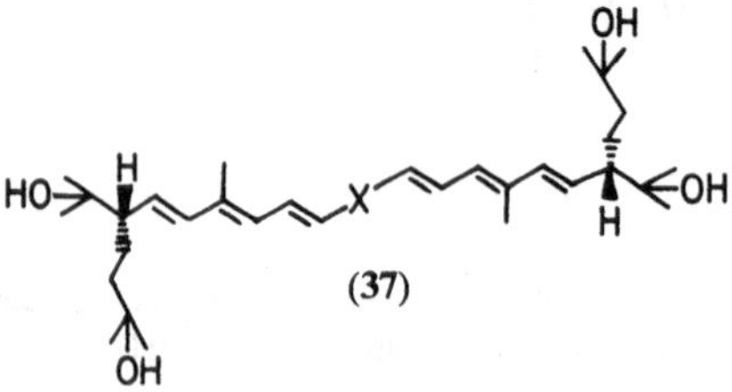

(37)

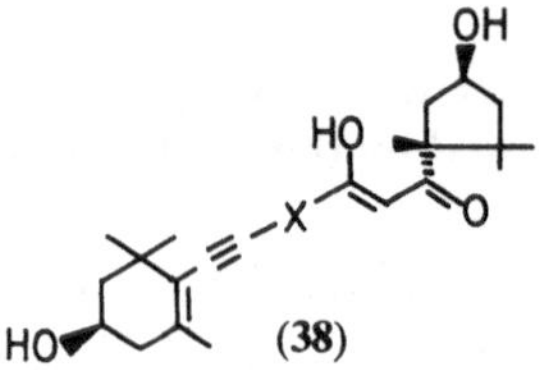

(38)

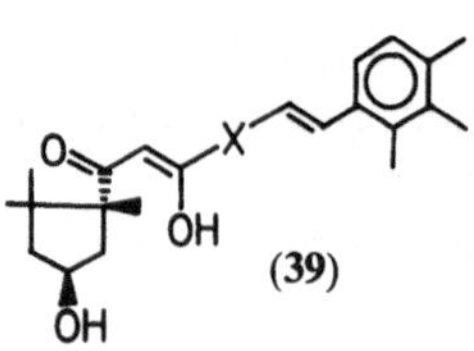

(39)

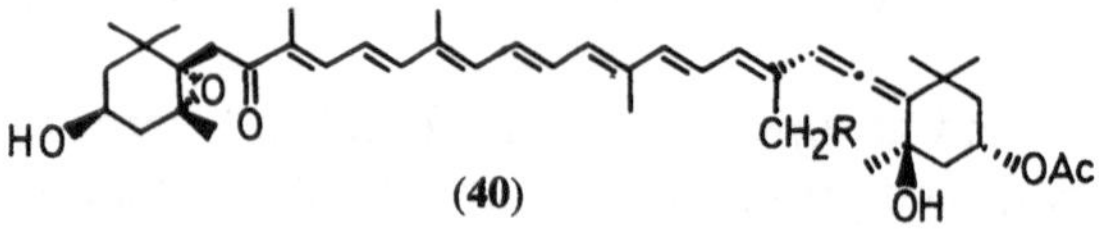

(40)

(*182*) is enhanced for mono-*cis* isomers, mostly for isomers with near-to-central *cis* bonds (*69, 89, 126*). Reversal of the Cotton effect has also been reported for 13- and 13′-mono-*cis* isomers of fucoxanthin [(**40**), R=H] and for its presumed 6′S isomer (*22*). For the monochiral

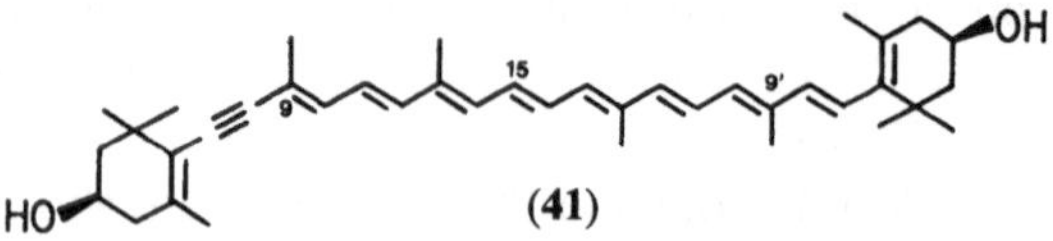

(41)

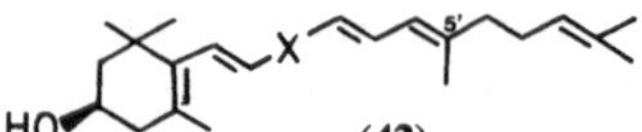

(42)

gazaniaxanthin considered to be the 5′-mono-*cis* isomer of rubixanthin (**42**) (*33*) with a terminal *cis* bond no change in sign of the Cotton effect is observed (*15*).

The sensitivity of CD spectra to steric alterations in the polyene chain is also reflected by the CD of in-chain substituted carotenoids such as loroxanthin (of unknown configuration), which is the 19′-hydroxy derivative of lutein (**14**) (*89*). In-chain substituents are known to influence the geometry of the polyene chain (*49, 54*).

Carotenoproteins such as the blue alloporin (*151*), in which (3*S*,3′*S*)-astaxanthin (**28**) is the prosthetic group, exhibit CD spectra similar in shape and bathochromically displaced relative to unbound (**28**). The carotenoprotein resulting from recombination of the apoprotein of alloporin with (3*S*,3′*S*)-7,8,7′,8′-tetradehydroastaxanthin (**34**) exhibits very similar chiroptical properties to those of alloporin (*151*).

The CD spectrum of the light-harvesting peridinin (**30**)-chlorophyll-protein complex from dinoflagellates (*86*) has recently been used together with fluorescence excitation spectra for interpretation of the molecular topology of this complex (*122*).

Lately the CD spectrum of the carotenoprotein α-crustacyanin which exhibits two maxima at 690 nm (+) and 583 nm (−), in phosphate buffer has been interpreted in terms of dipole-dipole coupling between the transition moments of the two carotenoid molecules on each subunit. The CD splitting corresponded to an exciton band width of 2800 cm$^{-1}$ which leads to an interchromophore distance of *ca.* 13 Å (*123a*).

With few exceptions (*e. g.* **36**) reliable CD spectra have not been obtained in the visible region. Although a theoretical interpretation of carotenoid CD spectra has been lacking, the method is indispensable for studies of absolute configuration on an empirical basis and requires only microgram samples. Even when considered merely as an identification criterion CD will gain increasing importance (*2, 8, 168*). In special cases the CD spectrum may reduce several structural alternatives to a minimum (*25, 43*).

Recently SNATZKE (*157a*) has shown that application of simple HMO theory (inherent chirality and exciton theory) gives a correct picture of the sign pattern of the Cotton effects observed for carotenoids such as zeaxanthin (**26**), lutein (**14**) and mono-*cis* derivatives thereof.

## 7. Linear Dichroism

Linear dichroic spectra of renierapurpurin-20-al (**23**) and other synthetic cross-conjugated carotenals have been analyzed and used to prove the same 13-*cis* configuration (*49*) previously assigned to the naturally occurring rhodopin-20-al (**5**) (*54, 152*).

## 8. High Pressure Liquid Chromatography

As the last physical method useful for stereochemical studies of carotenoids one should mention high pressure liquid chromatography. This was first applied to the carotenoid field in 1971 by STEWART and WHEATON (*159*) and subsequently shown to be superior for separation of geometrical isomers (*75, 76, 158*) and diastereomers (*46, 75*). Future developments on a preparative scale are anticipated.

## 9. Horeau Method

Turning now to chemical methods used for stereochemical studies of carotenoids the modified HOREAU method (*32*) has been successfully used to confirm the configuration of (2*R*)-β,β-caroten-2-ol [(**33**), R = H] (*38*) and to establish the configuration of aleuriaxanthin (**36**) (*34*). However, no conclusive results could be obtained for plectaniaxanthin [(**35**), R = R' = H] (*28*). In principle the HOREAU method is based on partial resolution of racemic and meso α-phenylbutyric anhydride by means of an optically active secondary alcohol, which reacts preferentially with one diastereomer. The unreacted anhydride is quantitatively determined by an appropriate procedure (*32, 68, 94*).

## 10. Oxidative Degradation

So far chiral centres in carotenoids have been restricted to the end groups. In cases where the chiroptical properties of the intact carotenoid are not significant, oxidative degradation may be resorted to. Ozone, zink permanganate (*26*) or nickel peroxide (*140*) are conveniently used (*26, 38, 40, 108*). Thus for example both fucoxanthin [(**40**), R = H] and peridinin (**30**) upon oxidative degradation provided the acetate of the allenic ketone (**3**) (*27, 108*) and 19'-hexanoyloxyfucoxanthin [(**40**), R = OCO(CH$_2$)$_4$CH$_3$] the corresponding hexanoyloxy derivative (*91*). The absolute configuration of lutein (**14**), first wrongly assigned as 3'-*S* (*79*), was elucidated in EUGSTER's laboratory (*40*) by means of oxidative degradations. Lutein (**14**) was converted to the 3'-didehydro derivative upon allylic oxidation with nickel peroxide and oxidative cleavage provided (6*R*)-3-oxo-α-ionone (**43**) which was identified by CD, thus demonstrating 6'*R*-chirality for lutein (**14**) (see Scheme 4). Lutein (**14**) dimethyl ether was degraded by photooxidation with Rose Bengal as sensitizer to provide 3-methoxy-α- and -β-ionone derivatives. The assignment for (3*R*)-methoxy-β-ionone (**44**) was based on application of

MILLS' rule (*67, 68*) and gave the stereochemistry at C-3 for lutein (**14**). Finally the configuration at C-3′ was proved by direct comparison, including GC and $^1$H NMR, of (3*R*,6*R*)-3-methoxy-α-ionone (**45**) obtained by degradation of lutein (**14**) dimethyl ether of unchanged configuration at C-3′ with the authentic compound and its 3*S* epimer.

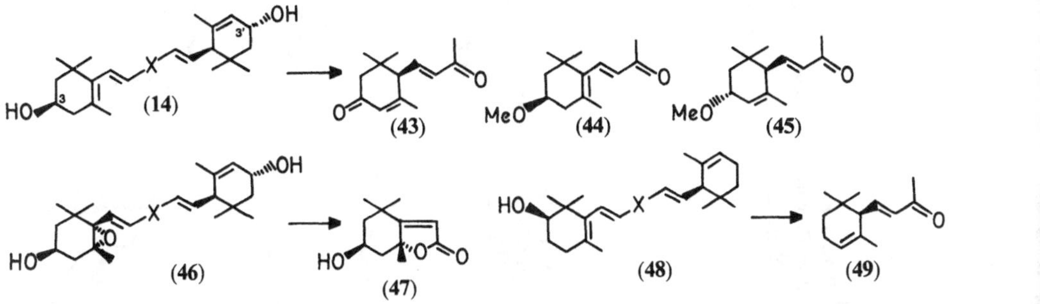

*Scheme 4*

Natural lutein epoxide (**46**) gave (−)-loliolide (**47**) after conversion to the diastereomeric furanoxides and oxidative degradation with nickel peroxide, thus proving the 3′*R*,6′*R*-configuration for (**46**) (*46*). Similar degradation of β,ε-caroten-3-ol (**48**) provided (6*R*)-α-ionone (**49**) (*38*).

Oxidative degradation of carotenoids has also been performed with the purpose of localizing *cis* double bonds in the polyene chain (*163*). Thus controlled alkaline permanganate oxidation to apocarotenals served to identify violeoxanthin as the 9-*cis* isomer of violaxanthin (*32*).

## 11. Partial Synthesis

In the carotenoid field functional group modifications or conversions involving no or minor modification of the carbon skeleton are classified as partial syntheses. The transformation of carotenoids with unknown absolute configuration to carotenoids of known chirality by reactions that retain the chiral centres intact will therefore be discussed in this section. Key conversions are summarized in Scheme 5.

Upon base catalyzed isomerization of the Δ4 double bond lutein (**14**) has been converted to optically inactive meso zeaxanthin (**50**) (*64*).

(3*R*,3′*R*)-Zeaxanthin (**26**), as diacetate, was transformed to the diacetate of eschscholtzxanthin (**27**) by dehydrogenation with N-bromosuccinimide (*68*).

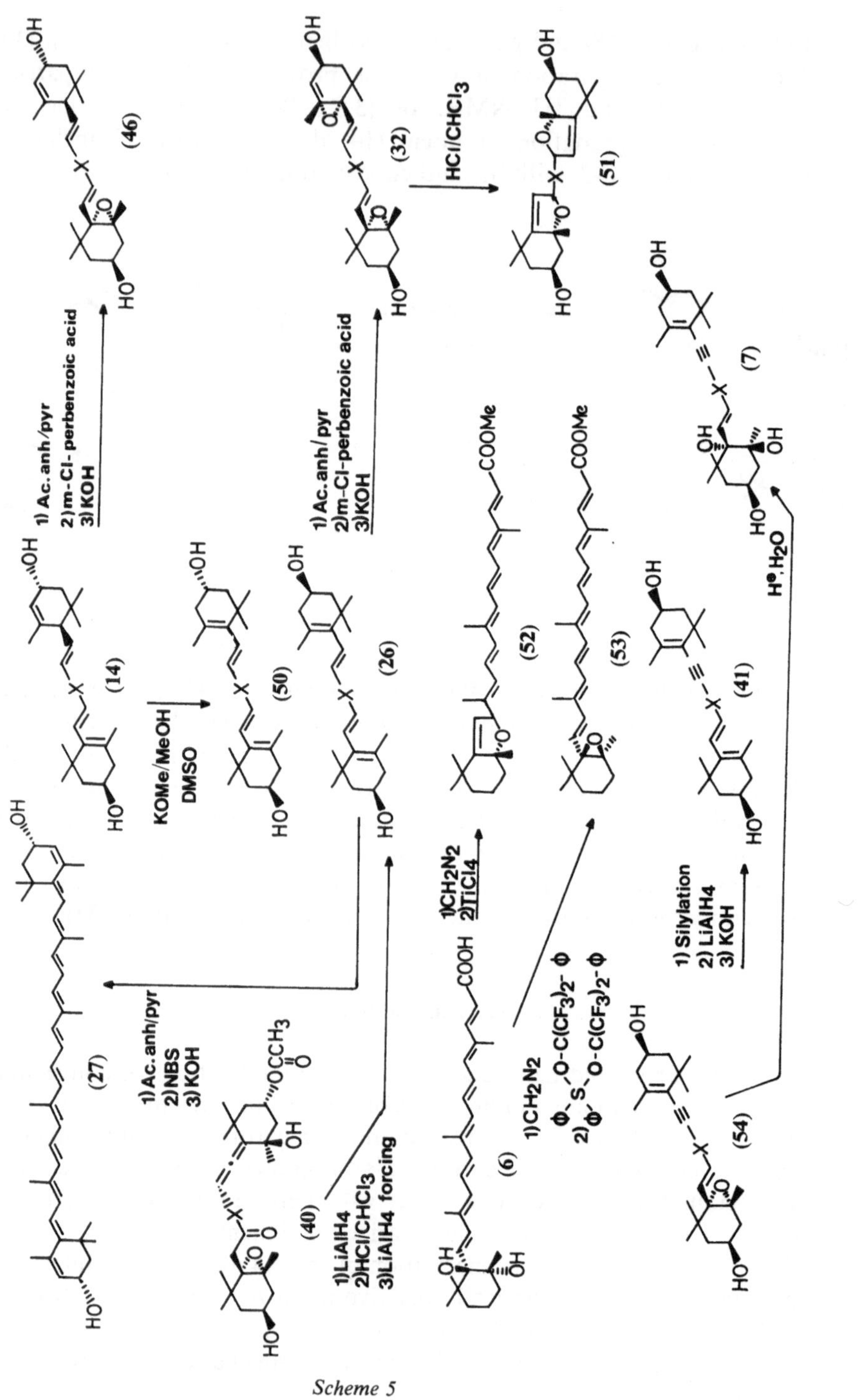

*Scheme 5*

Fucoxanthin (**40**) in a series of reactions involving lithium aluminium hydride reduction of the keto and acetate function to fucoxanthinol, allylic dehydration and subsequent epoxide rearrangement in acidic chloroform of fucoxanthinol to the epimeric fucochromes and further reaction with lithium aluminium hydride under forcing conditions, finally provided (3*R*,3′*R*) zeaxanthin (**26**) (*22, 26*).

Lutein (**14**), as diacetate, upon epoxidation with *m*-chlorperbenzoic acid gave a minor epoxide identical with lutein epoxide (**46**) diacetate besides the major 3,5-*cis* epoxide (*46*). In a similar way zeaxanthin (**26**) diacetate had previously been converted to violaxanthin (**32**) diacetate (*20*), the 3,5- and 3′,5′-*trans* compounds being obtained as minor products in addition to the major 3,5- and 3′,5′-*cis* derivatives, consistent with results for simple model compounds (*20*) and subsequent $^1$H NMR data (*78*).

A classical reaction in carotenoid chemistry is the acid-catalyzed rearrangement of 5,6-epoxides such as violaxanthin (**32**) to the corresponding 5,8-furanoxides (**51**) (*101, 110*). Retention of configuration at C-5 during this rearrangement is now documented (*46, 78*), whereas both C-8 epimers are formed from the C-8 carbonium ion intermediate.

Azafrin (**6**) methyl ester was converted to the corresponding furanoid oxide (**52**) with titanium tetrachloride in benzene-ether, whereas treatment with an appropriate sulfuran, requiring *trans* glycol as substrate, gave the epoxide (**53**) (*70*). De-epoxidation of diadinoxanthin [(**64**), as ditrimethylsilyl ether] to diatoxanthin (**4**) ditrimethylsilyl ether has been conducted with lithium aluminium hydride under forcing conditions (*42*). Heteroxanthin (**7**) and its C-5 epimer have also been prepared from diadinoxanthin (**54**) by treatment with aqueous acid (*42*).

Among other partial syntheses that have proved useful for stereochemical research may be mentioned that the failure of crustaxanthin (β,β-carotene-3,4,3′,4′-tetrol) to undergo dehydration on treatment with acidified chloroform, has been ascribed to the *trans* glycol arrangement (*141*). The conversion of plectaniaxanthin [(**35**), R = R′ = H] to its acetonide for comparison with a synthetic model (*153*) and the previously mentioned allylic oxidation of decaprenoxanthin (**8**) and lycoxanthin (**22**) to the corresponding aldehydes for $^1$H NMR analysis of the geometry of the terminal double bonds (*113, 156*) provide other examples. Whereas methylation of hydroxy groups by $CH_3I/Ag_2O$ or BaO (*40*) or $CH_3I/NaH$ (*144*) does not effect the geometry of the C-O bond of the carotenoid, methylation of the allylic hydroxy group in ε-rings with acidified methanol by an $S_N1$ mechanism results in racemization of the C-3 centre (*40, 75*). Racemization of an allylic hydroxyl in the 2-position, for instance in plectaniaxanthin [(**35**), R = R′ = H], has been achieved by oxidation followed by complex metal hydride reduction (*28*).

In principle the conversion of carotenoids with an allenic end group of the type present in neoxanthin (**1**) acetate to carotenoids with acetylenic end groups by HCl/CHCl$_3$ or by POCl$_3$ in pyridine (*41, 105*) or to products with chlorinated β-type end groups by HCl/CHCl$_3$ (*41, 105*) could be useful for stereochemical considerations.


## 12. Total Synthesis

In the absence of X-ray crystallographic analyses the total synthesis of chiral model carotenoids which are more acessible and better yet, of the optically active natural carotenoids for confirmation of configurational assignments represents a challenge.

Total syntheses of carotenoids until 1970 were expertly reviewed by Mayer and Isler (*133*) and up-dated in 1975 by Weedon (*176*). A complete account of total syntheses of optically active carotenoids is given elsewhere by Mayer (*131*).

In the present chapter recent total syntheses of chiral carotenoids as racemates which have been useful for assigning relative configuration by $^1$H NMR, will be discussed as well as syntheses of optically active model carotenoids and of optically active naturally occurring carotenoids or their enantiomers. Also included are recent syntheses related to naturally occurring carotenoids with Z-configuration in the polyene chain.


### a) Racemic Carotenoids

No allenic carotenoids have yet been prepared by total synthesis, but the allenic ketones (**55**) and (**56**) have been synthesized as racemates (*92*). In a recent preliminary note the total synthesis of optically inactive 2,6-*trans*-2',6'-*trans* decaprenoxanthin has been outlined. Natural deca-prenoxanthin (**8**) has $2R,6R,2'R,6'R$-configuration (*11*) with the 2,6(2',6')-substituents in a *cis* relationship.

By a stereochemically controlled synthesis the C$_{40}$ cyclopropane carotenoid precursor prephytoene alcohol [(**57**), as racemate] has been obtained (*47*). The total synthesis of azafrin [(**6**), as racemate] (*175, 176*), of lutein [(**14**), as racemate] and 3'-epilutein [(**58**), as racemate] (*176*) and of 3,6(3',6')-*cis* (**59**) and 3,6(3',6')-*trans* ε,ε-carotene-3,3'-diol (**60**), both as racemates (*176*), have been announced in symposium reviews. Experimental details have not yet been published. The $^1$H NMR data of racemic (**6**) were consistent with the *threo* structure of natural azafrin (**6**) that has been documented in other ways, while the $^1$H NMR properties of the models (**59**) and (**60**) were consistent with the previously established

3′,6′-*trans* configuration of natural lutein (**14**) (*176*), and have a bearing on the structure of tunaxanthin (**17**) (*152*). Structure (**58**) (3′-epilutein) has been suggested for calthaxanthin from *Caltha palustris* (*59*). The structure of aleuriaxanthin (**36**) (*14, 34*) has been confirmed by synthesis of the racemate (*119*).

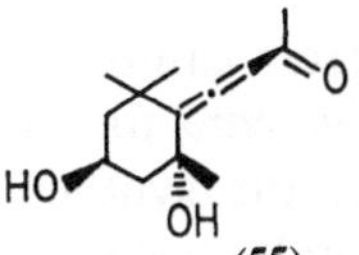

(55) 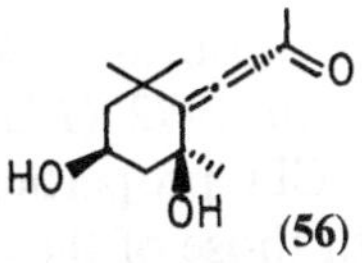    (56) 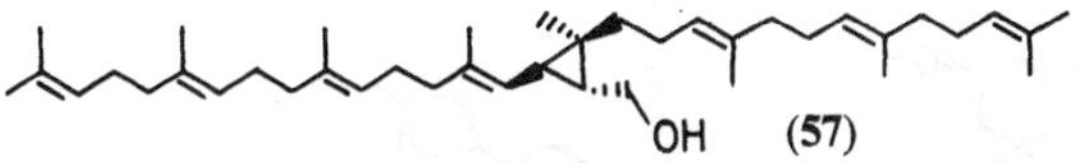    (57) 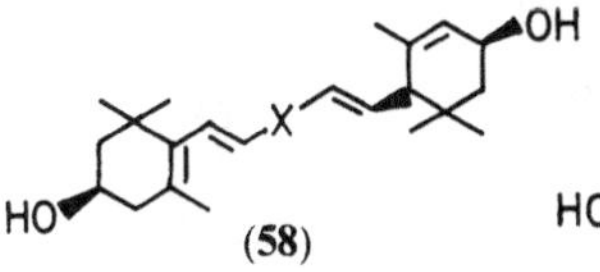    (58) 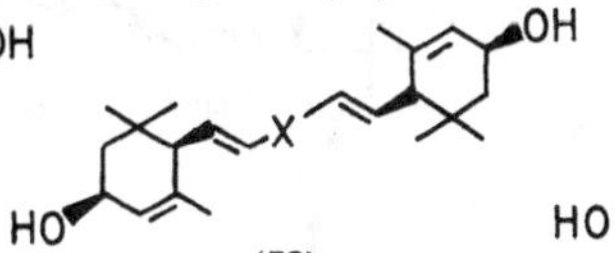    (59) 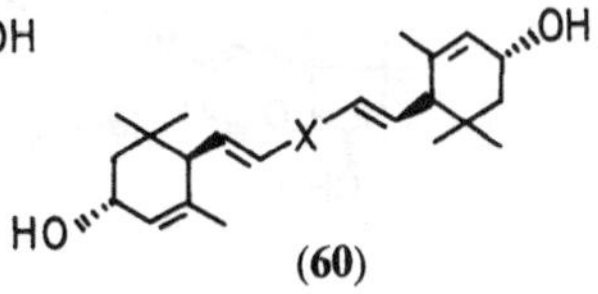    (60)

## b) Optically Active Model Carotenoids

For comparative ¹H NMR and CD studies with natural bicyclic $C_{50}$ carotenoids the $C_{42}$ carotenoids (**9**), (**10**), (**61**) and (**62**) have been synthesized from the appropriate irones (*5, 6, 10*). The $C_{50}$ carotene (**63**) has been prepared from *R*-lavandulyl acetate as a possible chiral model for bacterioruberin (**37**) (*104, 107*), while the $C_{45}$ carotene (**64**) and

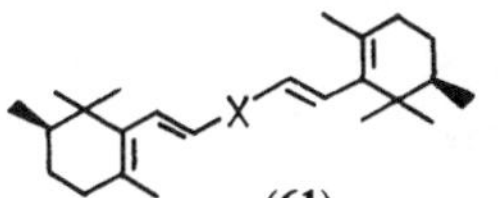

(61) 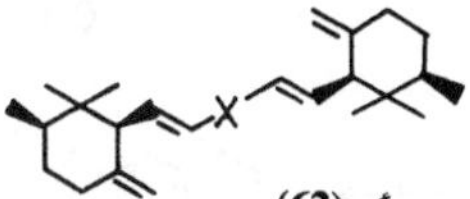    (62) 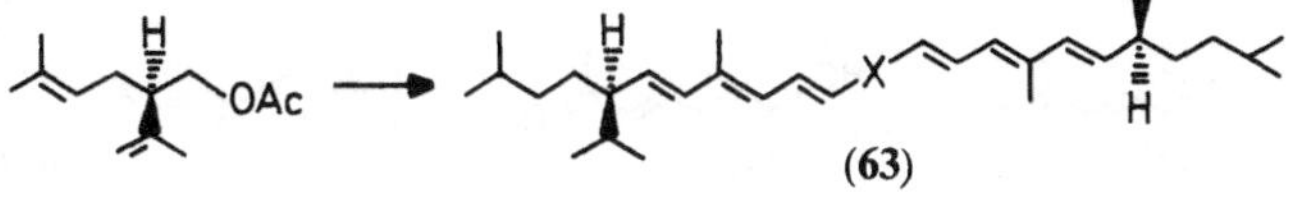    (63) 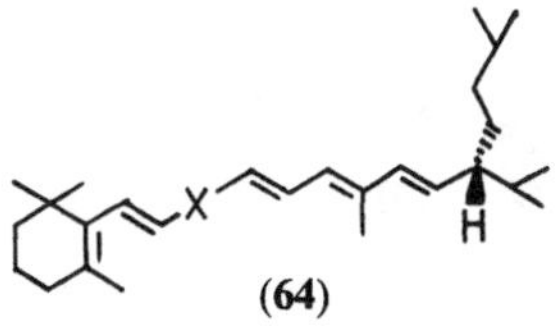    (64) 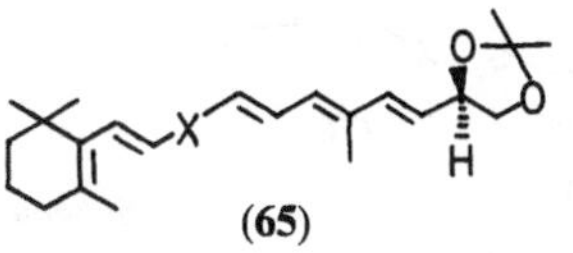    (65)

16′,17′-dinorplectaniaxanthin acetonide (65) were synthesized in connection with studies on the absolute configuration of plectaniaxanthin [(35), R=R′=H] (153).

## c) Optically Active Carotenoids

The total synthesis of optically active capsorubin (66) from camphoric acid by a lengthy route was outlined in 1972 (175) (Scheme 6), but details have not yet been published. CD properties of the synthetic capsorubin were in good agreement with those of the naturally-occurring material (175).

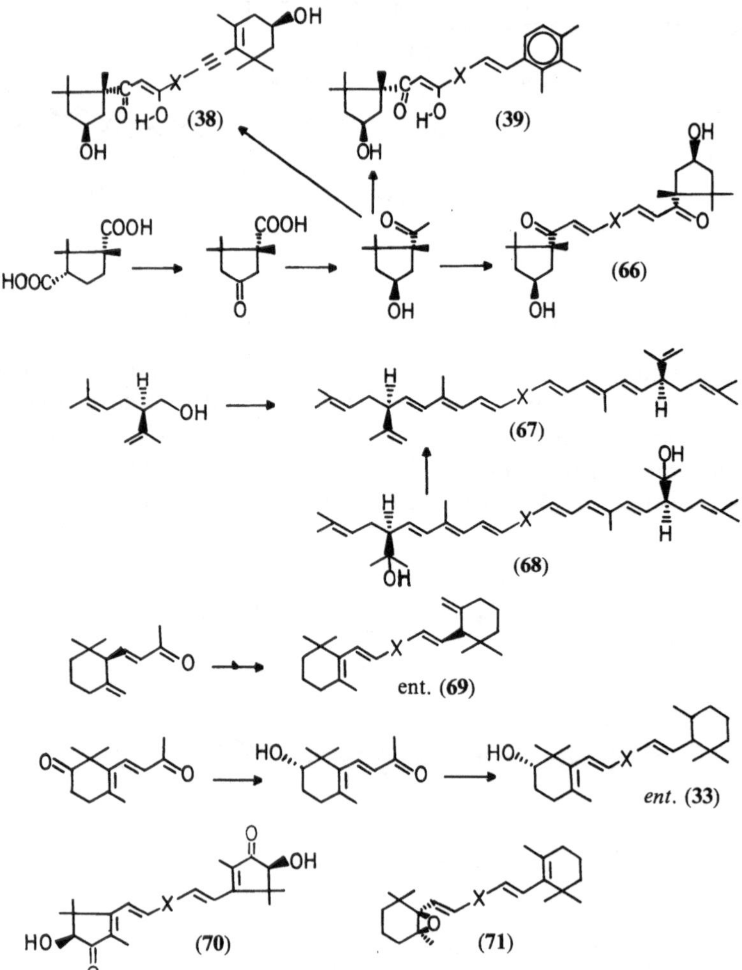

*Scheme 6*

(2S,2'S)-Tetraanhydrobacteriorubin (67), synthesized either from (R)-lavandulol or by dehydration of natural bacterioruberin (37) or bisanhydrobacterioruberin (68), had identical CD properties (104, 107). The total synthesis of (67) consequently permitted configurational assignment of the aliphatic $C_{50}$ carotenoids (37) and (68).

Optically active (6'R)-β,γ-carotene [ent-(69)], prepared from racemic γ-ionone partly resolved via its menthydrazone, had Cotton effect opposite to that of natural β,γ-carotene (69), which therefore has the 6'S-configuration (84).

(2S)-β,β-Caroten-2-ol [ent-(33)] has been prepared by a total synthesis (102) which involves regio- and enantioselective microbial reduction of the 2-oxo-β-ionone intermediate to form the chiral C-2 centre. The chirality at C-2 was defined after CD comparison of the synthetic $C_{40}$-ol with natural (2R)-β,β-caroten-2-ol [(33), R=H] of known configuration (38).

Recently total syntheses of the enolized β-diketones trikentriorhodin (39) and the 9-cis isomer of mytiloxanthin (38) with known stereochemistry at all chiral centres have been reported in brief (57). The route utilized the same intermediate methyl ketone used earlier in the total synthesis of capsorubin (66), although it was prepared from (+)-camphor by a different approach involving diastereoselective introduction of the chiral centre by hydroboration of an alkene intermediate. Synthetic (38) had chromatographic, IR, ¹H NMR and MS properties consistent with published data (3); the CD maxima appear to correspond to those of natural trikentriorhodin subsequently reported (39). However, the Cotton effect is very weak (39, 57). Comparative CD data for synthetic 9-cis (38) and the 9-cis isomer of natural mytiloxanthin were not given.

Total syntheses of (3S,3'S)-astaxanthin (28), (3R,3'R)-astaxanthin [ent-(28)] and (3S,3'R)-astaxanthin have been accomplished (69, 116, 131). Also recently synthesized in the Roche laboratories were optically pure (3R,3'R)-zeaxanthin (26) (132), (3S,3'S)-zeaxanthin [ent-(26)] and (3R,3'S)-zeaxanthin (50) (131), all-trans (3S,3'S)-7,8-didehydroastaxanthin (80) and all-trans (3S,3'S)-7,8,7',8'-tetradehydroastaxanthin (34) (23a) as well as both actinioerythrin enantiomers [(70) and ent-(70)] (138). This important achievement has been discussed by MAYER (131). Furthermore, optically pure all-trans (3R,3'R)-alloxanthin (31) has been obtained by total synthesis (153a). Recently (5S,6R)-5,6-epoxy-5,6-dihydro-β,β-carotene (71) and (5R,6R)-5,6-dihydro-β,β-carotene-5,6-diol (29) have been synthesized from azafrin (6) (71). The diol has not yet been encountered in nature, and correlations between the synthetic epoxide and naturally occurring mono- and diepoxides of β,β-carotene (162) remain to be carried out. This is also the case for the corresponding (5S,6R)-5,6-

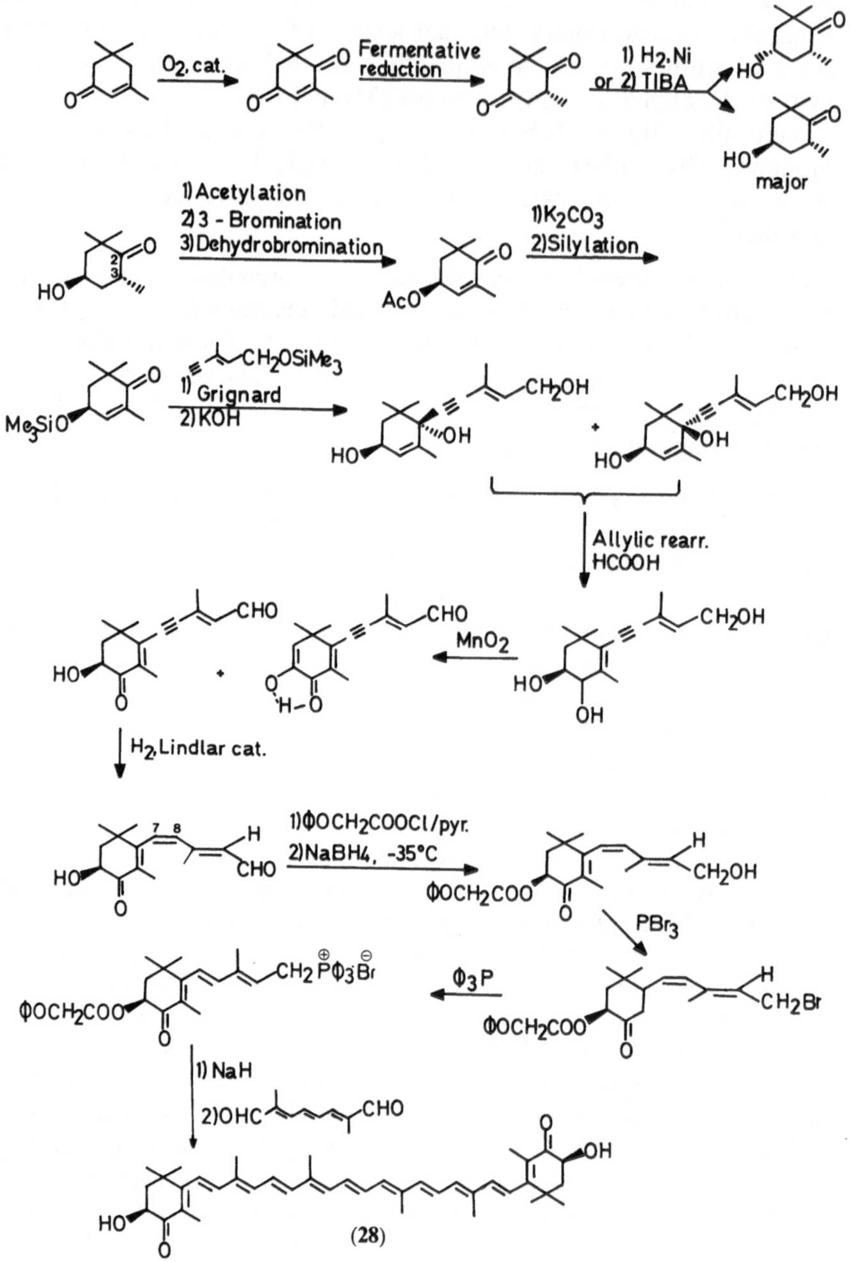

*Scheme 7*

epoxides of β,ε-carotene and β,ψ-carotene now synthetically available (*71 b*).

As an example of the state of the art the detailed route to (3*S*,3′*S*)-astaxanthin (**28**) used by KIENZLE and MAYER (*116*) is illustrated in Scheme 7.

A route to the key synthon (4*R*,6*R*)-4-hydroxy-2,2,6-trimethylcyclo-hexanone had previously been developed by LEUENBERGER *et al.* (*124*) starting from the readily available isophorone. The chiral centre was introduced by enantioselective reduction with bakers yeast, and regio-selective reduction of one keto group was achieved by chemical reduction with nickel catalyst or by triisobutylaluminium under conditions where the desired *trans* diastereomer was the major product.

The route from (4*R*,6*R*)-4-hydroxy-2,2,6-trimethylcyclohexanone to (3*S*,3′*S*)-astaxanthin (**28**) (*116*) was as outlined in Scheme 7. The *cis* con-figuration of the 7-double bond formed upon hydrogenation with Lindlar catalyst proved to be unexpectedly stable. However, spontaneous iso-merization to *trans* accompanied formation of the phosphonium salt with triphenyl phosphine. Wittig olefination with the $C_{10}$-dial, using sodium hydride as base, caused simultaneous loss of the phenoxyacetate protecting groups and provided the target molecule in a high yield.

### d) Z-Isomers

Certain carotenoids occur in nature with particular double bonds in the polyene chain in the *cis* (Z) configuration (*101, 182*) (see Part IV). Synthetic routes have been developed for regiospecific introduction of *cis* double bonds. Thus reaction of 4-hydroxy-3-methylbut-2-enolide (**72**) with allylic Wittig reagents gave polyene acids with complete retention of *cis* stereochemistry around the α,β-double bond and a high proportion of *cis*-configuration around the newly formed carbon-carbon double bonds (*146*). Using sensitized photochemical isomerizations LIU and co-workers have synthesized several sterically hindered isomers related to Vitamin A, including the 7-*cis*, 7,9-di-*cis* and 7,19,13-tri-*cis* isomers of retinal (**73**) (*130, 149, 150*). Several carotenoids with *cis*-configuration of the central Δ-15 bond have been obtained by selective hydrogenation of the required acetylenic intermediate in the presence of Lindlar catalyst (*133*). In the total synthesis of acetylenic carotenoids with triple bonds in 7 or 7,7′-positions, such as alloxanthin (**31**), by Wittig condensation of the appropriate acetylenic phosphonium salt with apo-carotenals the undesired formation of *cis* double bonds in the 9(9′) position has been a problem (*57, 96, 131, 173, 176*). Using a low-temperature KOH/*i*-propyl alcohol method for the Wittig condensation improved yields of the

9(9')-*trans* isomer may be obtained (*176*). The recent elucidation of the mechanism of the Wittig reaction (*23b*) may prove useful for optimizing reaction conditions so as to produce pure geometrical isomers.

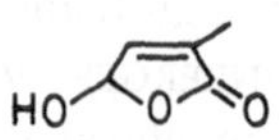

(72) 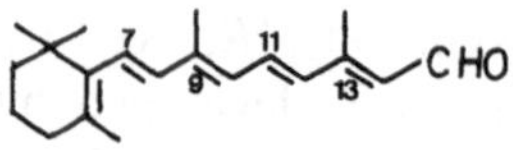          (73)

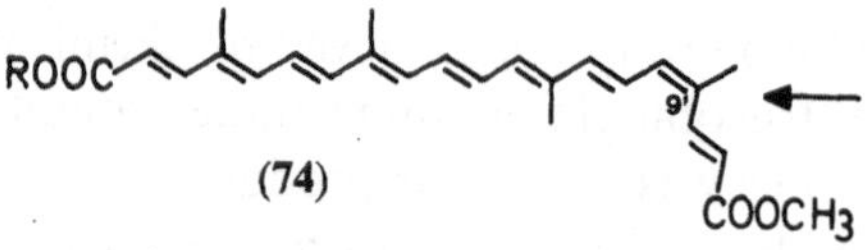

(74) 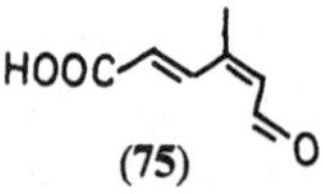          (75)

COOCH₃

Methyl natural bixin [(**74**), R = CH₃] with a 9'-*cis* (by C₄₀-carotenoid numbering) double bond has been synthesized by a stereochemically controlled synthesis with (**75**), obtained from the appropriate lactone as key intermediate (*145*). The configuration of natural bixin [(**74**), R = H] rested on ¹H NMR assignments (*19*).

Phytoene (**21**) occurs in nature in the all-*trans* and 15-*cis* form, as proved by comparative ¹H NMR studies with shorter synthetic models prepared under stereochemical control which possess four alternative geometrical arrangements for the triene chromophore (*EEE, EZE, EEZ, EZZ*) (*98, 122*).

Total syntheses of cross-conjugated caroten-20-als related to 13-*cis* rhodopin-20-al (**5**) have been carried out (*136, 137*). Functionalization at C-20 was effected by N-bromosuccinimide at the 8,8'-diapocarotene-8,8'-dial level (*138*). The synthetic cross-conjugated 20-carotenals showed the same preference for the 13-*cis* configuration as the naturally occurring carotenoids of this series (*78*). Manixanthin has been considered as the 9,9'-di-*cis* isomer of alloxanthin (**31**) (*139*). Synthetic, optically active 9,9'-di-*cis* (**31**) is now available (*153a*).

## IV. Geometrical Isomerism

According to the new IUPAC nomenclature for carotenoids (*98*) the *cis-trans* convention is still used to denote geometrical isomerism of the polyene chain. However, the *E/Z* designations may also be used, especially when the prefixes *cis* and *trans* might lead to ambiquity (*98*).

Geometrical isomerism in the carotenoid series is well covered in the earlier literature (*174, 181, 182*) through 1970. It has been known

*References, pp. 164—172*

for a long time that *cis*-carotenoids differ from the parent all-*trans* isomers in melting point, adsorption affinity, electronic spectra including the so-called *cis*-peak *etc.* Unmethylated double bonds, except the central one, have a sterically hindered *cis* configuration. Common methods for effecting *trans-cis* isomerization include thermal procedures and photochemical methods with and without catalytic amounts of iodine. Sterically hindered *cis* double bonds are not formed upon stereomutation. The generalization that the all-*trans* isomer is usually the naturally occurring stereoisomer and the thermodynamically most stable one still holds with some exceptions discussed below. Thus for bicyclic carotenoids the all-*trans* isomer usually predominates in iodine-catalyzed quasi-equilibrium mixtures, whereas this is not the case for aliphatic carotenoids with long chromophores such as spirilloxanthin (**76**) and bacterioruberin (**37**), which are also less sterically stable during laboratory manipulations (*125*).

During the last decade additional information about HPLC, ¹H NMR and CD properties of *cis*-carotenoids has been obtained, as already discussed in Part III.

Configurational assignment of the poly-*cis* carotenoids proneurosporene and pro-γ-carotene (*182*) is still lacking. Recent ¹H NMR and ¹³C-NMR evidence has suggested that prolycopene is 7,9,7′,9′-tetra-*cis*-ψ,ψ-carotene with two sterically hindered double bonds (7,7′). By a total synthesis designed to prepared this tetra-*cis*-lycopene several geometrical isomers were obtained, including one with HPLC properties corresponding to natural prolycopene (*135 a*).

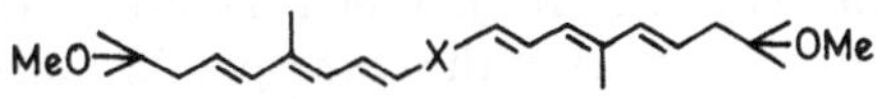

(**76**)

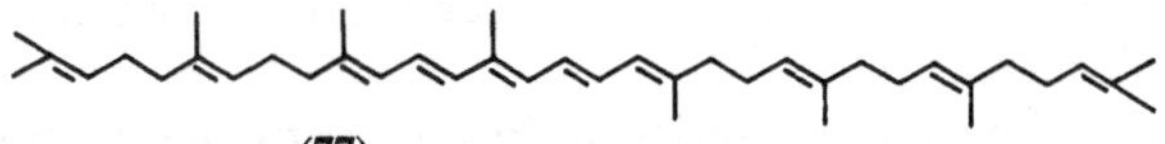

(**77**)

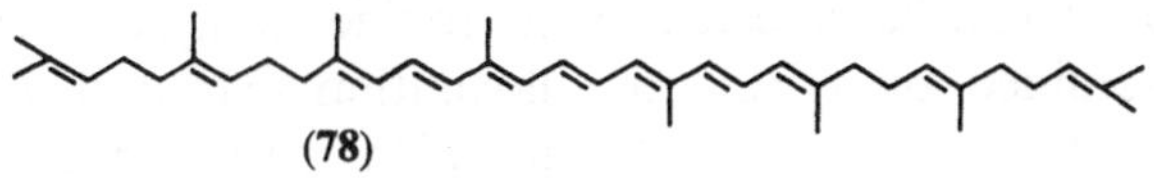

(**78**)

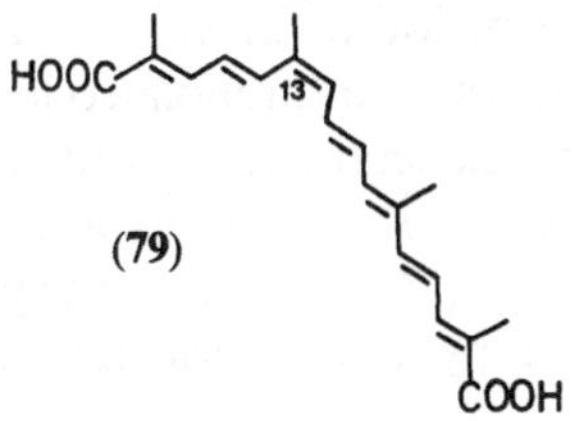

(**79**)

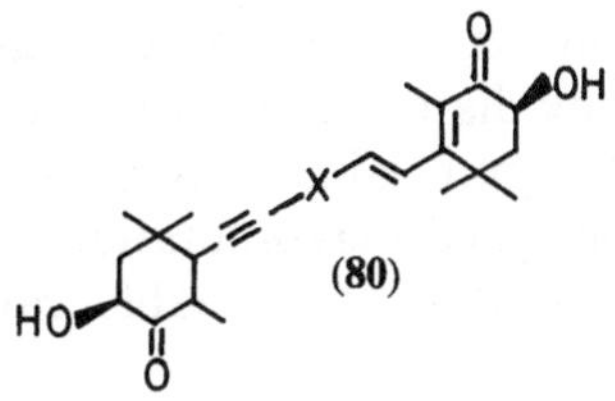

(**80**)

The geometry of most naturally-occurring mono-*cis* carotenoids has now been defined, frequently by approaches involving $^1$H NMR and synthesis. Thus the common mono-*cis*-isomer of phytoene (21) is 15-*cis* (*18, 62, 115*). The main phytofluene (77) and ζ-carotene (78) of natural occurrence are also probably 15-*cis* (*115, 174*). This preservation of *cis*-geometry during the sequential dehydrogenation of phytoene (21) to phytofluene (77) and ζ-carotene (78) in the biosynthesis of coloured carotenoids (*80*), at least in some organisms, is interesting.

The 9'-*cis* configuration of natural bixin [(74), R = H] (*19, 145*) has already been discussed. Also natural crocetin (79) is a mono-*cis* diapo-carotenoid for which the 13-*cis* configuration preferred in the early going is favoured by exclusion since the dimethyl esters of the 15-*cis* and 9-*cis* isomers have been synthesized and found to be different (*174*). The 5'-*cis* configuration of gazaniaxanthin [5'-*cis* (42)] also rests on an exclusion principle (*15, 33*), and positive $^{13}$C NMR evidence should be sought. Violeoxanthin is the 9-*cis* isomer of violaxanthin (32) as judged by $^1$H NMR and $^{13}$C NMR data (*136*). The 6-*S* isomer of fucoxanthin [(40), R = H] is claimed to be naturally occurring and also to be formed upon stereomutation of the all-*trans* isomer (*22*). The configurational assignment rests on $^1$H NMR arguments (*22*). It is, however, surprising that no 9'-*cis* isomer was detected in the stereomutation mixture of fuco-xanthin [(40), R = H] (*22*), especially since the corresponding 9-*cis* isomer of neoxanthin (1) is assumed to be common (*55*). Geometrical isomerism in the allenic carotenoid series should be further examined by FT $^1$H NMR, $^{13}$C NMR and CD.

The cross-conjugated carotenals of the rhodopin-20-al (5) series occur exclusively in the 13-*cis* configuration (*1, 49, 54*) and the all-*trans* isomer is not formed upon stereomutation (*1, 109*). When the cross-conjugated aldehyde is reduced, the all-*trans* isomers of the corresponding alcohol and acetate may be isolated (*49*). The preference for the 13-*cis* configuration in the cross-conjugated 20-al series is ascribed to a combination of steric and electronic factors (*49*).

In the 7,7'-diacetylenic series it has recently been shown that no all-*trans* isomer can be detected in the iodine catalyzed stereomutation mixtures examined by HPLC of alloxanthin (31) and 7,8,7',8'-tetra-dehydroastaxanthin (34) diacetate, in which the 9,9'-di-*cis* isomers are dominant (*76*). However, in the absence of iodine all-*trans* (31) and all-*trans* (34) diacetate are rather stable. In the monoacetylenic series exemplified by diatoxanthin (41) and the diacetate of 7,8-didehydroasta-xanthin (80), appreciable amounts of the all-*trans* isomers are present in the iodine catalyzed stereomutation mixture (*76, 89*). Again the preference for *cis*-configuration is probably caused by both steric and electronic factors.

*References, pp. 164—172*

## V. Established Chirality

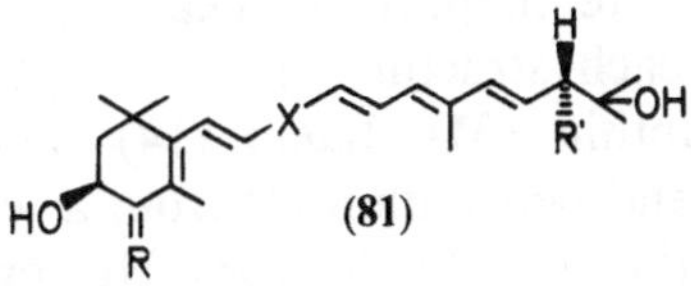

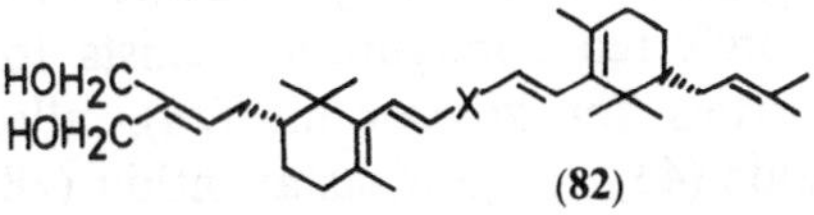

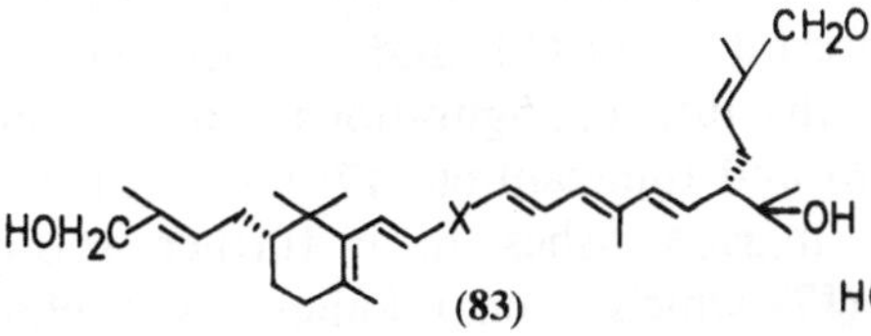

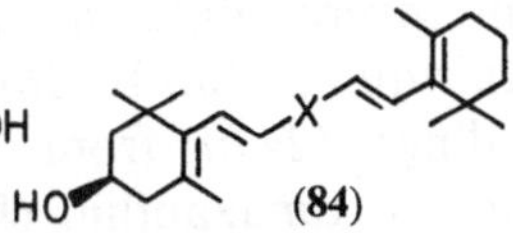

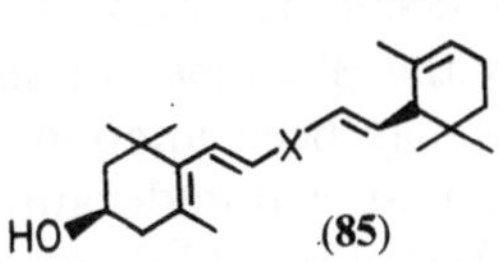

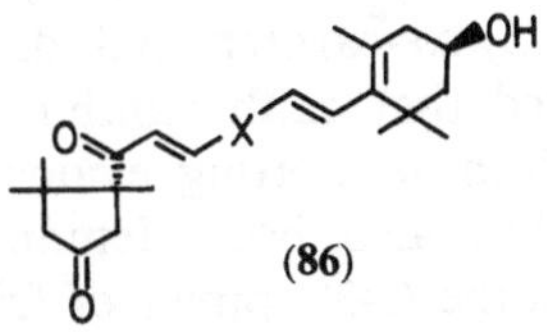

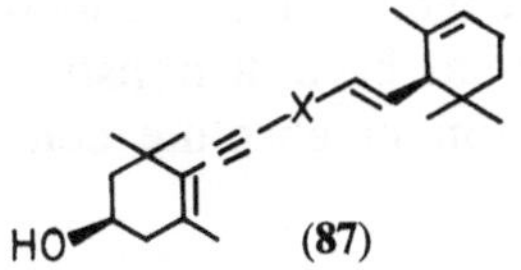

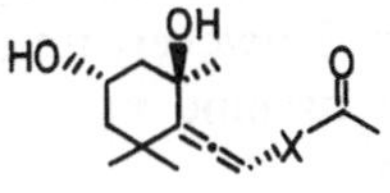

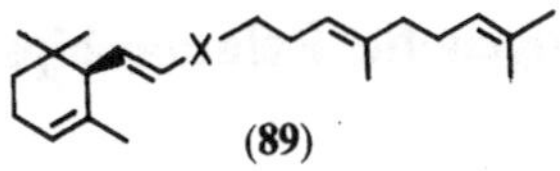

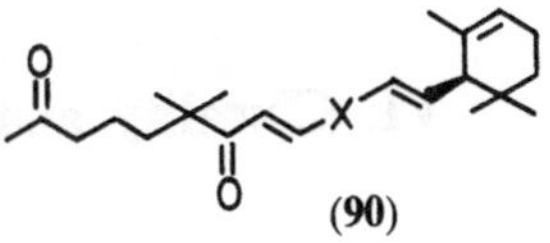

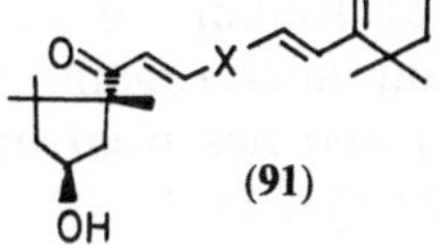

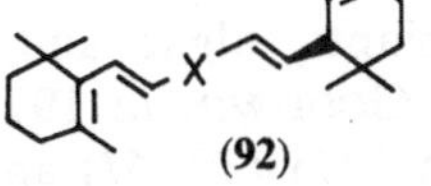

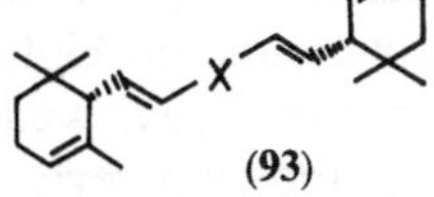

Having treated the methods used and their recent application to configurational assignments in the carotenoid series in Part III, a compilation will be given in Scheme 8 and in Table 1 which lists the chiral end groups so far encountered in natural carotenoids, the key evidence for the configurational assignment and the carotenoids in which the

(81)

(82)

(83)

(84)

(85)

(86)

(87)

(88)

(89)

(90)

(91)

(92)

(93)

particular end groups are documented. Listed are 5 aliphatic, 17 β-ring derived, 2 γ-ring derived, 6 ε-ring derived and 2 κ-ring derived end groups; altogether 32 chiral end groups of more or less well established absolute configuration. To this must now also be added *ent-*(**XXV**), since it was recently reported that ε,ε-carotene (**93**) isolated from chicken retina as well as from the green alga *Ulva lactua* has the 6*S*,6′*S* configuration (*60*). This includes nearly all chiral end groups so far encountered in carotenoids.

As concerns the possibility that carotenoids of identical constitution when produced by different types of organisms may have different absolute configurations, the early evidence seemed to support the idea that only one preferred configuration exists for each structure. The chiroptical properties for zeaxanthin (**26**), alloxanthin (**31**), lutein (**14**), lutein epoxide (**43**), (3*S*,3′*S*)-astaxanthin (**28**) and neoxanthin (**1**) from a wide variety of sources were consistent with this view (*2, 20, 168*). However, exceptions turned up subsequently. The occurrence of (3*R*,3′*R*)-asta-xanthin [*ent-*(**28**)] in yeast is well established (*12*) and ε,ε-carotene-3,3′-diol occurs in several different absolute configurations: chiriquixan-thin A (**15**), chiriquixanthin B (**16**) and tunaxanthin (**17**) (*24, 152*). Un-published findings (*24 b*) from marine fishes have further shown that in addition to tunaxanthin (**17**) which is enantiomeric with chiri-quixanthin A (**15**), *ent-*(**16**) (3*R*,6*S*,3′*R*,6′*S*) and the diastereomeric (3*S*,6*S*,3′*S*,6′*S*)-ε,ε-carotene-3,3′-diol also occur naturally. It has also been suggested that calthaxanthin (**58**) is the 3′-epimer of lutein (**14**) (*125*), for which supporting evidence now has been announced (*43 a*). Recently it has also been demonstrated that α-doradexanthin [(**96**), Scheme 14] is the C-3′ epimer of fritschiellaxanthin (*43 a*). Although no other examples have so far been demonstrated, it is desirable that CD and $^1$H NMR spectra should routinely be included in the identification criteria for carotenoids. However, until it has been demonstrated other-wise, it is practical to assume no variation in absolute configuration for a given constitution.

# VI. Chirality and Biogenetic Relationships

## 1. De novo Synthesis

Evidence regarding the biosynthetic routes leading to coloured carotenoids in higher plants, algae and certain microorganisms was thoroughly reviewed by Goodwin in 1971 (*80*) and has been updated in symposium reviews in 1975 (*30, 61*) and 1978 (*60, 147*).

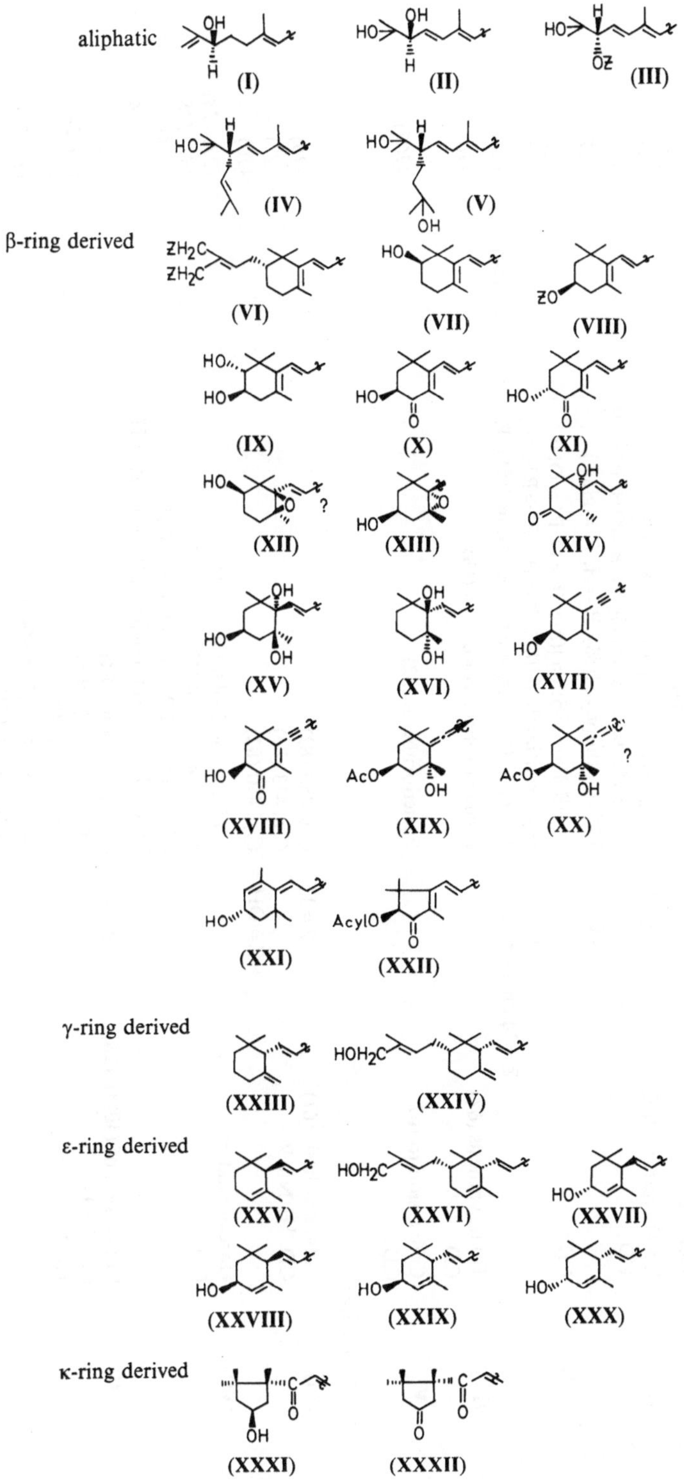

aliphatic

(I)    (II)    (III)

(IV)    (V)

β-ring derived

(VI)    (VII)    (VIII)

(IX)    (X)    (XI)

(XII)    (XIII)    (XIV)

(XV)    (XVI)    (XVII)

(XVIII)    (XIX)    (XX)

(XXI)    (XXII)

γ-ring derived

(XXIII)    (XXIV)

ε-ring derived

(XXV)    (XXVI)    (XXVII)

(XXVIII)    (XXIX)    (XXX)

κ-ring derived

(XXXI)    (XXXII)

*Scheme 8*

Table 1. *Chiral End Groups in Naturally Occurring Carotenoids*

| End group | Key information | | Established presence in | References |
|---|---|---|---|---|
| *Aliphatic* | | | | |
| (I) | Horeau analysis | | Aleuriaxanthin (36) | (33) |
| (II) | Total synthesis (65) | Z = H | Plectaniaxanthin [(35), R = R' = H] | (153) |
| | CD | | 2-OH-Plectaniaxanthin [(35), R = OH, R' = H] | (27, 153) |
| | | Z = Gluc. | Pheixanthophyll [(35), R = H, R' = gluc.] | (27) |
| (III) | CD | Z = H | 2'-OH-Flexixanthin [(81), R = O, R' = OH] | (27) |
| | | Z = Rham. | Myxoxanthophyll [(81), R = H₂, R' = Rham.] | (27) |
| | | Z = Methylpentoside | Myxol-2'-O-methyl-methylpentoside [(81), R = H, R' = OMe-pent.] | (27) |
| (IV) | Total synthesis (67) | | Bisanhydrobacterioruberin (68) | (107) |
| | CD | | | |
| (V) | Conversion to (67) | | Bacterioruberin (32) | (107) |
| | CD | | | |
| *β-Ring derived* | | | | |
| (VI) | Total synthesis (61) | Z = H | C. p. 450 (82) | (10) |
| | CD, ¹H NMR | | C. p. 473 (83) | (27) |
| | | Z = OH | C. p. 450 (82) | (10) |
| (VII) | Horeau analysis | | β,β-Carotene-2-ol [(33), R = H] | (37, 118) |
| | CD | | β,ε-Carotene-2-ol (48) | (37, 118) |
| | | | β,β-Carotene-2,2'-diol [(37), R = OH] | (118) |
| | | | 2-OH-Plectaniaxanthin [(35), R = OH, R' = H] | (82, 176) |
| (VIII) | Conversion of (40) to (26) | Z = H | β-Cryptoxanthin (84) | (20) |
| | CD, ORD | | α-Cryptoxanthin (85) | (20, 24) |
| | Total synthesis | | Rubixanthin (42) | (20) |
| | | | Gazaniaxanthin [5'-cis-(42)] | (15, 20) |
| | | | Diatoxanthin (41) | (20) |
| | | | Zeaxanthin (26) | (20, 92, 131, 132) |
| | | | Lutein (14) | (4, 39) |
| | | | Calthaxanthin (58) | (59, 43a) |

| | | Compound | Method | Ref. |
|---|---|---|---|---|
| | | Antheraxanthin [(26)-5,6-epoxide] | | (20) |
| | | Capsanthin (2) | | (20, 58) |
| | | Capsanthinone (86) | | (20) |
| | | Caloxanthin (24) | | (42) |
| | Z = Gluc. | Zeaxanthin [(26)-dirhamnoside] | | (88) |
| | Z = Acyl. | Physalien [(26)-dipalmitate] | | (162) |
| (IX) | | Caloxanthin (24) | CD | (42) |
| | | Nostoxanthin (25) | | (42) |
| (X) | | Flexixanthin [(81), R = O, R′ = H] | CD LAH-red. product | (27) |
| | | 2′-OH-Flexixanthin [(81), R = O, R′ = OH] | Total synthesis | (27) |
| | | Astaxanthin (28) | | (7, 116) |
| | | 7,8-Didehydroastaxanthin (80) | | (21, 23a) |
| | | α-Doradexanthin (96) | | (43a) |
| | | Fritschiellaxanthin [3′-epi-(96)] | | (43a) |
| (XI) | | (3R,3′R)-Astaxanthin [ent-(28)] | CD | (12) |
| (XII) | | 5,6-Epoxy-5,6-didehydro-β,β-caroten-2-ol [(33)-5,6-epoxide, R = H] | Partial synthesis from (33), R = H | (143) |
| (XIII) | | Diadinoxanthin (54) | Conversion to (VIII) | (41) |
| | | Antheraxanthin [(26)-5,6-epoxide] | Partial synthesis from (VIII) | (20, 78) |
| | | Lutein epoxide (46) | | (46) |
| | | Neoxanthin (1) | $^1$H NMR 5,8-furanoxide | (55, 92) |
| | | Dinoxanthin [(1)-3-acetate] | ORD, CD | (108) |
| | | Violaxanthin (32) | | (20, 78) |
| | | Fucoxanthin [(40), R = H] | | (22, 25) |
| | | Fucoxanthinol [Desacetyl-(40), R = H] | | (162) |
| | | 19-Hexanoyloxyfucoxanthin [(40), R = OCO(CH$_2$)$_4$CH$_3$] | | (91) |
| | | Peridinin (30) | | (108) |
| (XIV) | | Isomytiloxanthin (22) | $^1$H NMR | (114) |
| (XV) | | Heteroxanthin (7) | H-bonding Silylation CD Partial synthesis from (XVII) | (41) |
| (XVI) | | Azafrin (6) | H-bonding Conversion to (47) | (70) |

Table 1 (continued)

| End group | Key information | Established presence in | References |
|---|---|---|---|
| (XVII) | Hydrogenation to perhydro (VIII) ORD, CD | Alloxanthin (31) | (20) |
| | | Crocoxanthin (87) | (20) |
| | | Diatoxanthin (41) | (20) |
| | | Diadinoxanthin (54) | (41) |
| | | Mytiloxanthin (38) | (57, 114) |
| | | Isomytiloxanthin (22) | (114) |
| (XVIII) | CD LAH-red. product | 7,8-Didehydroastaxanthin (80) | (21, 23a) |
| | Total synthesis | 7,8,7',8'-Tetradehydroastaxanthin (34) | (21, 23a) |
| (XIX) | R = H | Neoxanthin (1) | (55, 92) |
| | Conversion to (3) (X-ray) | Fucoxanthinol [Desacetyl-(40), R = H] | (162) |
| | $^1$H NMR | Paracentrone (88) | (93, 174) |
| | R = Ac | Dinoxanthin [(1)-3-acetate] | (108) |
| | | Fucoxanthin [(40), R = H] | (23, 107) |
| | | 19'-Hexanoyloxyfucoxanthin [(40), R = OCO(CH$_2$)$_4$CH$_3$] | (91) |
| | | Peridinin (26) | (108) |
| (XX) | $^1$H NMR, CD | Minor isomer of fucoxanthin [6S-(40), R = H] | (22) |
| (XXI) | Partial synthesis from (26) $^1$H NMR, CD | Eschscholtzxanthin (27) | (8) |
| (XXII) | Total synthesis | Actinioerythrin (70) | (138, 139) |
| γ-Ring derived | | | |
| (XXIII) | Total synthesis | β,γ-Carotene (69) | (84) |
| (XXIV) | R = H | Sarcinaxanthin [(11), R = H' = H] | (90) |
| | Total synthesis (62) CD, $^1$H NMR | Sarcinaxanthin monoglucoside [(11), R = Gluc., R' = H] | (90) |
| | R = Gluc. | Sarcinaxanthin diglucoside [(11), R = R' = Gluc.] | (90) |

**ε-Ring derived**

| | Compound | Method | Ref. |
|---|---|---|---|
| (XXV) | β,ε-Carotene | Oxidative degradation | (20, 72, 165) |
| | ε,ψ-Carotene | | (34) |
| | α-Zeacarotene (89) | | (36) |
| | Crocoxanthin (87) | | (36) |
| | α-Cryptoxanthin (85) | Synthesis from (+)-α-ionone | (20, 24) |
| | Semi-α-carotenone (90) | | (44) |
| | ε-Carotene-8-al | CD, ORD | (36) |
| ent-(XXV) | (6S,6'S)-ε,ε-Carotene | CD | (60) |
| (XXVI) | Decaprenoxanthin (8) | Total synthesis (9) and racemate | (11, 56) |
| | | ¹H NMR, CD | |
| (XXVII) | Lutein (14) | Oxidative degradation | (4, 39) |
| | Lutein epoxide (46) | Conversion to ent-(VIII), Z = H | (46) |
| | Fritschiellaxanthin [3'-epi-(96)] | CD, ¹H NMR | (43a) |
| (XXVIII) | Calthaxanthin (58) | ¹H NMR, CD | (59, 43a) |
| | α-Doradexanthin (96) | | (43a) |
| (XXIX) | Tunaxanthin (17) | ¹H NMR, CD | (152) |
| | (3S,6S,3'S,6'S)-ε,ε-carotene-3,3'-diol | | (24a) |
| (XXX) | Tunaxanthin (17) | ¹H NMR, CD | (152) |
| | (3R,6S,3'R,6'S)-ε,ε-carotene-3,3'-diol [ent-(16)] | | (24a) |

**κ-Ring derived**

| | Compound | Method | Ref. |
|---|---|---|---|
| (XXXI) | Cryptocapsin (91) | Ozonolysis | (20) |
| | Capsanthin (2) | IR | (20, 58) |
| | Capsorubin (66) | ¹H NMR, ORD | (58, 74) |
| | Mytiloxanthin (38) | Total Synthesis | (57, 115) |
| | Trikentriorhodin (39)? | | (38, 57) |
| (XXXII) | Capsanthinone (86) | Partial synthesis from (2) | (25) |
| | | ORD | |

Most information is available for the early steps, common to caro-
tenogenic systems, which lead from mevalonic acid to the colourless
$C_{40}$ precursor phytoene (21) with subsequent dehydrogenation and
cyclization to coloured carotenes. Far less evidence has been obtained
for the later steps in carotenoid biosynthesis which include the introduction of oxygen functions. Chirality is usually introduced at a fairly
late stage and is generally connected with ε-, γ- and κ-rings, allenes,
$C_{45}$- and $C_{50}$-carotenoids and particular oxygen functions (secondary
alcohols and derivatives thereof, certain tertiary alcohols, epoxides
and glycosides). The large number of carotenoids produced by photosynthetic bateria is achiral (127); the greatest number of chiral centres
(six) has been encountered in algal carotenoids such as peridinin (30)
and fucoxanthin [(40), R = H].

When biosynthetic data are lacking mere knowledge of chirality may
provide indirect information about possible biogenetic relationships, as
illustrated by examples in the discussion below. In other cases established
chirality may render reexamination of presumed biosynthetic pathways
necessary.

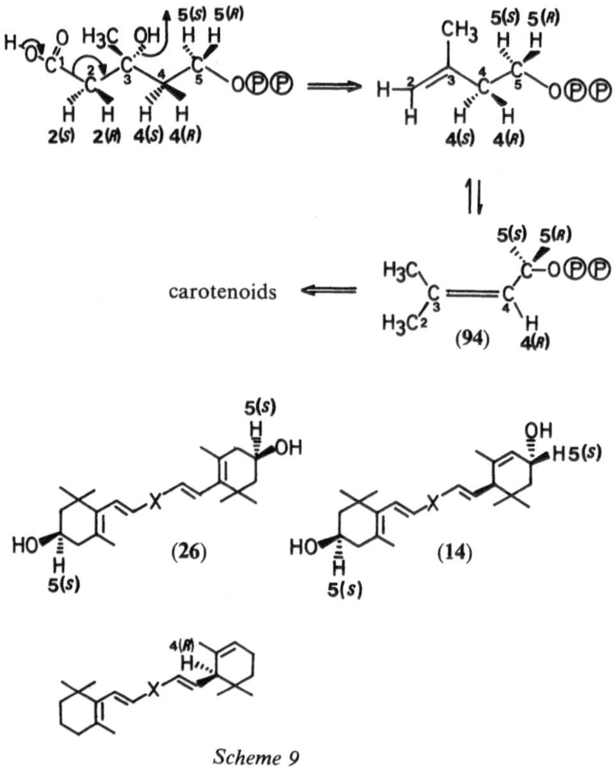

Scheme 9

The chirality of lutein (14) is now firmly established with the 3,3'-hydroxy functions in the β- and ε-rings possessing opposite absolute configuration (4, 40). The stereochemistry of the hydroxylation step in zeaxanthin (26) biosynthesis in a *Flavobacterium* sp. has been determined by using $(5R)$-[2-$^{14}$C, 5-$^3$H$_1$] mevalonate as substrate which demonstrated retention of the 5-*pro-S* hydrogen at C-3(3') (30), Scheme 9. Also in the case of lutein (14) it has been shown that the 5-*pro-S* hydrogen at C-3' is retained (81, 171) and β,ε-carotene (92) biosynthesized from $(4R)$-[2-$^{14}$C, 4-$^3$H$_1$] mevalonate retains the tritium at C-6 (82). Any mechanistic interpretation of the biosynthetic evidence must be consistent with the established chirality.

Eschscholtzxanthin (27) produced in Californian poppies from $(4R)$-[2-$^{14}$C, 4-$^3$H] mevalonate showed the same $^{14}$C:$^3$H ratio (8 : 6) as β,β-carotene (Scheme 10), and the biosynthetic pathway suggested *via* antheraxanthin [= the 5,6-epoxide of zeaxanthin (26)] (80, 178) is consistent with later findings on the absolute configuration of (27) (8), shown to be the same as in zeaxanthin (26).

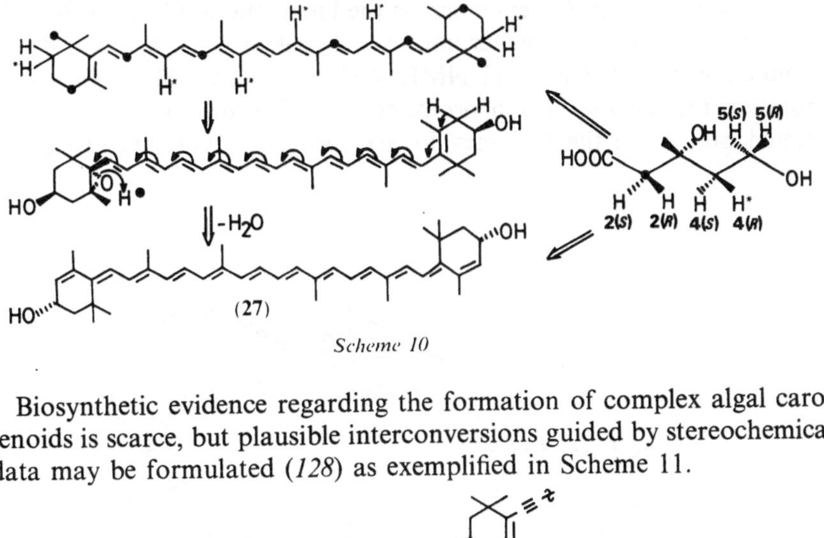

*Scheme 10*

Biosynthetic evidence regarding the formation of complex algal carotenoids is scarce, but plausible interconversions guided by stereochemical data may be formulated (128) as exemplified in Scheme 11.

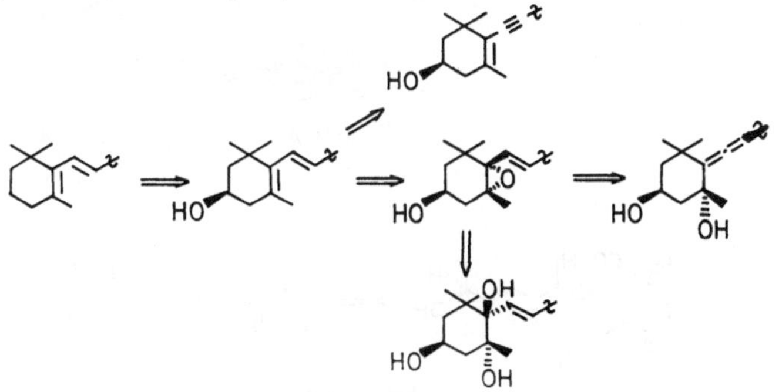

*Scheme 11*

The mechanism of cyclization has been a major problem in carotenoid biosynthesis for a long time. Inheritance studies with tomatoes (164)
and isotopic labelling studies (82, 83, 179) support the independent
formation of β- and ε-rings in the $C_{40}$-series. Since the chirality of β,γ-
carotene (69) is opposite to that of β,ε-carotene (92) different modes of
cyclization are indicated for ε- and γ-rings. It has been demonstrated
that the 17-methyl group (by carotenoid numbering) of trisporic acid
(95), (Scheme 12) was labelled by [2-$^{14}$C]-mevalonate. Since trisporic
acid (95) is further known to be formed from β,β-carotene this experi-
ment defines the biosynthetic origin of the gem-methyl groups in its
precursor and suggests that folding A is required for the acyclic $C_{40}$
precursor formed via dimethylallyl pyrophosphate [(94), Scheme 9] prior
to cyclization to β,β-carotene. The same folding A would, however
result in the wrong enantiomer of β,ε-carotene. Hence it was suggested
by BuLock et al. (45) that the opposite folding B is required for the
biosynthesis of the ε-ring. In order to accommodate the stereochemical
result for β,γ-carotene (69) folding A may than be postulated for the
cyclization to γ-rings. Recent work on the biosynthesis of zeaxanthin (26)
from lycopene (94) in a bacterium, carried out in $D_2O$, has revealed in
conjunction with MS and $^1H$ NMR studies the (2$S$,2'$S$)-[2,2-$^2H_2$]-con-
figuration for the zeaxanthin produced (31). One inconsistency may be
pointed out. Since the C-2 mevalonate derived gem-methyl group and

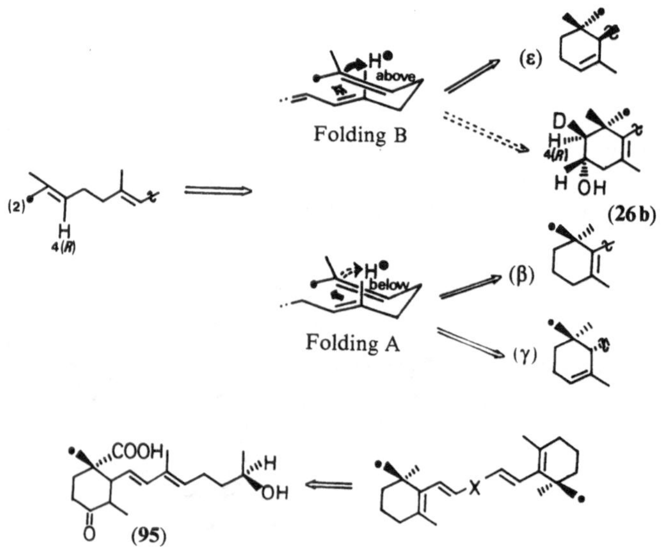

Scheme 12

the 4-*pro*(*R*) mevalonate derived hydrogen are *cis* in the acyclic precursor (*cf.* **93**) models reveal that a simple *trans* addition type cyclization mechanism leading to that absolute configuration at C-2 in zeaxanthin would result in the steric arrangement of the *gem*-methyl groups given in (**26b**), Scheme 12. This is not in agreement with the conclusion reached from the trisporic acid evidence. Subsequent work with the same *Flavobacterium* sp. by the Liverpool group (*30a*) using [2-$^{13}$C]-mevalonate as substrate has now indeed confirmed the expected geometry (**26b**) for zeaxanthin formed *via* the aliphatic [16-$^{13}$C]-lycopene (the 16-methyl group is *trans* to the carbon chain of lycopene) on the basis of $^{13}$C-NMR evidence. The folding involved for β-ring formation in the bacterial (*30a*) and fungal system (*45*) therefore appears to be different.

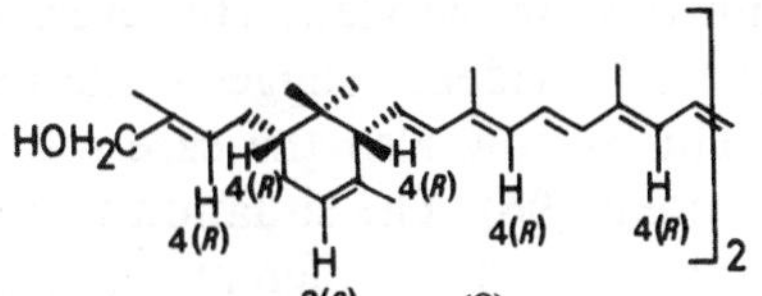

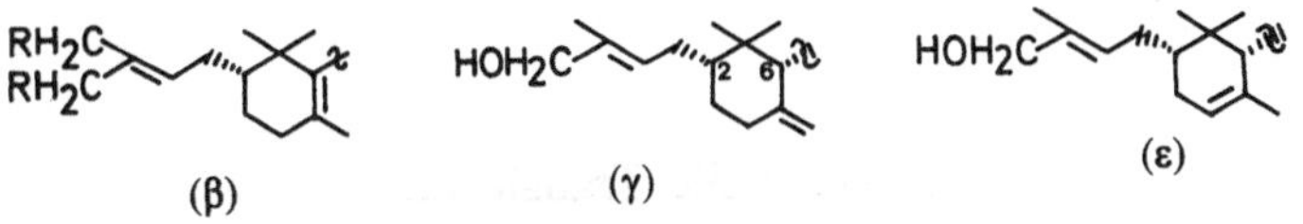

*Scheme 13*

In the above discussion only chair folding and *E*-configuration of Δ5 of the acyclic precursor has been considered. A thorough theoretical treatment of the biological cyclization reaction was recently made by Eugster (*71b*) who, after considering the eight steric alternatives by chair ($C_1$ or $C_2$), or boat ($B_1$ or $B_2$) foldings, the geometry of the Δ1 and Δ5 double bonds (*E* or *Z*) of the acyclic precursor and the prochirality of the ene surface concluded that the stereochemical information is not yet conclusive as to the cyclization mechanism for any carotenoid belonging to the β, γ- or ε-series. For zeaxanthin (**26**) the biosynthetic evidence (*30a, 31*) now appears to be compatible with folding B = $C_2$ (*E,E*) folding in Eugster's terminology.

In the bicyclic $C_{50}$-series recent biosynthetic evidence from a cell-free system has defined the origin of the H-2,6,10,14,2',6',10',14' and the olefinic proton of the two isopropylidene groups of decaprenoxanthin (**8**) as the 4-*pro*(*R*) hydrogen of mevalonic acid. The 2-*pro*(*R*) and 5-*pro*(*S*) hydrogens were also retained in the polyene chain in the same way as in the $C_{40}$-series, and the 2-*pro*(*S*) hydrogen of mevalonate was retained as H-4,4' in (**8**) (*73*) (Scheme 13). Were the origin of the geminal 16,17-methyl groups also known, all stereochemical details required to formulate the isopentenylation/cyclization reaction for decaprenoxanthin (**8**) would be available, because the chirality of (**8**) is already established (*11*). Present evidence is compatible with folding A (Scheme 12) = $C_1$ (*E,E*) folding. Since $C_{50}$-carotenoids with ε-, β- and γ-rings have identical configuration at C-2(2') and those with ε- and γ-rings have identical chirality at C-6(6') the same folding is likely to be involved in the cyclization leading to their formation.

It has been suggested that mirror image processes may be involved in the cyclization of $C_{40}$ and $C_{50}$ carotenoids (*73*), and available bio-synthetic and stereochemical evidence for zeaxanthin (**26**) and $C_{50}$-caro-tenoids is consistent with this simple view. However, the trisporic acid (*45*) and zeaxanthin (*30a, 31*) evidence suggests alternative foldings for formation of a β-ring. This is now also the case for the ε-ring in β,ε-carotene (**92**) and ε,ε-carotene (**96**). The situation therefore seems to be complicated by different situations in different types of organisms and additional evidence is required before a clearer picture can emerge.

## 2. Metabolic Transformations

It is generally assumed that animals do not have the capability to carry out *de novo* carotenoid synthesis. However, they occassionally have the ability to modify structurally carotenoids obtained through their diet. The absolute configuration of the carotenoids involved may throw light on such metabolic transformations.

For instance, it has been claimed on the basis of isotopic labelling experiments that lutein (**14**) was converted to (3*S*,3'*S*)-astaxanthin (**28**) in fish (*95, 111, 112*). The route proposed which did not involve an epi-merization at C-3' is not compatible with the absolute configurations later established for (**14**) and (**28**). However, recent evidence for the 3'-*S* chirality of α-doradexanthin (**96**) from goldfish (*43a*) has led to the plausible suggestion that the biosynthetic conversion of lutein (**14**) proceeds *via* 3'-didehydrolutein → (3'*S*)-epilutein (**58**) → (3'*S*)-α-doradexanthin (**96**) → (3*S*,3'*S*)-astaxanthin (**28**). The reported conver-sion of alloxanthin (**31**) to 7,8,7',8'-tetradehydroastaxanthin (**34**) in gold-fish (*85*) is compatible with stereochemical data. But this would also be true for a conversion of (3*S*,3'*S*)-astaxanthin (**28**) to its 7,8,7',8'-tetra-dehydroderivative (**28**).

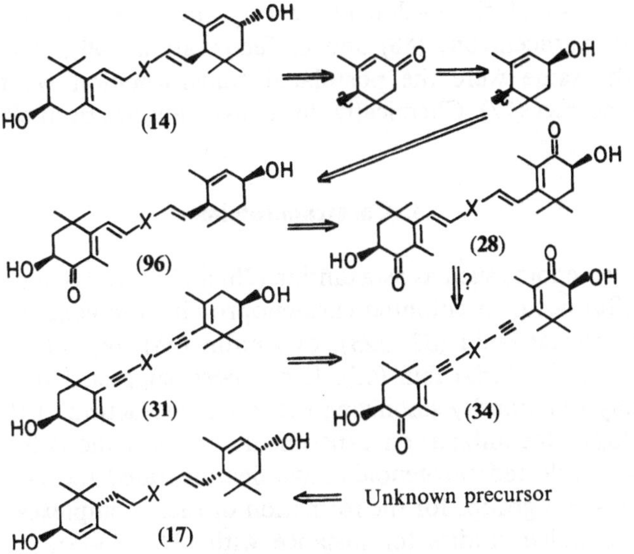

*Scheme 14*

Tunaxanthin (**17**) is a challenge in biogenetic respects. The 6S,6'S-chirality is opposite to that of those $C_{40}$ carotenoids which have ε-rings and are known to be synthesized by primary producers. However, the recent establishment of 6S,6'S-configuration for ε,ε-carotene (**93**) from a green alga (*60*) is of great interest in this context.

Were mytiloxanthin (**38**) and trikentriorhodin (**39**) formed by pinacolic rearrangement from dietary carotenoids with the common 3-hydroxy-5,6-epoxy β-ring system as has been hypothesized (*114*), the

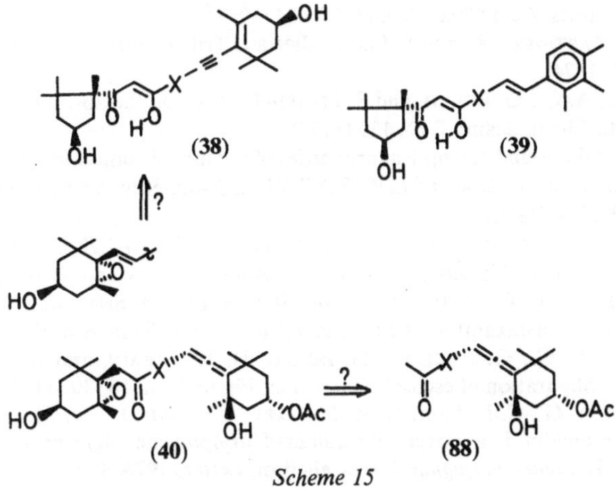

*Scheme 15*

chirality at C-3 of the cyclopentane ring should be maintained. The chirality of paracentrone (88) and of fucoxanthin [(40), R = H] should also be the same were the postulated transformation *via* the 3-keto derivative correct (77). Chemically this conversion has been effected (93).

## 3. Carotenoproteins

Ketocarotenoids such as astaxanthin (28) and its acetylenic derivatives (34) and (80) occur in coloured carotenoprotein complexes examples of which are crustacyanin (52, 123), ovoverdin (48), ovorubin (50), alloporin (151) and asteriarubin (65). It has been suggested that the carotenoids may function by stabilizing the tertiary structure of the proteins involved (51). Recombination experiments between the colourless apoproteins and selected carotenoids have demonstrated the need for conjugated 4,4'-keto groups for the formation of these complexes (151, 180). However, detailed studies for instance with (3S,3'S)-astaxanthin (28), its 3R,3'R enantiomer and the mesoform for elucidation of the steric requirements for the complexing carotenoid have not yet been reported.

The molecular topology of the peridinin (30)-chlorophyll-protein complex (122) and of crustacyanin (123a) has recently been studied and future progress in this field is anticipated.

## References

1. Aasen, A. J., and S. Liaaen-Jensen: Cross-conjugated carotenals. Acta Chem. Scand. 21, 2185 (1967).
2. Aasen, A. J., S. Liaaen-Jensen, and G. Borch: The chirality of zeaxanthin from different sources. Acta Chem. Scand. 25, 404 (1971).
3. Aguilar-Martinez, M., and S. Liaaen-Jensen: Trikentriorhodin. Acta Chem. Scand. B 28, 1247 (1974).
4. Andrewes, A. G., G. Borch, and S. Liaaen-Jensen: On the absolute configuration of lutein. Acta Chem. Scand. B 28, 139 (1974).
5. — — — Synthesis and chiroptical properties of the model compounds (2R,6R,2'R,6'R)-2,2'-dimethyl-ε,ε-carotene and (2R,6S,2'R,6'S)-2,2'-dimethyl-ε,ε-carotene. Acta Chem. Scand. B 30, 214 (1976).
6. — — — Synthesis of the model compounds (2R,6S,2'R,6'S)-2,2'-dimethyl-γ,γ-carotene and (2R,2'R,6'S)-2,2'dimethyl-β,γ-carotene. Acta Chem. Scand. B 31, 212 (1977).
7. Andrewes, A. G., G. Borch, S. Liaaen-Jensen, and G. Snatzke: On the absolute configuration of astaxanthin and actinioerythrin. Acta Chem. Scand. B 28, 730 (1974).
8. Andrewes, A. G., G. Englert, G. Borch, H. H. Strain, and S. Liaaen-Jensen: Absolute configuration of eschscholtzxanthin. Phytochem. 18, 303 (1979).
9. Andrewes, A. G., C. L. Jenkins, M. P. Starr, J. Shepherd, and M. Hope: Structure of xanthomonadin I, a novel dibrominated arylpolyene pigment produced by the bacterium *Xanthomonas juglandis*. Tetrahedron Letters 1976, 4023.

10. ANDREWES, A. G., S. LIAAEN-JENSEN, and G. BORCH: Synthesis of (2R,2'R)-2,2'-dimethyl-β,β-carotene and absolute configuration of (2R,2'R)-(4-hydroxymethyl-2-butenyl)-2'-(3-methyl-2-butenyl)-β,β-carotene. Acta Chem. Scand. B 28, 737 (1974).

11. ANDREWES, A. G., S. LIAAEN-JENSEN, and O. B. WEEKS: Absolute configuration of decaprenoxanthin. Acta Chem. Scand. B 29, 884 (1975).

12. ANDREWES, A. G., and M. P. STARR: (3R,3'R)-Astaxanthin from the yeast *Phaffia rhodozyma.* Phytochem. 15, 1009 (1976).

13. ARPIN, N., J.-L. FIASSON, S. NORGÅRD, G. BORCH, and S. LIAAEN-JENSEN: C₅₀-carotenoids from *Arthrobacter glacialis.* Acta Chem. Scand. B 29, 921 (1975).

14. ARPIN, N., H. KJØSEN, G. W. FRANCIS, and S. LIAAEN-JENSEN: The constitution of aleuriaxanthin. Phytochem. 12, 2751 (1973).

15. ARPIN, N., and S. LIAAEN-JENSEN: Rubixanthin and gazaniaxanthin. Phytochem. 8, 185 (1969).

16. ARPIN, N., S. LIAAEN-JENSEN, and M. TROUILLOUD: Isolation of decaprenoxanthin mono- and diglucoside from an *Arthrobacter* sp. Acta Chem. Scand. 26, 2524 (1972).

17. ARPIN, N., W. A. SVEC, and S. LIAAEN-JENSEN: New fucoxanthin-related carotenoids from *Coccolithus huxleyi.* Phytochem. 15, 529 (1976).

18. BARLOW, L., and G. PATTENDEN: Synthesis of poly-Z-isomers of 2,6,11,15-tetramethylhexadeca-2,6,8,10,14-pentane, a C₂₀ analogue of phytoene. Re-examination of the stereochemistry of a new isomer of phytoene from *Rhodospirillum rubrum.* J. Chem. Soc. Perkin I 1976, 1029.

19. BARBER, M. S., A. HARDISSON, L. M. JACKMAN, and B. C. L. WEEDON: Stereochemistry of the bixins. J. Chem. Soc. 1961, 1625.

20. BARTLETT, L., W. KLYNE, W. P. MOSE, P. M. SCOPES, G. GALASKO, A. K. MALLAMS, and B. C. L. WEEDON: Optical rotatory dispersion of carotenoids. J. Chem. Soc (C) 1969, 2527.

21. BERGER, R., G. BORCH, and S. LIAAEN-JENSEN: Chirality of asterinic acid. Acta Chem. Scand. B 31, 243 (1977).

22. BERNHARDT, K., G. P. MOSS, G. TOTH, and B. C. L. WEEDON: Stereoisomers of fucoxanthin. Tetrahedron Letters 1974, 3099.

23. — — — — — Absolute configuration of fucoxanthin. Tetrahedron Letters 1976, 115.

23a. BERNHARD, K., F. KIENZLE, H. MAYER, and R. K. MÜLLER: The synthesis of (3S,3'S)-asterinic acid. Abstracts Fifth Internat. IUPAC Symp. Carotenoids 1978, 4.

23b. BESTMAN, H. J.: Synthesis of polyenes via phosphonium ylids. Pure and Applied Chem. 51, 515 (1979).

24. BINGHAM, A., H. S. MOSHER, and A. G. ANDREWES: Epimeric 3,3'-dihydroxy-ε,ε-carotene from the skin of the yellow Costa Rican frog, *Atelopus chiriquiensis.* Chem. Commun. 1977, 96.

24a. BINGHAM, A., D. W. WILKIE, and H. S. MOSHER: Tunaxanthin: Occurrence and absolute stereochemistry. Comp. Biochem. Physiol. 62B, 489 (1979).

25. BODEA, C., A. G. ANDREWES, G. BORCH, and S. LIAAEN-JENSEN: Physoxanthin. Phytochem. 17, 2038 (1978).

26. BONNETT, R., A. K. MALLAMS, A. A. SPARK, J. L. TEE, B. C. L. WEEDON, and A. McCORMICK: Structure and reactions of fucoxanthin. J. Chem. Soc. (C) 1969, 429.

27. BORCH, G., S. NORGÅRD, and S. LIAAEN-JENSEN: Circular dichroism and relative configuration of C₅₀-carotenoids. Acta Chem. Scand. 25, 402 (1971).

28. BORCH, G., H. RØNNEBERG, R. BERGER, N. ARPIN, R. BUCHECKER, S. HERTZBERG, M. HALLENSTVET, K. L. SIMPSON, and S. LIAAEN-JENSEN: On the chirality of 2'-oxygenated carotenoids related to plectaniaxanthin. Phytochem. To be published.

29. BREMSER, W., and J. PAUST: Die ¹³C-NMR-Spectren von β-Carotin und die Ladungsverteilung in der Polyenkette von Apocarotinalen. Org. Magn. Res. 6, 433 (1974).

30. Britton, G.: Later reactions of carotenoid biosynthesis. Pure and Applied Chem. 47, 223 (1976).

30a. Britton, G., T. W. Goodwin, W. J. S. Lockley, N. J. Patel, and G. Englert: Stereochemistry of the cyclization reaction in carotenoid biosynthesis: Studies with stable isotopes. Abstracts Fifth Internat. IUPAC Symp. Carotenoids 1978, 5.

31. Britton. G.. W. J. S. Lockley, N. J. Patel, T. W. Goodwin, and G. Englert: Use of deuterium labelling to elucidate the stereochemistry of the initial step of the cyclization reaction in zeaxanthin biosynthesis in a Flavobacterium. J. C. S. Chem. Commun. 1977, 665.

32. Brooks, C. J. W., and J. D. Gilbert: Absolute configuration of secondary alcohols. A gas chromatographic modification of Horeau's method. J. C. S. Chem. Commun. 1973, 194.

33. Brown, B. O., and B. C. L. Weedon: Rubixanthin and gazaniaxanthin. J. C. S. Chem. Commun. 1968, 382.

34. Buchecker, R., N. Arpin, and S. Liaaen-Jensen: Absolute configuration of aleuriaxanthin. Phytochem. 15, 1013 (1976).

35. Buchecker, R., and C. H. Eugster: Absolute Konfiguration von δ-Carotin aus der Tomatenmutante Del/Del 65-3-54-5. Helv. Chim. Acta 54, 327 (1971).

36. — — Absolute Konfiguration von Picrocrocin. Helv. Chim. Acta 56, 1121 (1973).

37. — — Absolute Konfiguration von α-Zeacarotin, α-Apo-8-carotinal, und α-Apo-8-carotinol. Helv. Chim. Acta 56, 1124 (1973).

38. Buchecker, R., C. H. Eugster, H. Kjøsen, and S. Liaaen-Jensen: Absolute configuration of β,ε-caroten-2-ol, β,β-caroten-2-ol and β,β-carotene-2,2'-diol. Acta Chem. Scand. B 28, 449 (1974).

39. Buchecker, R., C. H. Eugster, and C. Litchfield: Carotinoide aus marinen Schwämmen (Porifera): Isolierung und Struktur von sieben Hauptcarotinoiden aus Agelas schmidtii. Helv. Chim. Acta 60, 2780 (1977).

40. Buchecker, R., P. Hamm, and C. H. Eugster: Absolute Konfiguration von Xanthophyll (Lutein). Helv. Chim. Acta 57, 631 (1974).

41. Buchecker, R., and S. Liaaen-Jensen: Reaktionen an Allencarotinoiden. Helv. Chim. Acta 58, 89 (1975).

42. — — Absolute configuration of heteroxanthin and diadinoxanthin. Phytochem. 16, 729 (1977).

43. Buchecker, R., S. Liaaen-Jensen, G. Borch, and H. W. Siegelman: Carotenoids of Anacystis nidulans. Structures of caloxanthin and nostoxanthin. Phytochem. 15, 1015 (1976).

43a. Buchecker, R., A. Weber, and C. H. Eugster: Stereochemistry of α-doradexanthin and related compounds. Biochemical implications. Abstracts Fifth Internat. IUPAC Symp. Carotenoids 1978, 6.

44. Buchecker, R., H. Yokoyama, and C. H. Eugster: Absolute Konfiguration von Semi-α-carotinon aus Murraya exotica. Helv. Chim. Acta 53, 1210 (1970).

45. Bu'Lock, J. D., D. J. Austin, G. Snatzke, and L. Hruban: Absolute configuration of trisporic acids and the stereochemistry of cyclization in β-carotene biosynthesis. J. C. S. Chem. Commun. 1970, 255.

46. Cadosch, H., and C. H. Eugster: Beweis der absoluten Konfiguration der Xanthophyll-5,6-epoxide (Luteinepoxide). Helv. Chim. Acta 57, 1472 (1974).

46a. Cadosch, H., U. Vögeli, P. Rüedi, and C. H. Eugster: Über die Carotinoide Flavoxanthin und Crysanthemaxanthin: $^1$H-NMR, $^{13}$C-NMR und Massen-Spektren und absoluten Konfiguration. Helv. Chim. Acta 61, 783 (1978).

47. Campbell, R. V. M., L. Crombie, D. A. R. Findley, R. W. King, G. Pattenden, and D. A. Whiting: Synthesis of (±)-presqualene alcohol, (±)-prephytoene alcohol and structurally related compounds. J. Chem. Soc. Perkin I 1975, 897.

48. CECCALDI, H. J., D. F. CHEESMAN, and P. F. ZAGALSKY: Quelques propriétés et caractéristiques de l'ovoverdine. Soc. Biol. Marseille 1966, 582.
49. CHAE, C., P.-S. SONG, J. E. JOHANSEN, and S. LIAAEN-JENSEN: Linear dichroic spectra of cross-conjugated carotenals and configurations of in-chain substituted carotenoids. J. Amer. Chem. Soc. 99, 5609 (1977).
50. CHEESMAN, D. F.: Ovorubin, a chromoprotein from the eggs of the gastropod mollusc Pomacea canaliculata. Roy. Soc. Proc. B 149, 571 (1958).
51. CHEESMAN, D. F., W. L. LEE, and P. F. ZAGALSKY: Carotenoproteins in invertebrates. Biol. Rev. 42, 132 (1967).
52. CHEESMAN, D. F., P. F. ZAGALSKY, and H. J. CECCALDI: Purification and properties of crustacyanin. Roy. Soc. Proc. B 164, 130 (1966).
53. CHENG, J. Y., M. DON-PAUL, and N. J. ANTIA: Isolation of an unusually stable cis-isomer of alloxanthin from a bleached autolyzed culture of Chroomonas salina grown heterotropically on glycerol. Observation on cis-trans isomerization of alloxanthin. Z. Protozool. 21, 761 (1974).
54. CHIN, A., and P. S. SONG: Electronic spectra of carotenoids: A theoretical analysis of the electronic spectrum of rhodopinal. J. Mol. Spectrosc. 52, 224 (1974).
55. CHOLNOKY, L., K. GYØRGYFI, A. RONAI, J. SZABOLCS, G. TÓTH, G. GALASKO, A. K. MALLAMS, E. S. WAIGHT, and B. C. L. WEEDON: Structure of neoxanthin (folia-xanthin). J. Chem. Soc. (C) 1969, 1256.
56. CHOPRA, A. K., B. P. S. KHAMBAY, H. MADDEN, G. P. MOSS, and B. C. L. WEEDON: Synthesis of $C_{50}$ carotenoids; the structure of decaprenoxanthin. J. C. S. Chem. Commun. 1977, 357.
57. CHOPRA, A. K., G. P. MOSS, and B. C. L. WEEDON: Synthesis of the enolic β-diketone carotenoids mytiloxanthin and trikentriorhodin. J. C. S. Chem. Commun. 1977, 467.
58. COOPER, R. D. G., L. M. JACKMAN, and B. C. L. WEEDON: Stereochemistry of capso-rubin and synthesis of optically inactive epimers. J. Chem. Soc. (C) 1962, 215.
59. DABBAGH, A. G., and K. EGGER: Calthaxanthin — Ein Stereoisomer des Lutein aus Caltha palustris. Z. Pflanzenphysiol. 72, 177 (1974).
60. DAVIES, B. H.: Solved and unsolved problems of carotenoid formation. Pure and Applied Chem. 51, 623 (1979).
61. DAVIES, B. H., and R. F. TAYLOR: Carotenoid biosynthesis — the early steps. Pure and Applied Chem. 47, 211 (1976).
62. DAVIES, J. B., L. M. JACKMAN, P. T. SIDDONS, and B. C. L. WEEDON: The structure and synthesis of phytoene, phytofluene, ζ-carotene and neurosporene. J. Chem. Soc. (C) 1966, 2154.
63. DE VILLE, T. E., M. B. HURSTHOUSE, S. W. RUSSELL, and B. C. L. WEEDON: Stereo-chemistry of allenes. J. C. S. Chem. Commun. 1969, 754.
64. — — — — Absolute configuration of carotenoids. J. C. S. Chem. Commun. 1969, 1311.
65. ELGSAETER, A., J. D. TAUBER, and S. LIAAEN-JENSEN: Carotenoid distribution and caro-tenoprotein of Asterias rubens. Biochim. Biophys. Acta 530, 402 (1978); 531, 357 (1978).
66. ELIEL, E. L.: Stereochemistry of carbon compounds, p. 411. New York: Mc Graw-Hill. 1962.
67. — Stereochemie der Kohlenstoffverbindungen, pp. 87, 136. Weinheim: Verlag Chemie. 1966.
68. ENGLERT, G.: A carbon-13 NMR study of cis-trans isomeric vitamins A, carotenoids and related compounds. Helv. Chim. Acta 58, 2367 (1975).
69. ENGLERT, G., F. KIENZLE, and K. NOACK: $^1$H NMR-, $^{13}$C-NMR-, UV- und CD-Daten von synthetischen (3S,3'S)-Astaxanthin, seinen 15-cis Isomeren und einigen analogen Verbindungen. Helv. Chim. Acta 60, 1209 (1977).
70. ESCHENMOSER, W., and C. H. EUGSTER: Absolute Konfiguration von Azafrin. Helv. Chim. Acta 58, 1722 (1975).

168                                    S. Liaaen-Jensen:

71. Eschenmoser, W., and C. H. Eugster: Synthese und Chiralität von (5S,6R)-5,6-
    Epoxy-5,6-dihydro-β,β-carotin und (5R,6R)-5,6-Dihydro-β,β-carotin-5,6-diol, einen
    Carotinoid mit ungewöhnlichen Eigenschaften. Helv. Chim. Acta 61, 822 (1978).
71a. — — Synthesis of optically active carotenoid glycols and epoxides. Abstract Fifth
    Internat. IUPAC Symp. Carotenoids 1978, 14.
71b. Eugster, C. H.: Characterization, chemistry and stereochemistry of carotenoids.
    Pure and Applied Chem. 51, 463 (1979).
72. Eugster, C. H., R. Buchecker, C. Tscharner, G. Uhde, and G. Ohloff: Bestimmung
    der Chiralität der enantiomeren α-Cyclogeraniumsäuren, α-Cyclogeraniale, α-Jonone,
    γ-Jonone, α-Carotine, ε-Carotine und verwandte Verbindungen durch chemische Ver-
    knüpfungsreaktionen. Helv. Chim. Acta 52, 1729 (1969).
73. Fahey, D. P., and B. V. Milborrow: The stereochemistry of biosynthesis of
    decaprenoxanthin in a cell-free system. Phytochem. 17, 2077 (1978).
74. Faigle, H., and P. Karrer: Die Konfiguration des natürlichen (+)-Capsanthins und
    des natürlichen Capsorubins. Helv. Chim. Acta 44, 1904 (1961).
75. Fiksdahl, A., J. T. Mortensen, and S. Liaaen-Jensen: High-pressure liquid chro-
    matography of carotenoids. J. Chromatography 157, 111 (1978).
76. Fiksdahl, A., J. D. Tauber, S. Liaaen-Jensen, G. Saucy, and G. F. Weber: Steric
    stability of acetylenic carotenoids. Acta Chem. Scand. 33B, 192 (1979).
77. Galasko, G., J. Hora, T. P. Toube, B. C. L. Weedon, D. André, M. Barbier,
    E. Lederer, and V. R. Villanueva: Allenic carotenoids in sea urchins. J. Chem.
    Soc. (C) 1969, 1264.
78. Goodfellow, D., G. P. Moss, J. Szabolcs, G. Tóth, and B. C. L. Weedon:
    Configuration of carotenoid epoxides. Tetrahedron Letters 1973, 3925.
79. Goodfellow, D., G. P. Moss, and B. C. L. Weedon: The absolute configuration of
    lutein. J. C. S. Chem. Commun. 1970, 1578.
80. Goodwin, T. W.: Biosynthesis. In: Carotenoids (Isler, O., ed.), Chap. VII. Basel:
    Birkhäuser. 1971.
81. — Biosynthesis of carotenoids and plant triterpenes. Biochem. J. 123, 293 (1971).
82. Goodwin, T. W., and R. J. H. Williams: A mechanism for the cyclization of an
    acyclic precursor to form β-carotene. Biochem. J. 94, 5 C (1965).
83. — — A mechanism for the biosynthesis of α-carotene. Biochem. J. 97, 28 C (1965).
83a. Granger, G., B. Maudinas, R. Herbec, and J. Villoutreix: ¹H and ¹³C NMR
    spectra of cis and trans phytoene isomers. J. Magn. Res. 10, 43 (1973).
84. Hallenstvet, M., R. Buchecker, G. Borch, and S. Liaaen-Jensen: Absolute
    configuration of β,γ-carotene and biosynthetic implications. Phytochem. 16, 583 (1977).
85. Hata, M., and M. Hata: Metabolism of injested cynthiaxanthin. Tokohu J. Agr.
    Res. 21, 183 (1970).
86. Haxo, F. T., J. H. Kycia, G. F. Somers, A. Bennett, and H. W. Siegelman:
    Peridinin-chlorophyll a-proteins of the dinoflagellate Amphidinium carterae (Plymouth
    450). Plant Physiol. 57, 297 (1976).
87. Hemmer, E., and S. Liaaen-Jensen: Spectroscopic properties of methyl triacetyl α-
    and β-L-rhamnosides. Acta Chem. Scand. 24, 3019 (1970).
88. Hertzberg, S., G. Borch, and S. Liaaen-Jensen: Absolute configuration of zea-
    xanthin dirhamnoside. Arch. Mikrobiol. 110, 95 (1976).
89. — — — CD spectra of mono-cis carotenoids. Acta Chem. Scand. 33B, 42 (1979).
90. Hertzberg, S., and S. Liaaen-Jensen: Absolute configuration of sarcinaxanthin
    and sarcinaxanthin mono-β-D-glucoside. Isolation of sarcinaxanthin diglucoside.
    Acta Chem. Scand. B 31, 215 (1977).
91. Hertzberg, S., T. Mortensen, G. Borch, H. W. Siegelman, and S. Liaaen-Jensen:
    On the absolute configuration of 19'-hexanoyloxyfucoxanthin. Phytochem. 16, 587
    (1977).

92. HLUBUCEK, J. R., J. HORA, W. S. RUSSELL, T. P. TOUBE, and B. C. L. WEEDON: Stereochemistry and synthesis of the allenic end groups. Absolute configuration of zeaxanthin. J. Chem. Soc. Perkin I **1974**, 848.

93. HORA, J., P. T. TOUBE, and B. C. L. WEEDON: Conversion of fucoxanthin into paracentrone. J. Chem. Soc. (C) **1970**, 241.

94. HOREAU, A.: Principe et application d'une nouvelle methode de determination de configuration dite "par dedoublement partiel". Tetrahedron Letters **1961**, 506.

95. HSU, W.-J., D. B. RODRIGUEZ, and C. O. CHICHESTER: The conversion of $^{14}$C-lutein and $^{14}$C-β-carotene in goldfish. Int. J. Biochem. **3**, 333 (1971).

95a. HURSTHOUSE, M. B., and G. P. MOSS: X-Ray studies on the stereochemistry of carotenoid end groups. Abstracts Fifth Internat. IUPAC Symp. Carotenoids **1978**, 24.

96. IKE, T., J. INANAGO, A. NAKANO, N. OKUKADO, and M. YAMAGUCHI: Total synthesis of natural acetylenic analogues of isorenieratene and renieratene. Bull. Chem. Soc. Jap. **47**, 350 (1974).

97. International Union of Pure and Applied Chemistry: Plenary lectures presented at the Third International Symposium On Carotenoids Other Than Vitamin A. Pure and Applied Chem. **35**, 1 (1973).

98. — Nomenclature of carotenoids (Rules approved 1974). London: Butterworths. 1974.

99. — Main lectures presented at the Fourth International Symposium on Carotenoids. Pure and Applied Chem. **47**, 97 (1976).

100. — Main lectures presented at the Fifth International Symposium on Carotenoids. Pure and Applied Chem. **51**, 435 (1979).

101. ISLER, O. (ed.): Carotenoids. Basel: Birkhäuser. 1971.

102. ITO, M., R. MASAHARA, and K. TSUKIDA: Synthesis of (2S)-β,β-caroten-2-ol. Tetrahedron Letters **1977**, 2767.

103. JAUTELAT, M., J. B. GRUTZNER, and J. D. ROBERTS: Natural-abundance $^{13}$C nuclear magnetic resonance spectra of terpenes and carotenes. Proc. Nat. Acad. Sci. **65**, 288 (1970).

104. JOHANSEN, J. E.: Chemical studies of selected algal and bacterial carotenoids. Trondheim: Univ. Trondheim. 1977.

105. JOHANSEN, J. E., and S. LIAAEN-JENSEN: Chemical reactions of allenic carotenoids. Acta Chem. Scand. **B 28**, 949 (1974).

106. — — Reactions of diapocarotenals with N-bromosuccinimide and synthetic applications. Acta Chem. Scand. **B 29**, 315 (1975).

107. — — Total synthesis of bacterioruberin derivatives. Absolute configuration of bacterioruberin. Tetrahedron Letters **1976**, 955.

108. — — Studies on the absolute configuration of peridinin and dinoxanthin. In: Marine Natural Products Chemistry (FAULKNER, D. J., and W. H. FENICAL, eds.), p. 225. New York: Plenum. 1977.

109. — — Total synthesis of cross-conjugated carotenals. Tetrahedron **33**, 381 (1977).

110. KARRER, P., and E. JUCKER: Carotenoids. New York: Elsevier. 1950.

111. KATAYAMA, T., K. HIRATA, H. YOKOYAMA, and C. O. CHICHESTER: The carotenoids of sea breams. Bull. Jap. Soc. Scient. Fish. **36**, 709 (1970).

112. KATAYAMA, T., H. YOKOYAMA, and C. O. CHICHESTER: The structures of α-doradexanthin and β-doradexanthin. Int. J. Biochem. **1**, 438 (1970).

113. KELLY, M., S. A. ANDRESEN, and S. LIAAEN-JENSEN: The stereochemistry of lycoxanthin and lycophyll. Acta Chem. Scand. **25**, 1607 (1971).

114. KHARE, A., G. P. MOSS, and B. C. L. WEEDON: Mytiloxanthin and isomytiloxanthin, two novel acetylenic carotenoids. Tetrahedron Letters **1973**, 3921.

115. KHATOON, N., D. E. LOEBER, T. P. TOUBE, and B. C. L. WEEDON: Stereochemistry of phytoene. J. C. S. Chem. Commun. **1972**, 996.

116. KIENZLE, F., and H. J. MAYER: Synthese von (3S,3'S)-Astaxanthin. Helv. Chim. Acta **61**, 2609 (1978).

*117.* Kjøsen, H., and S. Liaaen-Jensen: Application of the tris (dipivalomethanato)europium (III) nuclear magnetic shift reagent to carotenoids. Acta Chem. Scand. **26**, 2185 (1972).

*118.* Kjøsen, H., N. Arpin, and S. Liaaen-Jensen: The carotenoids of *Trentepohlia iolithus*. Isolation of β,β-caroten-2-ol, β,ε-caroten-2-ol and β,ε-carotene-2,2'-diol. Acta Chem. Scand. **26**, 3053 (1972).

*119.* Kjøsen, H., and S. Liaaen-Jensen: Total synthesis of aleuriaxanthin. Acta Chem. Scand. **27**, 2495 (1973).

*120.* Kleinig, H., W. Heumann, W. Meister, and G. Englert: New carotenoids from *Rhizobium lupini*. Helv. Chim. Acta **60**, 254 (1977).

*121.* Kleinig, H., W. Meister, and G. Englert: The effect of nicotine on the carotenoid pattern of *Rhizobium lupini*. Arch. Microbiol. **119**, 71 (1978).

*122.* Koka, P., and P.-S. Song: The chromophore topography and binding environment of peridinin-chlorophyll *a* — protein complexes from marine dinoflagellate algae. Biochim. Biophys. Acta **495**, 220 (1977).

*123.* Kuhn, R., and H. Kühn: Crustacyanin, ein Chromoproteid aus Hummerpanzer. European J. Biochem. **2**, 349 (1967).

*123 a.* Lee, T. Y.: Spectroscopic characterization of crustacyanin. Abstracts Fifth Internat. IUPAC Symp. Carotenoids **1978**, 32.

*123 b.* Lee, W. L. (ed.): Carotenoproteins in animal coloration. Stroudsbourg: Dowden, Hutchinson and Ross. 1977.

*124.* Leuenberger, H. G. H., W. Bogoth, E. Widmer, and R. Zell: Synthese der chiralen Schlüsselverbindung (4*R*,6*R*)-4-hydroxy-2,2,6-trimethylcyclohexanon. Helv. Chim. Acta **59**, 1832 (1976).

*125.* Liaaen-Jensen, S.: The constitution of some bacterial carotenoids and their bearing on biosynthetic problems. Kgl. Norske Vit. Selsk. Skr. No. 8, 1962.

*126.* — New Structures. Pure and Applied Chem. **47**, 129 (1976).

*127.* — Chemistry of carotenoid pigments. In: Photosynthetic Bacteria (Clayton, R. K., and W. R. Sistrom, eds.), Chap. 12. New York: Plenum. 1979.

*128.* — Marine Carotenoids. In: Marine natural products. Chemical and biological prospects (Scheuer, P., ed.), Vol. 2, Chap. 1. New York: Academic Press. 1979.

*129.* Liu, I.-S., T. H. Lee, H. Yokoyama, K. L. Simpson, and C. O. Chichester: Isolation and identification of 2-hydroxy-plectaniaxanthin from *Rhodotorula aurantiaca*. Phytochem. **12**, 2953 (1973).

*130.* Liu, R. S. H., A. E. Asato, and M. Denny: New geometric isomers of vitamin A and carotenoids. 5,7-*cis*-3-dehydroretinal and 7-*cis*-3-dehydro-$C_{18}$-ketone from direct irradiation of the *trans* isomers in polar solvents. J. Amer. Chem. Soc. **99**, 895 (1977).

*131.* Mayer, H.: Synthesis of optically active carotenoids and related compounds. Pure and Applied Chem. **51**, 535 (1979).

*132.* Mayer, H., W. Boguth, H. G. W. Leuenberger, E. Widmer, and R. Zell: The synthesis of all-*trans*-(3*R*,3'*R*)-zeaxanthin. Abstr. Fourth Int. IUPAC Symp. Carot. Berne **1975**, 43.

*133.* Mayer, H., and O. Isler: Total synthesis. In: Carotenoids (Isler, O., ed.), Chap. 6. Basel: Birkhäuser. 1971.

*134.* Mills, J. A.: Correlations between monocyclic and polycyclic unsaturated compounds from molecular rotation differences. J. Chem. Soc. **1952**, 4976.

*135.* Moss, G. P.: Carbon-13 NMR spectra of carotenoids. Pure and Applied Chem. **47**, 97 (1976).

*135 a.* — Physico-chemical and synthetic studies on carotenoids. Pure and Applied Chem. **51**, 507 (1979).

*136.* Moss, G. P., J. Szabolcs, G. Tóth, and B. C. L. Weedon: The stereochemistry of the carotenoid violeoxanthin. Acta Chim. Acad. Sci. Hung. **87**, 301 (1975).

137. MÜLLER, H., and P. KARRER: Methyläther von Carotinoiden. Konfiguration des Azafrins. Helv. Chim. Acta **48**, 291 (1965).

138. MÜLLER, R. K., H. J. MAYER, K. NOACK, and J. J. DALY: Total synthesis of (3S,3'S)- and (3R,3'R)-actinioerythrin. To be published.

139. MÜLLER, R. K., H. J. MAYER, K. NOACK, J. J. DALY, D. J. TAUBER, and S. LIAAEN-JENSEN: Absolute configuration of actinioerythrin. Helv. Chim. Acta **61**, 2881 (1978).

140. NAKAGAWA, K., R. KONATA, and T. NAKATA: Oxidation with nickel peroxides. I. Oxidation of alcohols. J. Org. Chem. **27**, 1597 (1962).

141. NICOARA, E., G. ILLYES, M. SUTEO, and C. BODEA: The conformation of crustaxanthin. Rev. roum. chim. **12**, 547 (1967).

142. NOGGLE, J. H., and R. E. SCHIRMER: The nuclear Overhauser effects and chemical applications. New York: Academic Press. 1971.

143. NYBRAATEN, G., and S. LIAAEN-JENSEN: New carotenoid epoxides from *Trentepohlia iolithus*. Acta Chem. Scand. **B 28**, 483 (1974).

144. — — Improved O-methylation of carotenoids. Acta Chem. Scand. **B 24**, 584 (1974).

145. PATTENDEN, G., J. E. WAY, and B. C. L. WEEDON: Synthesis of methyl natural bixin. J. Chem. Soc. (C) **1970**, 235.

146. PATTENDEN, G., and B. C. L. WEEDON: Synthesis of *cis*- and di-*cis*-polyenes by reactions of the Wittig type. J. Chem. Soc. (C) **1968**, 1984.

147. PORTER, J. W.: Enzymatic synthesis of carotenes. Pure and Applied Chem. In press.

148. PUNTERVOLD, O., and S. LIAAEN-JENSEN: Synthesis of lycopen-20-al and rhodopin-20(20')-al. Acta Chem. Scand. **B 28**, 1096 (1974).

149. RAMAMURTHY, V., and R. S. H. LIU: 7-*cis* Isomers of retinal via 7-*cis*- and 7,9-di-*cis*-β-$C_{18}$-tetraene ketones. Tetrahedron **31**, 201 (1975).

150. RAMAMURTHY, V., G. TUSTIN, C. C. YAU, and R. S. H. LIU: Preparation of sterically hindered geometric isomers of 7-*cis*-β-ionyl and β-ionylidene derivatives in the vitamin A series. Tetrahedron **31**, 193 (1975).

151. RØNNEBERG, H., G. BORCH, D. L. FOX, and S. LIAAEN-JENSEN: Alloporin — a new carotenoprotein. Comp. Biochem. Physiol. **62B**, 309 (1979).

152. RØNNEBERG, H., G. BORCH, S. LIAAEN-JENSEN, H. MATSUTAKA, and T. MATSUNO: Tunaxanthin. Acta Chem. Scand. **32B**, 621 (1978).

153. RØNNEBERG, H., and S. LIAAEN-JENSEN: On the absolute configuration of plectaniaxanthin. Acta Chem. Scand. To be published.

153a. SAUCY, G., G. WEBER, and J. GUTZWILLER: Total synthesis of all-*trans* and 9,9'-di-*cis*-alloxanthin. Helv. Chim. Acta. To be published.

154. SCHMIDT, K.: Biosynthesis of carotenoids. In: Photosynthetic bacteria (CLAYTON, R. K., and W. R. SISTROM, eds.), Chap. 39. New York: Plenum. 1979.

155. SCHMIDT, K., G. W. FRANCIS, and S. LIAAEN-JENSEN: New carotenoid glucosides and remarkable $C_{43}$-carotenoid artefacts of cross-conjugated carotenals. Acta Chem. Scand. **25**, 2476 (1971).

156. SCHWIETER, U., and S. LIAAEN-JENSEN: Stereochemistry of the terminal double bonds of *dehydrogenans*-P 439. Acta Chem. Scand. **23**, 1057 (1969).

157. SCOPES, P. M.: Application of the chiroptical techniques to the study of natural products. Fortschr. Chem. Org. Naturstoffe **32**, 167 (1975).

157a. SNATZKE, G., F. SNATZKE, and S. LIAAEN-JENSEN: Circular dichroism: Carotenoid polyenes. To be published.

158. STEWART, I.: Provitamin A and carotenoid content of citrus juices. J. Agr. Food Chem. **25**, 1132 (1977).

159. STEWART, I., and T. A. WHEATON: Continuous flow separation of carotenoids by liquid chromatography. J. Chromatog. **55**, 325 (1971).

160. STRAIN, H. H., W. A. SVEC, K. AITZETMÜLLER, M.-C. GRANDOLFO, J. J. KATZ, H. KJØSEN, S. NORGÅRD, S. LIAAEN-JENSEN, F. T. HAXO, P. WEGFAHRT, and H.

RAPOPORT: The structure of peridinin — the characteristic dinoflagellate carotenoid. J. Amer. Chem. Soc. **43**, 1823 (1971).

161. STRAIN, H. H., W. A. SVEC, P. WEGFAHRT, M. RAPOPORT, F. T. HAXO, S. NORGÅRD, H. KJØSEN, and S. LIAAEN-JENSEN: Structural studies on peridinin. Part 1. Structure elucidation. Acta Chem. Scand. **330**, 109 (1976).
162. STRAUB, O.: Key to carotenoids. Lists of natural carotenoids. Basel: Birkhäuser. 1976.
163. SZABOLCS, J.: Some studies on the stereochemistry of carotenoids. Pure and Applied Chem. **47**, 147 (1976).
164. TOMES, M. L.: Competitive effect of the β- and δ-carotene genes on α- or β-ionone ring formation in the tomato. Genetics **1967**, 227.
165. TSCHARNER, C., C. H. EUGSTER, and P. KARRER: Synthese der optisch aktiven α-Carotine. Helv. Chim. Acta **40**, 1676 (1957).
166. — — — Synthese des (+)- und des (−)-ε-Carotine. Helv. Chim. Acta **41**, 32 (1958).
167. UEDA, I., and W. NOWACKI: Crystal structure of capsanthin di-*p*-bromobenzoate. Z. Krist. **140**, 190 (1974).
168. VEERMAN, A., G. BORCH, R. PEDERSEN, and S. LIAAEN-JENSEN: Chirality of astaxanthin of different biosynthetic origin. Acta Chem. Scand. **B 29**, 525 (1975).
169. VETTER, W., G. ENGLERT, N. RIGASSI, and U. SCHWIETER: Spectroscopic methods. In: Carotenoids (ISLER, O., ed.), Chap. IV. Basel: Birkhäuser. 1971.
170. VÖGELI, V., W. ESCHENMOSER, and C. H. EUGSTER: Strukturbestimmung von O-Methylazafrin-methylester durch $^{13}$C-NMR-Spektroskopie. Helv. Chim. Acta **58**, 2044 (1975).
171. WALTON, T. J., G. BRITTON, and T. W. GOODWIN: Biosynthesis of xanthophyll in higher plants. Stereochemistry of hydroxylation at C-3. Biochem. J. **112**, 383 (1969).
172. WEEDON, B. C. L.: Spectroscopic methods for structural elucidation of carotenoids. Fortschr. Chem. Org. Naturstoffe **27**, 81 (1969).
173. — Allenic and acetylenic carotenoids. Rev. Pure Appl. Chem. (Australia) **20**, 51 (1970).
174. — Stereochemistry. In: Carotenoids (ISLER, O., ed.), Chap. 5. Basel: Birkhäuser. 1971.
175. — Some recent studies on carotenoids and related compounds. Pure and Applied Chem. **35**, 113 (1973).
176. — Synthesis of carotenoids and related polyenes. Pure and Applied Chem. **47**, 161 (1976).
177. WEHRLI, F. W., and T. WIRTHLIN: Interpretation of carbon-13 NMR spectra. London: Heyden. 1976.
178. WILLIAMS, R. J. H., G. BRITTON, and T. W. GOODWIN: A possible mechanism for the biosynthesis of eschscholtzxanthin. Biochim. Biophys. Acta **124**, 197 (1966).
179. — — — The biosynthesis of cyclic carotenes. Biochem. J. **105**, 99 (1967).
180. ZAGALSKY, P. F.: Carotenoid-protein complexes. Pure and Applied Chem. **47**, 103 (1976).
181. ZECHMEISTER, L.: *Cis-trans* isomeric carotenoid pigments. Fortschr. Chem. Org. Naturstoffe **18**, 223 (1960).
182. — *Cis-trans* isomeric carotenoids, vitamins A and arylpolyenes. Wien: Springer. 1962.

*(Received August 14, 1978)*

# Chemistry and Biochemistry of γ-Glutamyl Derivatives from Plants Including Mushrooms (Basidiomycetes)

By T. KASAI* and P. O. LARSEN, Chemistry Department, Royal Veterinary and Agricultural University, Copenhagen, Denmark

## Contents

* Present Address: Department of Agricultural Chemistry, Faculty of Agriculture, Hokkaido University, Sapporo, Japan.

# I. Introduction

More than 70 γ-glutamyl derivatives of amino acids and amines
have been isolated from plants, including mushrooms (Basidiomycetes),
during the last twenty years, and it has been shown that these compounds
often predominate in the fraction of free amino acids and other amphoteric
plant constituents giving positive ninhydrin reactions. The present review
describes the structures and distribution of these compounds and the
methods used for their isolation and identification. Special emphasis is
laid on the differences in chemical behaviour of γ-glutamyl derivatives and
α-dipeptides with glutamic acid as the amino- and as the carboxy-
terminal. Furthermore, what little is known about the biochemistry,
especially the enzymology, of γ-glutamyl derivatives in plants is de-

scribed and compared with the available information on glutamine and glutathione.

Present knowledge regarding the distribution and biochemistry of γ-glutamyl derivatives in animals and microorganisms is dealt with only for the purpose of comparison, mostly by reference to recent reviews. Chemical synthesis is treated only briefly by referring to a number of reviews on peptide synthesis and with special treatment of compounds which present unusual problems.

A short review of γ-glutamyl derivatives in plants has appeared in 1962 (*330*), and a general review of naturally occurring peptides published in 1966 (*364*) includes a list of 25 γ-glutamyl peptides from plants.

The present review covers the literature appearing up to the end of July 1978.

## II. Nomenclature

According to official nomenclature recommendations (*393*), γ-glutamyl derivatives of amino acids should be named as γ-glutamyl followed by the name of the parent amino acid (for example γ-glutamylvaline). This system is employed throughout the present review. Previously, most γ-glutamyl derivatives of amino acids in the indexes of Chemical Abstracts were registered as glutamine derivatives [e. g. N-(1-carboxyethyl)glutamine for γ-glutamylalanine; when no superscript is given to N, this signifies substitution at $N^5$. Substitution at the nitrogen of the α-amino group is indicated with $N^2$], but from 1972 the indexes of Chemical Abstracts list glutamyl derivatives under the parent amino acid, preceded by N-γ-glutamyl (example: valine, N-γ-glutamyl).

In a number of cases the parent amino acid is a non-protein amino acid with a trivial name. Most of these have been retained, but systematic names for such amino acids are given in Tables 2, 3 and 4.

γ-Glutamyl derivatives of amines are registered in the indexes of Chemical Abstracts as glutamine derivatives (e. g. N-ethylglutamine) and this practise is followed here, although with indication of the nitrogen which is substituted (e. g. $N^5$-ethylglutamine).

Some γ-glutamyl derivatives have historically acquired trivial names. These names are indicated in Tables 2—6, and in most cases are used in the text. However, Tables 2—6 also give alternative systematic or semi-systematic names.

In agreement with common usage and with official nomenclature the word peptide is used in a broad sense to include all compounds with amide bonds between amino acids.

The following abbreviations are used in the formulae:

γ-Glu for

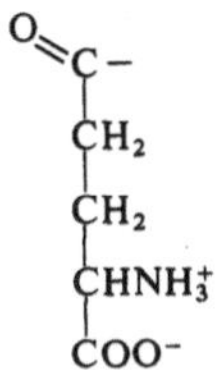

$$O = C -$$
$$| CH_2$$
$$| CH_2$$
$$| CHNH_3^+$$
$$| COO^-$$

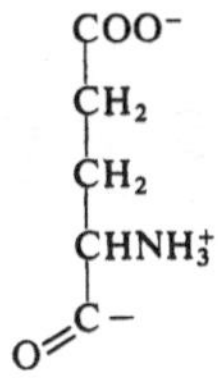

γ-L-Glu for

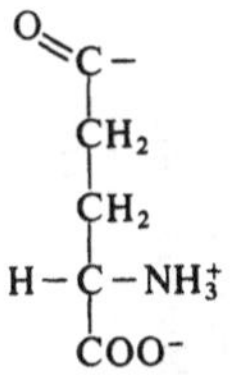

$$O = C -$$
$$| CH_2$$
$$| CH_2$$
$$H - C - NH_3^+$$
$$| COO^-$$

a-Glu for

$$COO^-$$
$$| CH_2$$
$$| CH_2$$
$$| CHNH_3^+$$
$$O = C -$$

a-L-Glu for

$$COO^-$$
$$| CH_2$$
$$| CH_2$$
$$H - C - NH_3^+$$
$$O = C -$$

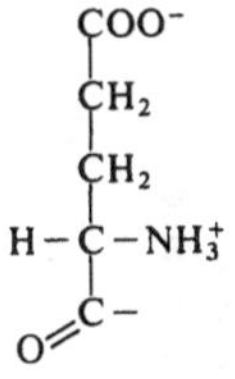

# III. Structures and Distribution of γ-Glutamyl Derivatives from Plants

γ-Glutamyl derivatives have so far been identified in many different plants distributed throughout the plant kingdom. Although some patterns of distribution are apparent, it has proved impossible to make a systematic division and classification of these compounds according to their occurrence. It would, of course, be possible to list the compounds under the individual species. This will, however, not be attempted here, firstly because the compounds identified reflect to a certain extent the analytical methods employed (see Section III.5 and IV), and secondly because few systematic investigations have so far been made of the occurrence of γ-glutamyl derivatives in botanically well-defined groups.

On the other hand, a structural classification is possible and has been implemented in the following by dividing the compounds into six groups. It must, however, be stressed that in many cases compounds from different groups have been found together in the same plant.

## 1. γ-Glutamyl Derivatives of Protein Amino Acids

The γ-glutamyl derivatives of protein amino acids so far identified in plants are listed in Table 1. The table gives information on the character of the isolated material and attempts in this manner to give an indication of the reliability of the identification. In many cases no assignment of the configuration of the amino acids has been made but L-configurations are assumed. However, this cannot be taken for granted, especially

Table 1. *γ-Glutamyl Derivatives of Protein Amino Acids Identified in Plants*

| Compound | Plant species | Plant family | Character of isolate, comments | Configuration of Glutamic acid moiety | Configuration of The second amino acid | Ref. | Chemical synthesis of γ-glutamyl derivative Method | Chemical synthesis of γ-glutamyl derivative Ref. |
|---|---|---|---|---|---|---|---|---|
| γ-L-Glutamylglycine | *Tulipa gesneriana*, fruit capsules | Liliaceae | Evap. residue, tentative identification | | | (*53*) | 2 | (*191*) |
| | *Pisum sativum*, seedlings | Leguminosae | Evap. residue | | | (*294*) | | |
| | *Arachis hypogea*, seed flour | Leguminosae | Evap. residue | | | (*317*) | | |
| γ-L-Glutamyl-L-alanine | *Iris tingitana*, leaves | Iridaceae | Cryst. | L | L | (*231*) | 2 | (*279*) |
| | *Iris tingitana*, bulbs and roots | Iridaceae | Paper chromatography | | | (*231*) | | |
| | *Tulipa gesneriana*, fruit capsules | Liliaceae | Evap. residue, tentative identification | | | (*53*) | | |
| | *Billia hippocastanum*, seeds | Hippocastanaceae | Evap. residue | | | (*57*) | | |
| | *Fagus silvatica*, seeds | Fagaceae | Evap. residue | | | (*177*) | | |
| γ-L-Glutamyl-L-valine | *Allium cepa*, bulbs | Alliaceae | Cryst. | L | L | (*356*) | 2 | (*279*) |
| | *Iris tingitana*, leaves | Iridaceae | Evap. residue | | | (*231*) | 1 | (*231*) |

| Compound | Source | Family | Method | Config | Ref | No. | Ref |
|---|---|---|---|---|---|---|---|
| | Tulipa gesneriana, fruit capsules | Liliaceae | Evap. residue tentative identification | | (53) | | |
| | Billia hippocastanum, seeds | Hippocastanaceae | Evap. residue | | (57) | | |
| | Fagus silvatica, seeds | Fagaceae | Evap. residue | | (177) | | |
| γ-L-Glutamyl-L-leucine | Phaseolus lunatus, seed meal | Leguminosae | Evap. residue | | (277) | 2 | (279) |
| | Allium cepa, bulbs | Alliaceae | Cryst. | | (356) | 1 | (234) |
| | Phaseolus vulgaris, seeds | Leguminosae | Cryst. | L | (234) | | |
| | Tulipa gesneriana, fruit capsules | Liliaceae | Evap. residue tentative identification | | (53) | | |
| | Phaseolus radiatus var. typicus, seeds | Leguminosae | Cryst. | L | (140) | | |
| | Fagus silvatica, seeds | Fagaceae | Mix. with γ-glutamyl-isoleucine | L | (177) | | |
| | Vigna mungo, seeds | Leguminosae | ? | | (256) | | |
| | Fagus silvatica var. purpurea, seedlings | Fagaceae | Mix. with γ-glutamyl-isoleucine | L | (129) | | |
| γ-L-Glutamyl-L-isoleucine | Allium cepa, bulbs | Alliaceae | Cryst. | L | (356) | | Variation (109) of method 1 |

Table 1 (*continued*)

| Compound | Plant species | Plant family | Character of isolate, comments | Configuration of Glutamic acid moiety | The second amino acid | Ref. | Chemical synthesis of γ-glutamyl derivative Method | Ref. |
|---|---|---|---|---|---|---|---|---|
| | *Fagus silvatica*, seeds | Fagaceae | Mix. with γ-glutamyl-leucine | | | (177) | | |
| | *Fagus silvatica*, var. *purpurea*, seedlings | Fagaceae | Mix. with γ-glutamyl-leucine | L | L | (129) | | |
| γ-L-Glutamyl-L-serine | *Tulipa gesneriana*, fruit capsules | Liliaceae | Evap. residue, tentative identification | | | (53) | 1 | (49) |
| γ-L-Glutamyl-L-threonine | *Tulipa gesneriana*, fruit capsules | Liliaceae | Evap. residue, tentative identification | | | (53) | 3 | (131) |
| | *Billia hippocastanum*, seeds | Hippocastanaceae | Evap. residue | | | (57) | | |
| γ-L-Glutamyl-L-aspartic acid | *Glycine max*, seedlings | Leguminosae | Evap. residue | | | (107) | 2 | (191) |
| | *Acacia georginae*, seeds | Leguminosae | | | | (108) | | |
| | *Billia hippocastanum*, seeds | Hippocastanaceae | Cryst. | | | (57) | | |

| Compound | Source | Family | Isolation | | | Ref. | No. | Ref. |
|---|---|---|---|---|---|---|---|---|
| γ-L-Glutamyl-L-glutamic acid | *Trifolium repens* var. *giganteum*, seeds | Leguminosae | Cryst. | L | L | (145) | | |
| | *Vicia faba*, seeds | Leguminosae | Evap. residue | | | (143) | | |
| | *Acacia georginae*, seeds | Leguminosae | | | | (108) | 2 | (191) |
| | *Billia hippocastanum*, seeds | Hippocastanaceae | Cryst. | | | (57) | | |
| | *Nicotiana tabacum*, cells, suspension culture | Solanaceae | Cryst. | L | L | (125) | | |
| | *Fagus silvatica*, seeds | Fagaceae | Evap. residue | | | (177) | | |
| γ-L-Glutamyl-L-asparagine | *Acacia georginae*, seeds | Leguminosae | | | | (108) | 3 | (131) |
| | *Billia hippocastanum*, seeds | Hippocastanaceae | Cryst. | | | (57) | | |
| γ-L-Glutamyl-L-glutamine | *Dactylis glomerata*, tillering nodes and roots | Gramineae | Cryst. | L | L | (31) | 1 | (282) |
| γ-L-Glutamyl-L-cysteine | *Triticum aestivum*, germ | Gramineae | Lyoph. as S-carboxymethyl deriv. | L | L | (334) | 4 | (307) |
| | | | | | | | 3 | (84) |
| N,N'-bis-(γ-L-Glutamyl)-L,L-cystine | *Allium sativum*, | Alliaceae | Evap. residue tentative identification | | | (319) | 3 | (84) |

Table 1 *(continued)*

| Compound | Plant species | Plant family | Character of isolate, comments | Configuration of Glutamic acid moiety | Configuration of The second amino acid | Ref. | Chemical synthesis of γ-glutamyl derivative Method | Chemical synthesis Ref. |
|---|---|---|---|---|---|---|---|---|
| | *Allium schoenoprasum*, seeds | Alliaceae | Cryst. | L | L | *(211)* | | |
| γ-L-Glutamyl-L-methionine | *Allium cepa*, bulbs | Alliaceae | Cryst. | L | | *(356)* | 1 | *(234)* |
| | *Phaseolus vulgaris*, seeds | Leguminosae | Cryst. | | | *(234)* | | |
| | *Phaseolus aureus*, seedlings | Leguminosae | Paper chromatography | | | *(24)* | | |
| | *Phaseolus radiatus* var. *typicus*, seeds | Leguminosae | Cryst. | L | L | *(140)* | | |
| | *Vigna mungo*, seeds | Leguminosae | ? | | | *(256)* | | |
| | *Boletus edulis* | Boletaceae | Evap. residue | | | *(337)* | | |
| γ-L-Glutamyl-L-arginine | *Allium cepa*, bulbs | Alliaceae | Evap. residue, powder | L | L | *(216)* | | |
| γ-L-Glutamyl-L-phenylalanine | *Allium cepa*, bulbs | Alliaceae | Cryst. | L | L | *(354)* | 1 | *(230)* |
| | *Allium sativum*, bulbs | Alliaceae | Cryst. | | | *(319)* | | |
| | *Lupinus albus*, | Leguminosae | Cryst. | L | L | *(378)* | | |

| Species | Family | Method | | | Ref. |
|---|---|---|---|---|---|
| L. angustifolius, seeds | Leguminosae | Cryst. | L | L | (230) |
| Glycine max, seeds | Leguminosae | Evap. residue, tentative identification | | L | (53) |
| Tulipa gesneriana, fruit capsules | Liliaceae | Paper chromatography | | | (43) |
| Astragalus, seeds | Leguminosae | Cryst. | | | (189) |
| Aubrietia deltoidea, seeds | Cruciferae | Cryst. | L | L | (264) |
| Lotus corniculatus, seeds | Leguminosae | Paper chromatography | | | (265) |
| Lotus species, seeds | Leguminosae | Evap. residue | | | (140) |
| Phaseolus radiatus var. typicus, seeds | Leguminosae | Cryst. | L | L | (145) |
| Trifolium repens var. giganteum, seeds | Leguminosae | | | | (38) |
| Dolichos sericeus, seeds | Leguminosae | | | | (39) |
| Macrotyloma, Dolichos, Pseudovigna sp., seeds | Leguminosae | Paper chromatography. ion-exchange chromatography | L | L | (143) |
| Vicia faba, seeds | Leguminosae | Evap. residue | | L | (317) |
| Arachis hypogea, seed flour | Leguminosae | Evap. residue | | L | (177) |
| Fagus silvatica, seeds | Fagaceae | Cryst. | | L | |

Table 1 (*continued*)

| Compound | Plant species | Plant family | Character of isolate, comments | Configuration of Glutamic acid moiety | The second amino acid | Ref. | Chemical synthesis of γ-glutamyl derivative Method | Ref. |
|---|---|---|---|---|---|---|---|---|
| | *Fagus* species, seeds | Fagaceae | Paper chromatography | | | (129) | | |
| γ-L-Glutamyl-L-tyrosine | *Lupinus albus*, *L. angustifolius*, seeds | Leguminosae | Cryst. | L | L | (378) | 2 | (191) |
| | *Glycine max*, seeds | Leguminosae | Cryst. | L | L | (230) | 1 | (230) |
| | *Aubrietia deltoidea*, seeds | Cruciferae | Cryst. | L | L | (189) | | |
| | *Astragalus* sp., seeds | Leguminosae | Paper chromatography | | | (43) | | |
| | *Lotus corniculatus*, seeds | Leguminosae | Cryst. | L | L | (265) | | |
| | *Lotus* species, seeds | Leguminosae | Paper chromatography | | | (265) | | |
| | *Phaseolus radiatus* var. *typicus*, seeds | Leguminosae | Evap. residue | | | (140) | | |
| | *Trifolium repens* var. *giganteum*, seeds | Leguminosae | Cryst. | L | L | (145) | | |
| | *Fagus silvatica*, seeds | Fagaceae | Evap. residue | | | (177) | | |

| | | | | | | |
|---|---|---|---|---|---|---|
| | Vicia faba, seeds | Leguminosae | Evap. residue | L | L | (143) |
| | Sophora secundiflora, seeds | Leguminosae | | | | (113) |
| | Arachis hypogea, seed flour | Leguminosae | Evap. residue | L | | (317) |
| γ-L-Glutamyl-L-tryptophan | Fagus silvatica var. purpurea, seeds | Fagaceae | Cryst. | L | L | (129) 3 (129) |

since γ-L-glutamyl-D-alanine has been found in pea seedlings (*66, 67*) and in exudates from roots of pea seedlings (*342*). The names of the plants are in most cases the correct Latin names, although trivial names may have been used in the original publications. Table 1 also gives information on chemical syntheses (cf. Section VI).

As is apparent from the Table, only three of the twenty possible γ-glutamyl derivatives of the protein amino acids have so far not been found in plants. Two of these, γ-glutamylhistidine and γ-glutamyllysine (two possible isomers), are derivatives of basic amino acids and are therefore neutral. Their absence from the list may simply reflect the fact that isolation of neutral amino acids and γ-glutamyl derivatives from plants is more difficult than the isolation of acidic derivatives. The third missing compound is γ-glutamylproline, which contains a secondary amino group in the parent amino acid and therefore differs in principle from all the others. However, the isolation of γ-glutamylpipecolic acid from *Gleditsia caspica* (*36*) (cf. Table 2), demonstrates that plants are able to produce such derivatives.

## 2. γ-Glutamyl Derivatives of Non-Protein Amino Acids Which Do not Contain Sulphur or Selenium

The γ-glutamyl derivatives of this type which have so far been identified in plants are listed in Table 2. Again, information is given in the table on the character of the isolate and on the determination of configuration.

Nearly all the compounds listed in Table 2 occur together with the free non-protein amino acid; the only exception being γ-L-glutamyl-(*R*)-3-amino-3-phenylpropanoic acid. The (*R,S*)-system is employed consistently for amino acids in which the amino group is not in an α-position to a carboxyl group, and for all amino acids with more than one chiral center, even when this was not done in the original publications. Table 2 also contains references to the occurrence of the free amino acids, in which references to structure elucidation of these amino acids can generally be found. For the last two compounds in the Table the absolute configurations indicated require special comment. The configurations of the two naturally occurring 2-(carboxycyclopropyl)glycines have been determined, but the configuration reported is in disagreement with the experiments on which they are based (*61*). The assignment of the (2*S*,1′*S*, 2′*S*)-configuration in Table 2 is based on a reinterpretation of the original results. The configuration of the naturally occurring 3-(methylenecyclopropyl)alanine has likewise been determined, but again the configuration reported is in disagreement with the experiments (*18*). As before,

Table 2. γ-Glutamyl Derivatives of Non-Protein Amino Acids, Not Containing Sulphur or Selenium. Identified in Plants

| Compound | Formula | Plant species | Plant family | Character of isolate, comments | Configuration of Glutamic acid moiety | The second amino acid | Ref. | Occurrence of second amino acid | Chemical synthesis of γ-glutamyl derivative Method | Ref. |
|---|---|---|---|---|---|---|---|---|---|---|
| γ-L-Glutamyl-D-alanine | | Pisum sativum, seedlings | Leguminosae | Cryst. | L | D | (66) | For D-amino acids in plants see (155, 156) | 1 | (66) |
| | | Pisum sativum, seedlings | Leguminosae | Evap. residue | | | (295, 349) | | | |
| | | Pisum sativum, exudate from root tips of seedlings | Leguminosae | Evap. residue | | | (342) | | | |
| γ-Glutamyl-3-[(amino-carboxyl)-amino]alanine (γ-Glutamyl-albizziine) | 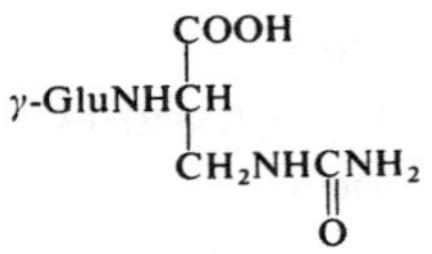 $\gamma\text{-GluNHCH}$ with $COOH$ and $CH_2NHCNH_2 {=}O$ | Acacia georginae, seeds | Leguminosae | Cryst. | | | (108) | Many sp. of Leguminosae including A. georginae (306) | | |
| γ-Glutamyl-homoserine | | Pisum sativum, seedlings | Leguminosae | Evap. residue | | | (294) | Ubiquitous | 1 | (294) |

Table 2 *(continued)*

| Compound | Formula | Plant species | Plant family | Character of isolate, comments | Configuration of Glutamic acid moiety | The second amino acid | Ref. | Occurrence of second amino acid | Chemical synthesis of γ-glutamyl derivative — Method | Ref. |
|---|---|---|---|---|---|---|---|---|---|---|
| γ-L-Glutamyl-β-cyano-L-alanine | 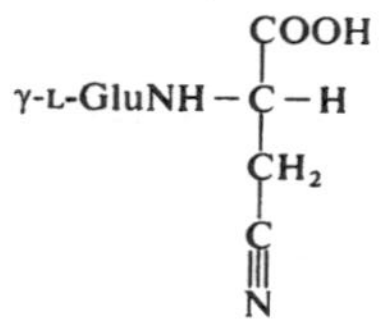 γ-L-GluNH—C—H, COOH, CH₂, C≡N | *Vicia sativa*, seeds | Leguminosae | Cryst. | L | L | (276) | | 1 | (275) |
| | | *V. sativa*, seedlings *Vicia, Lathyrus* sp., seeds | Leguminosae | Ion-exchange chromatography | | | (275) | | | |
| | | *Lathyrus sylvestris*, seeds and seedlings | Leguminosae | Ion exchange chromatography | | | (274) | | | |
| | | *Vicia sativa*, seeds | Leguminosae | Cryst. | | | (339) | *Vicia,* including *V. sativa* (13) | | |
| | | *Vicia* sp., seeds | Leguminosae | Paper chromatography | | | (13) | | | |
| γ-L-Glutamyl-L-2-aminohex-4-ynoic acid | 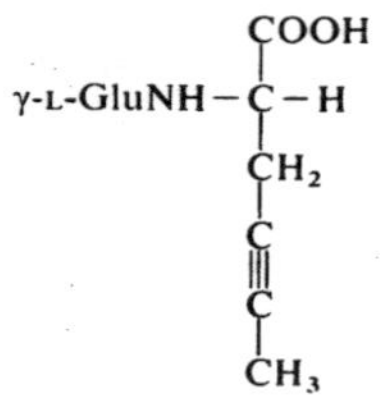 γ-L-GluNH—C—H, COOH, CH₂, C≡C, CH₃ | *Tricholomopsis rutilans* | Tricholoma-taceae | Cryst. | L | L | (242) | *Tricholomopsis rutilans* (93) | | |

| Compound | Structure | Source | Family | Method | Config. | Ref. | Other occurrence | Ref. |
|---|---|---|---|---|---|---|---|---|
| γ-L-Glutamyl-(2S,3S)-2-amino-3-hydroxyhex-4-ynoic acid | γ-L-GluNH–C(COOH)–H, HO–C–H, C≡C–CH₃ | Tricholomopsis rutilans | Tricholoma-taceae | Cryst. | L | 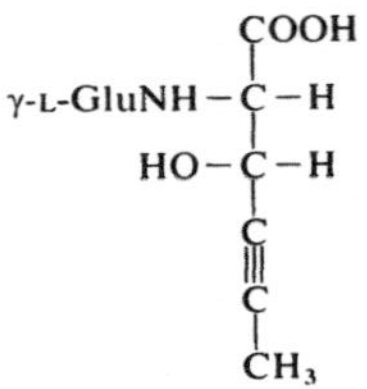 2S, 3S (242) | Tricholomopsis rutilans (241) | |
| γ-Glutamyl-2-amino-3-methylenepenta-noic acid (γ-Glutamyl-β-methylene-norvaline) | 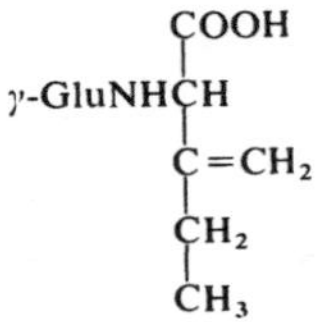 γ-GluNHCH(COOH)–C(=CH₂)–CH₂–CH₃ | Philadelphus sp., leaves | Philadelpha-ceae | Cryst. | | (35) | Philadelphus sp. (35); Lactarius helvus (194) | |
| γ-L-Glutamyl-(E)-L-2-amino-4-methylhex-4-enoic acid | 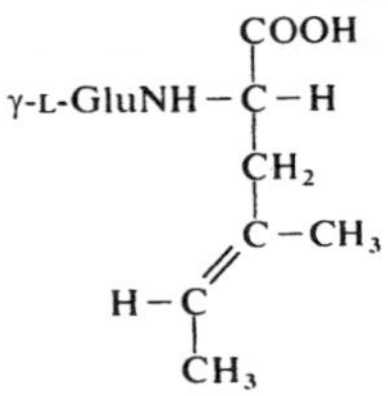 γ-L-GluNH–C(COOH)–H, CH₂, C–CH₃, H–C–CH₃ | Aesculus california, seeds | Hippocasta-naceae | Evap. residue | L | E-L (59) | Aesculus california (59) | |
| γ-L-Glutamyl-β-alanine | | Lunaria annua, seeds | Cruciferae | Cryst. | L | (186) | Ubiquitous | 2 (255) |
| | | Iris tingitana, bulbs | Iridaceae | Cryst. | L | (233) | | 1 (233) |
| | | Iris tingitana, leaves and roots | Iridaceae | Paper chromato-graphy | | (231) | | |
| | | Vicia sativa, young pods | Leguminosae | Cryst. | L | (266) | | |

Table 2 *(continued)*

| Compound | Formula | Plant species | Plant family | Character of isolate, comments | Configuration of Glutamic acid moiety | Configuration of The second amino acid | Ref. | Occurrence of second amino acid | Chemical synthesis of γ-glutamyl derivative: Method | Chemical synthesis of γ-glutamyl derivative: Ref. |
|---|---|---|---|---|---|---|---|---|---|---|
| γ-L-Glutamyl-γ-aminobutyric acid | | *Lunaria annua*, seeds | Cruciferae | Cryst. | | | *(187)* | Ubiquitous | 1 | *(187)* |
| | | *Vicia sativa*, young pods. | Leguminosae | Cryst. | L | | *(266)* | | 5 | *(50)* |
| | | *Beta vulgaris*, beets. | Chenopodiaceae | Cryst. | L | | *(54)* | | | |
| γ-L-Glutamyl-(*R*)-β-aminoisobutyric acid | γ-L-$GluNHCH_2-C(COOH)(CH_3)-H$ | *Iris tingitana*, bulbs | Iridaceae | Cryst. | L | *R* | *(232)* | *Iris tingitana* *(6)* | 2 | *(123)* |
| | | *Iris tingitana*, leaves and roots | Iridaceae | Paper chromatography | | | *(231)* | | | |
| | | *Lunaria annua*, seeds. | Cruciferae | Cryst. | L | *R* | *(186)* | *Lunaria annua* *(186, 187)* | | |

| Compound | Structure | Source | Family | Method | Config. | Ref. | Occurrence | Synthesis |
|---|---|---|---|---|---|---|---|---|
| Linatine (1-[(N-γ-L-glutamyl)amino]-D-proline) Chemical Abstracts: 1-[(4-amino-4-carboxy-1-oxobutyl)amino]-D-proline | 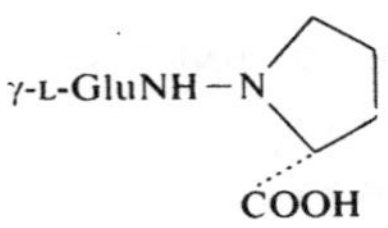 γ-L-GluNH–N (pyrrolidine) ⋯COOH | Linum usitatissinum. seeds | Linaceae | Evap. residue | L ... D | (158) | Linum usitatissinum (245) | 2 (158) See Section VI.2. Trifluoro-acetyl-L-glutamic acid α-ethylester +1-amino-D-proline benzylester (dicyclohexyl-carbo-diimide) (245) |
| γ-L-Glutamyl-L-pipecolic acid | 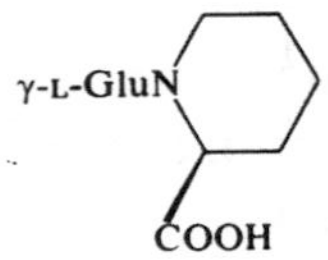 γ-L-GluN (piperidine) ⋯COOH | Gleditsia caspica, seeds | Leguminosae | Cryst. | L L | (36) | Ubiquitous | |
| γ-L-Glutamyl-β-pyrazol-1-yl-L-alanine (γ-L-Glutamyl-3-(1-pyrazolyl)-L-alanine) | 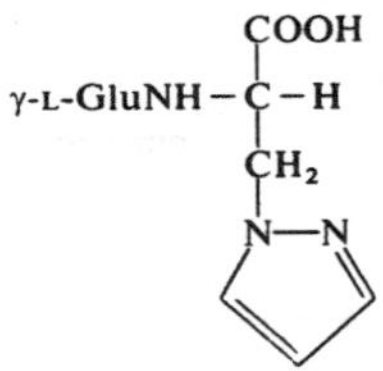 γ-L-GluNH–C–H (COOH, CH₂, pyrazole) | Cucumis sativus, seeds | Cucurbitaceae | Cryst. | L L | (41) | Cucurbitaceae including Cucumis sativus (42) | |
| | | Cucurbitaceae, seeds | Cucurbitaceae | Paper chromatography | | (42) | | |

Table 2 (continued)

| Compound | Formula | Plant species | Plant family | Character of isolate, comments | Configuration of Glutamic acid moiety | The second amino acid | Ref. | Occurrence of second amino acid | Chemical synthesis of γ-glutamyl derivative | |
|---|---|---|---|---|---|---|---|---|---|---|
| | | | | | | | | | Method | Ref. |
| γ-L-Glutamyl-3-(2-amino-4-pyrimidinyl)-L-alanine (γ-L-Glutamyl-L-lathyrine) | 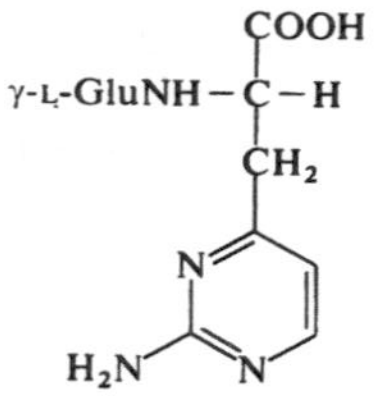 γ-L-GluNH–C–H, COOH, CH₂, pyrimidine ring with H₂N | Lathyrus japonicus, seeds | Leguminosae | Cryst. | L | L | (248) | Lathyrus, sp. (12) | | |
| γ-L-Glutamyl-3-(1-uracil)-L-alanine (γ-Glutamylwillardiine, (γ-L-Glutamyl-3-(3,4-dihydro-2,4-dioxy-1-(2H)pyrimidinyl)-L-alanine) | 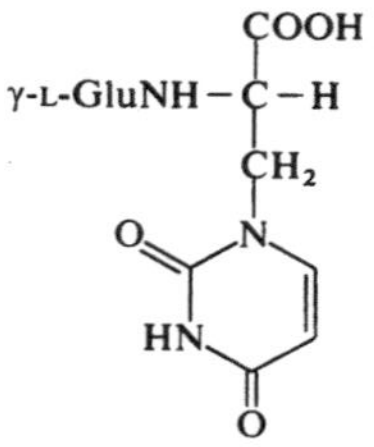 γ-L-GluNH–C–H, COOH, CH₂, uracil ring | Fagus silvatica, seeds | Fagaceae | Cryst. | L | L | (176) | Acacia sp. (306) Pisum sativum (184) | 1 | (309) |
| | | Fagus species, seeds | Fagaceae | Paper chromatography | | | (129) | Fagus sp. incl. F. silvatica (129) | | |
| γ-L-Glutamyl-(R)-3-amino-3-phenylpropanoic acid (γ-L-Glutamyl-β-L-phenyl-β-alanine) | 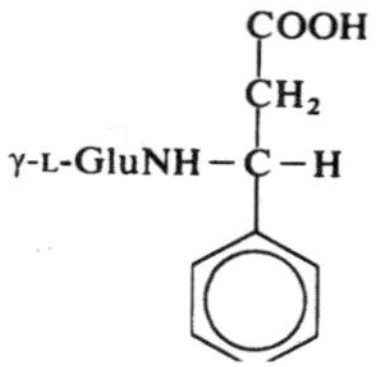 γ-L-GluNH–C–H, COOH, CH₂, phenyl ring | Phaseolus angularis, seeds | Leguminosae | Cryst. | L | R | (170) | Not found in plants | 1 | (171) |

| Compound | Structure | Source | Family | Method | L | Configuration (ref.) | Species (ref.) |
|---|---|---|---|---|---|---|---|
| γ-L-Glutamyl-(2S,1'S)-2-(methylene-cyclopropyl)-glycine (γ-L-Glutamyl-α-L-(methylene-cyclopropyl)-glycine) | 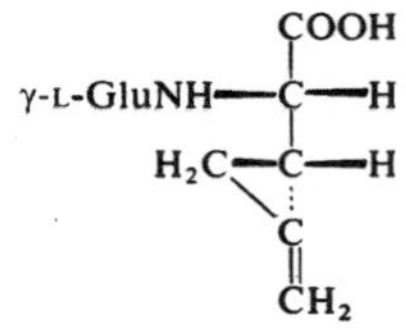 | *Billia hippo-castanum,* seeds | Hippo-castanaceae | Cryst. | L | 2S,1'S (57) | *Litchi chinensis* (77) |
| | | *Acer pseudo-platanus,* seeds | Aceraceae | Cryst. | L | 2S,1'S (57) | *Billia hippo-castanum* (47) |
| | | *Acer* species, seeds | Aceraceae | Paper chromatography | | (56) | *Acer,* including *A. pseudo-platanus* (56) |
| γ-L-Glutamyl-(2S,1'S,2'S)-(carboxycyclo-propyl)-glycine (γ-L-Glutamyl-*trans*-α-L-carboxycyclo-propyl)-glycine) | 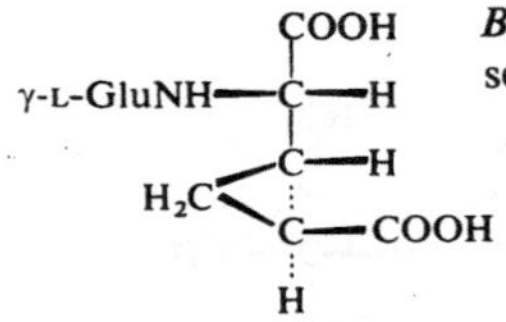 | *Blighia sapida,* seeds | Sapindaceae | Cryst. | L | 2S,1'S (60) 2'S | *Blighia sapida* (61) |

Table 2 (continued)

| Compound | Formula | Plant species | Plant family | Character of isolate, comments | Configuration of Glutamic acid moiety | The second amino acid moiety | Ref. | Occurrence of second amino acid | Chemical synthesis of γ-glutamyl derivative — Method Ref. |
|---|---|---|---|---|---|---|---|---|---|
| γ-L-Glutamyl-(2S,1'R)-3-methylenecyclopropyl)alanine (Hypoglycine B) | 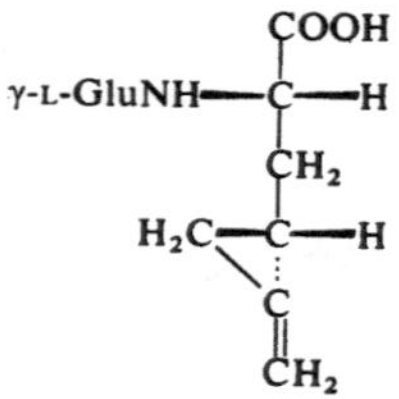 γ-L-GluNH— (COOH—C—H / CH₂ / C…H / C=CH₂ H₂C) | Blighia sapida, unripe fruits | Sapindaceae | Cryst. | L | 2S,1'R | (60, 91, 102, 119) | Blighia sapida (92, 102) | N-trityl-L-glutamic acid (90) |
| γ-L-Glutamyl-hypoglycine A | | Billia hippocastanum, seeds | Hippocastanaceae | Evap. residue | | | (47) | Billia hippocastanum (47) | trimethylamine salt + hypoglycine A methylester, Dicyclohexylcarbodiimide 0.2N NaOH, 50% AcOH |
| | | Acer pseudoplatanus, fruits | Aceraceae | Cryst. | | | (57) | Acer incl. A. Pseudoplatanus (56) | N-Trifluoroacetyl-L-glutamic acid (120) |

| | | | | |
|---|---|---|---|---|
| *Acer* species. seeds | Aceraceae | Paper chromatography | (56) | α-ethyl ester + iso-butyl-cloro-formate + (2S, 1′R)-3-(methyl-ene-cyclo-propyl)-alanine followed by re-moval of protect-ing groups with base |

Table 3. *γ-Glutamyl Tripeptides, Not Containing Sulphur or Selenium, Identified in Plants*

| Compound | Formula | Plant species | Plant family | Character of isolate, comments | Configuration of: Glutamic acid moiety | The second amino acid | The third amino acid | Ref. | Chemical synthesis of γ-glutamyl derivative Method | Ref. |
|---|---|---|---|---|---|---|---|---|---|---|
| γ-Glutamylvalyl-glutamic acid | 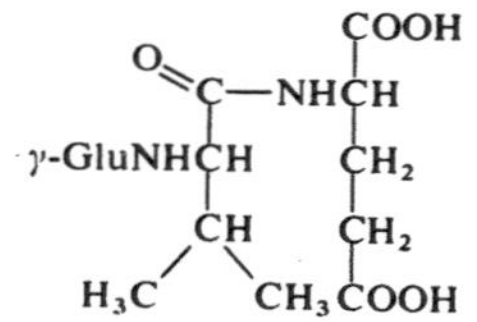 [structure] γ-GluNHCH; COOH, CH₂, CH₂, COOH; O=C−NHCH, CH, H₃C, CH₃COOH | *Juncus conglomeratus*, fresh plant parts | Juncaceae | Evap. residue | | | | (350) | | |
| γ-L-Glutamyl-L-phenylalanyl-L-willardiine | 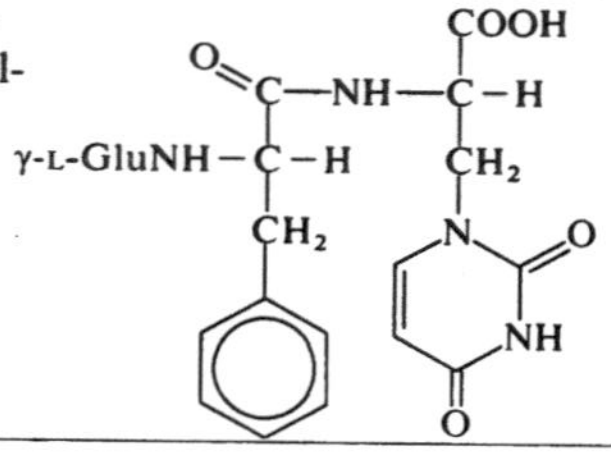 [structure] γ-L-GluNH; COOH, C−H, CH₂; O=C−NH−C−H, CH₂ (phenyl); willardiine ring | *Fagus silvatica*, seeds | Fagaceae | Cryst. | L | L | L | (176) For configuration see (309) | 1 | (309) |
| γ-L-Glutamyl-γ-L-glutamyl-L-phenylalanine | 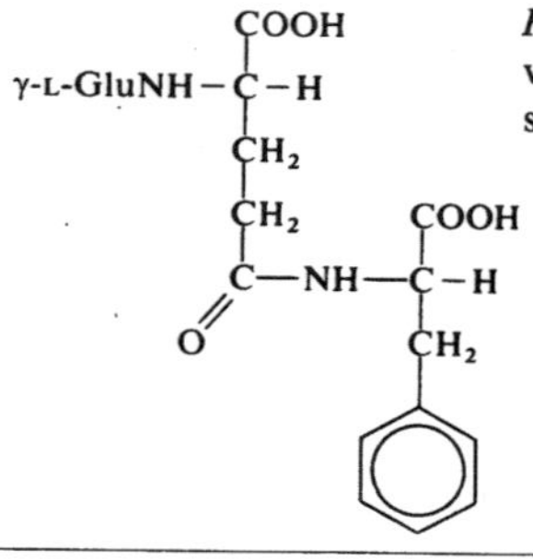 [structure] γ-L-GluNH−C−H; COOH, CH₂, CH₂, C=O, NH−C−H; COOH, CH₂ (phenyl) | *Fagus silvatica* var. *purpurea*, seeds | Fagaceae | Cryst. | L | L | L | (129) | 4 | (129) |

assignment of the $(2S,1'R)$-configuration in Table 2 is based on a reinterpretation of the original results.

Table 2 lists 13 γ-glutamyl derivatives of α-L-amino acids, one derivative of an imino acids, four derivatives of β- or γ-amino acids, one derivative of a compound with an $H_2N$-N-grouping, and one derivative of an α-D-amino acid. For comparison it may be mentioned that in addition to the protein amino acids about 130 L-α-amino acids without sulphur or selenium, about 35 imino acids, and about 8 β-, γ- or δ-amino acids are known which occur in higher plants and mushrooms. Thus the possibilities for new γ-glutamyl derivatives are considerable.

Those γ-glutamyl tripeptides which do not contain sulphur or selenium and which have so far been identified in plants are listed in Table 3. Table 2 and 3 also contain information on chemical syntheses (cf. Section VI).

### 3. γ-Glutamyl Derivatives of Non-Protein Amino Acids Containing Sulphur or Selenium

The γ-glutamyl derivatives belonging to this category which have so far been identified in plants are listed in Table 4. The table lists 16 derivatives of sulphur-containing and two derivatives of selenium-containing amino acids. For comparison, about 25 sulphur-containing and about eight selenium-containing amino acids are known to occur in plants.

As before, the table contains information on the occurrence of the free parent amino acids in plants. In most cases the free amino acids occur together with the γ-glutamyl derivative.

The assignment of structure are in many cases based on structure determination on the parent amino acids. In many cases configurations have not been determined, particularly at asymmetric sulphur atoms. However, most of the cysteine sulphoxides of established configuration which have been isolated from plants have the $S$-configuration at the $S$-atom (*10, 75*), a notable exception being the γ-glutamyl-S-(prop-1-enyl)cysteine sulphoxide from *Santalum album*. In the case of the γ-glutamyl-S-(prop-1-enyl)cysteine from *Allium* no information is available on the configuration at the double bond. It is, however, reasonable to assume the same configuration as for the corresponding sulphoxide.

Those γ-glutamyl tripeptides which contain a sulphur-containing amino acid and which have so far been identified in plants are listed in Table 5. Tables 4 and 5 also contain information on chemical syntheses (cf. Section VI).

Table 4. *γ-Glutamyl Derivatives of Non-Protein, Sulphur- or Selenium-Containing Amino Acids, Identified in Plants*

| Compound | Formula | Plant species | Plant family | Character of isolate, comments | Configuration of Glutamic acid moiety | Configuration of The second amino acid | Ref. | Occurrence of 2nd amino acid | Chemical synthesis of γ-glutamyl derivative: Method | Chemical synthesis of γ-glutamyl derivative: Ref. |
|---|---|---|---|---|---|---|---|---|---|---|
| γ-L-Glutamyl-S-methyl-L-cysteine | | *Phaseolus lunatus*, seeds | Leguminosae | Cryst. | L | L | (*277*) | *Phaseolus lunatus* (*277*) | Glutathione→S-Me-Glutathione Carboxypeptidase ⟶ γ-L-Glu-S-Me-L-cysteine | (*391*) |
| | | *Phaseolus vulgaris*, seeds | Leguminosae | Cryst. | L | L | (*391*) | *Phaseolus vulgaris* (*332*) | 1 | (*246*) |
| | | *Phaseolus aureus* seedlings | Leguminosae | Paper chromatography | | | (*24*) | *Allium sativum* (*101*) | | |
| | | *Allium sativum*, bulbs | Alliaceae | Evap. residue | | | (*319*) | *Allium cepa* (*353*) | | |
| | | *Astragalus* sp., seeds | Leguminosae | Paper chromatography | | | (*43*) | | | |

| Compound | Source | Family | Method | Ref. | Biosynthesis/Related |
|---|---|---|---|---|---|
| γ-Glutamyl-S-methyl-cysteine-sulphoxide | Hippophae rhamnoides, seeds | Elaeagnaceae | Evap. residue | (278) | Hippophae rhamnoides (278) |
| | Astragalus bisulcatus, various stages and parts | Leguminosae | Ion-exchange chromatography | (240) | Astragalus sp. (240) |
| | Vigna radiata, seeds | Leguminosae | Cryst. | (256) | |
| | Allium cepa, bulbs | Alliaceae | | (356) | |
| | Phaseolus vulgaris, seeds | Leguminosae | Probably produced from γ-glutamyl-S-methyl-cysteine during isolation | (391) | Phaseolus vulgaris (390) Gluta-thione→S-Me-Gluta-thione Carboxy-peptidase ↑ γ-L-Glu-S-Me-L-cysteine H₂O₂ → (391) |
| Chemical Abstracts: N-L-γ-Glutamyl-3-(methyl-sulfinyl)-L-alanine | Allium sativum, bulbs | Alliaceae | | (319) | Allium cepa (26) |
| | Astragalus sp., seeds | Leguminosae | Paper chromatography | (43) | Allium sativum (319) |

Table 4 (continued)

| Compound | Formula | Plant species | Plant family | Character of isolate, comments | Configuration of Glutamic acid moiety | The second amino acid | Ref. | Occurrence of 2nd amino acid | Chemical synthesis of γ-glutamyl derivative Method | Ref. |
|---|---|---|---|---|---|---|---|---|---|---|
| | | Phaseolus vulgaris, seeds | Leguminosae | Ion-exchange chromatography | | | (390) | Crucifer sp. (229) | | |
| | | Vigna radiata, seeds | Leguminosae | | | | (256) | | | |
| | | Phaseolus lunatus, seeds | Leguminosae | Paper chromatography | | | (277) | | | |
| γ-L-Glutamyl-S-(prop-1-enyl)-L-cysteine | | Allium schoeno-prasum, seeds | Alliaceae | Cryst. | L | L | (210, 359) | Allium cepa (312) Allium sativum (314) | | |
| γ-Glutamyl-S-(prop-1-enyl)-cysteine sulphoxide | γ-L-GluNH–C–H COOH / CH₂ / SO / H–C–C–H / CH₃ | Allium cepa, bulbs | Alliaceae | Powder | L | E-(2R, SS) | (357, 358) | Allium cepa (312) | | |
| Chemical Abstracts: E-N-L-γ-glutamyl-3-(1-propenyl-sulfinyl)-L-alanine | | Santalum album, leaves | Santalaceae | Cryst. | L | E-(2R, SR) | (183) | Allium porum (20) | | |
| | | Allium schoeno-prasum seeds | Alliaceae | | | | (348) | | | |

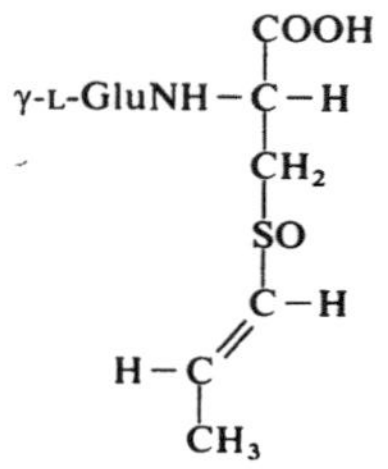

| Compound | Structure | Family | Source | State | Config. | Config. | Ref. | Species | No. | Ref. | Synthesis/Notes |
|---|---|---|---|---|---|---|---|---|---|---|---|
| γ-L-Glutamyl-S-allyl-L-cysteine | | Alliaceae | Allium sativum, bulbs | Cryst. | L | L | (320, 360) | Allium sativum (322) | 4 | (320) | |
| | | Alliaceae | Allium schoeno-prasum, seeds | | | | (348) | Allium porum (20) | | | |
| γ-L-Glutamyl-S-propyl-L-cysteine | | Alliaceae | Allium sativum bulbs | | | | (351) | Allium cepa (7) | | | |
| | | Alliaceae | Allium schoeno-prasum, seeds | | | | (211) | | | | |
| γ-L-Glutamyl-S-(2-carboxy-propyl)-L-cysteine | γ-L-GluNH—C—H with COOH, CH₂, S, CH₂, CHCH₃, COOH | Alliaceae | Allium cepa, bulbs | Cryst. | L | L(R) No configuration known for side chain | (216) | Acacia millefolia (71) | | (216) | Glutathione + Br–CH₂CHCOOH (—CH₃) → S-(2-carboxy-n-propyl)-glutathione; Carboxypeptidase → γ-L-Glu-S-(2-carboxy-n-propyl)-L-cysteine |

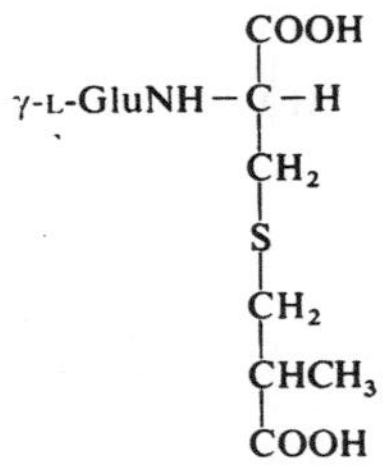

Table 4 *(continued)*

| Compound | Formula | Plant species | Plant family | Character of isolate, comments | Configuration of Glutamic acid moiety | The second amino acid | Ref. | Occurrence of 2nd amino acid | Chemical synthesis of γ-glutamyl derivative — Method | Ref. |
|---|---|---|---|---|---|---|---|---|---|---|
| γ-L-Glutamyl-S-allyl-mercapto-L-cysteine | 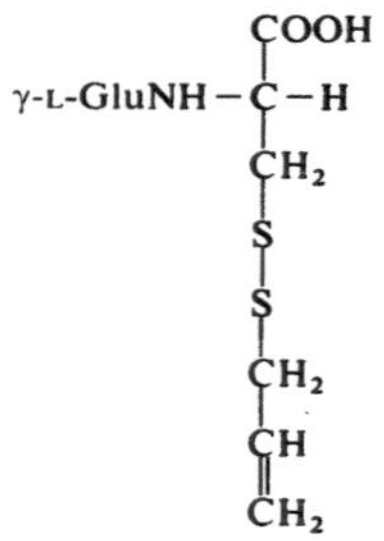 γ-L-GluNH—C—H; COOH; CH₂; S—S—CH₂; CH=CH₂ | *Allium sativum*, bulbs | Alliaceae | Cryst. | L | L | (315) | *Allium sativum* (315) | | |
| γ-Glutamyl-marasmine (γ-L-glutamyl-3-(methyl-thiomethyl-sulphinyl)-L-alanine) | 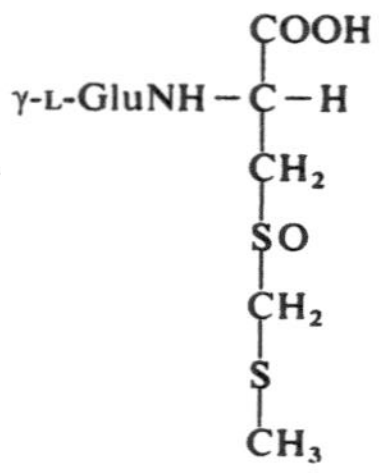 γ-L-GluNH—C—H; COOH; CH₂; SO; CH₂; S—CH₃ | *Marasmius* species | Tricholoma-taceae | Cryst. | L | L(R) No configuration known for chiral center at S | (72) | Not recorded from nature | | |

| Compound | Structure | Family | Source | Method | Config. | Reference / Occurrence |
|---|---|---|---|---|---|---|
| Lentinic acid (2-(γ-glutamyl-amino)-4,6,8,10-pentaoxo-4,6,8,10-tetrathia-undecanoic acid) | γ-L-GluNH–C–H, COOH, $CH_2SCH_2SCH_2SCH_2SCH_3$ (pentaoxo) | Tricholoma-taceae | 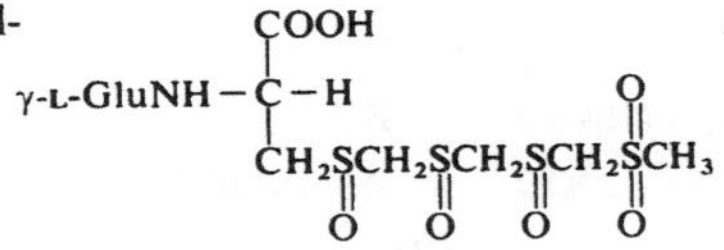 *Micromphale perforans, Collybia hariolorum, Lentinus edodes* | | L(R) No configuration known for chiral centers at S | *(100)* Not recorded from nature |
| | | | *Lentinus edodes* | | | (384) |
| γ-Glutamyl-djenkolic acid (N-Glutamyl-S-[[(2-amino-2-carboxy-ethyl)thio]methyl]-cysteine) | γ-L-GluNH–CH, COOH, $CH_2SCH_2SCH_2$, COO⁻, CHNH₃⁺ | Leguminosae | 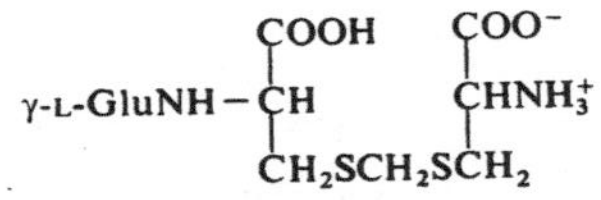 *Acacia georginae*, seeds | Evap. residue | L | *(306)* *Acacia* sp. including *A. georginae (306)* |
| | | Leguminosae | *Acacia* sp., seeds | Paper chromatography | | *(306)* |
| γ-Glutamyl-djenkolic acid sulphoxide (position of glutamyl residue/sulphoxide group not determined) | | Leguminosae | *Acacia georginae*, seeds | | | *(108)* *Acacia* sp. including *A. georginae (306)* |

Table 4 *(continued)*

| Compound | Formula | Plant species · Plant family | Character of isolate, comments | Configuration of Glutamic acid moiety | The second amino acid | Ref. | Occurrence of 2nd amino acid | Chemical synthesis of γ-glutamyl derivative — Method | Ref. |
|---|---|---|---|---|---|---|---|---|---|
| N,N-bis-(γ-Glutamyl)-3,3'-(1-methyl-ethylene-1,2-dithio)-dialanine 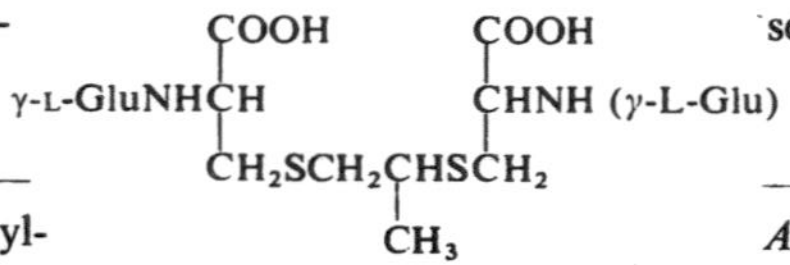 | γ-L-GluNHCH(COOH)CH₂SCH₂CHSCH₂ — CH₃ — COOH — CHNH (γ-L-Glu) | *Allium schoeno-prasum* seeds — Alliaceae | Cryst. | L | | (211, 212) | Not recorded from nature | | |
| N-γ-Glutamyl-3-[(3-Amino-3-carboxy-propyl)-thio]alanine and N-γ-Glutamyl-2-amino-4-[(2-amino-2-carboxyethyl)-thio]butanoic acid (the isomeric γ-glutamyl-cystathionines) | 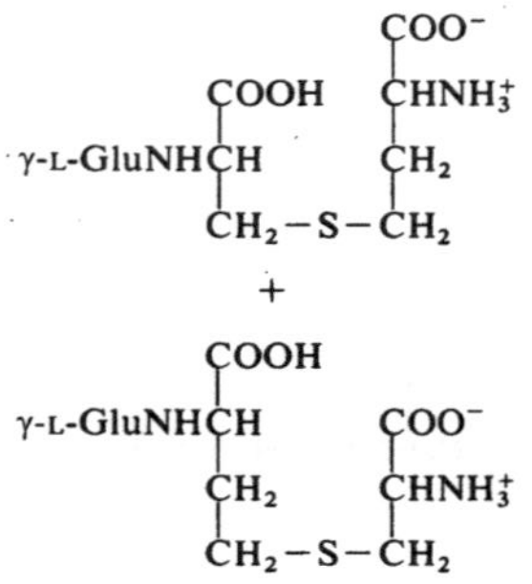 γ-L-GluNHCH(COOH)CH₂—S—CH₂(COO⁻)(CHNH₃⁺)CH₂ + γ-L-GluNHCH(COOH)CH₂—S—CH₂(COO⁻)(CHNH₃⁺)CH₂ | *Astragalus pectinatus,* seeds — Leguminosae | Mix. with corresponding Se-derivatives | | | (239) | *Astragalus pectinatus* (238); *Boletus erythropus* (114); *Neptunia amplexicaulis* (258) | | |

| Name | Structure | Source | Family | State | Configuration | Reference |
|---|---|---|---|---|---|---|
| methionine sulphoxide (N-L-γ-Glutamyl-2-L-amino-4-(methyl-sulphinyl)-butanoic acid) | 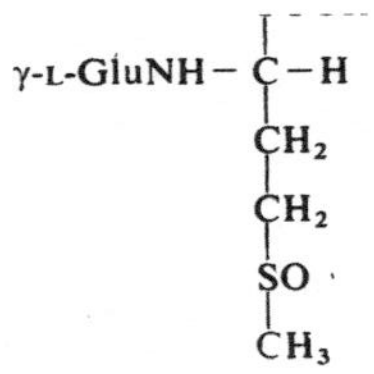 γ-L-GluNH—CH with CH₂ CH₂ SO CH₃ | *radiatus* var. *typicus*, seeds | | residue | | *vulgaris* (390) |
| | | *Fagus silvatica*, seeds | Fagaceae | Evap. residue Probably produced from γ-glutamyl-methionine during isolation | Configuration not known for chiral S-center | (177) *Malus* spurs (15) |
| | | *Vigna mungo*, seeds | Leguminosae | Cryst. | | (256) |
| γ-Glutamyl-methionine sulphone | 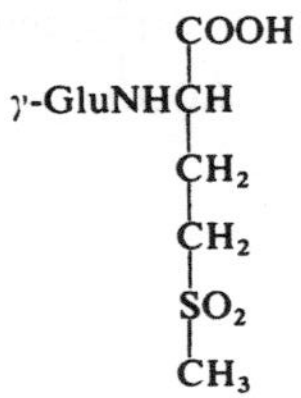 γ-GluNHCH with COOH CH₂ CH₂ SO₂ CH₃ | *Boletus edulis* | Boletaceae | Evap. residue | Not identified in plants | (337) |
| γ-L-Glutamyl-Se-methyl-seleno-L-cysteine | 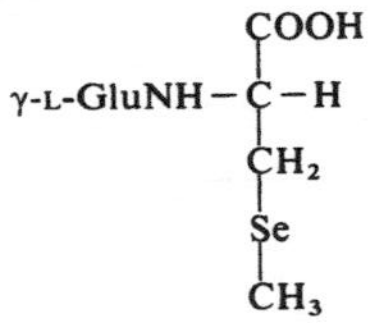 γ-L-GluNH—CH with COOH CH₂ Se CH₃ | *Astragalus bisulcatus*, seeds | Leguminosae | Cryst. | L  L | *Astragalus bisulcatus*, (338) (237) |
| | | *Astragalus bisulcatus*, various stages and parts | Leguminosae | Cryst. | | (240) |

Table 4 *(continued)*

| Compound | Formula | Plant species | Plant family | Character of isolate, comments | Configuration of Gluta-mic acid moiety | The second amino acid | Ref. | Occurrence of 2nd amino acid | Chemical synthesis of γ-glutamyl derivative — Method | Ref. |
|---|---|---|---|---|---|---|---|---|---|---|
| N-γ-Glutamyl-2-amino-4-[(2-amino-2-carboxyethyl)-seleno]butanoic acid and | γ-GluNHCH (COOH) — CH$_2$ — CH$_2$—Se—CH$_2$ — CH (COO$^-$) CHNH$_3^+$ | *Astragalus pectinatus*, seeds | Leguminosae | Mixture with the corresponding S-derivatives | | | (239) | *Astragalus pectinatus* (238) | | |
| | | | | | | | | *Lecythis ollaria* (150) | | |
| N-γ-Glutamyl-3-[(3-amino-3-carboxypropyl)seleno]alanine (The isomeric γ-glutamyl-selenocystathionines) | γ-GluNHCH (COOH) — CH$_2$ — CH$_2$—Se—CH$_2$ — CH (COO$^-$) CHNH$_3^+$ | | | | | | | *Neptunia amplexicaulis* (258) | | |

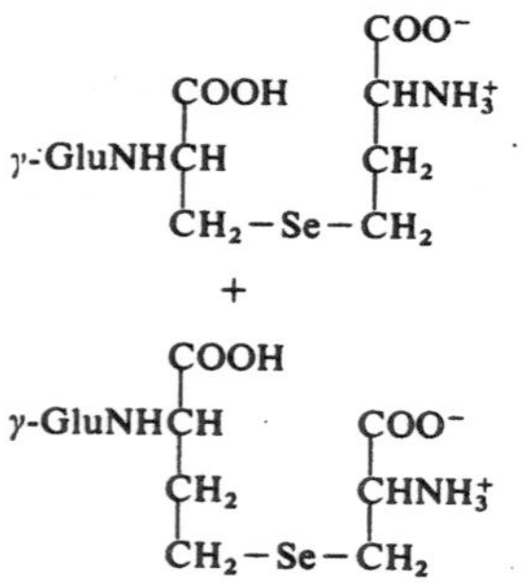

Table 5. *S-Containing γ-Glutamyl Tripeptides (Except Glutathione), Identified in Plants*

| Compound | Formula | Plant species | Plant family | Character of isolate, comments | Configuration of: Glutamic acid moiety | The second amino acid | The third amino acid | Ref. | Chemical synthesis of γ-glutamyl derivative Method Ref. |
|---|---|---|---|---|---|---|---|---|---|
| Homoglutathione (γ-L-Glutamyl-L-cysteinyl-β-alanine) | (structure) | *Phaseolus aureus,* seedlings | Leguminosae | Cryst. (oxidized form) | L | L | | (24, 25) | |
| γ-Glutamyl-S-(prop-1-enyl)-cysteinyl-S-(prop-1-enyl)-cysteine sulphoxide | 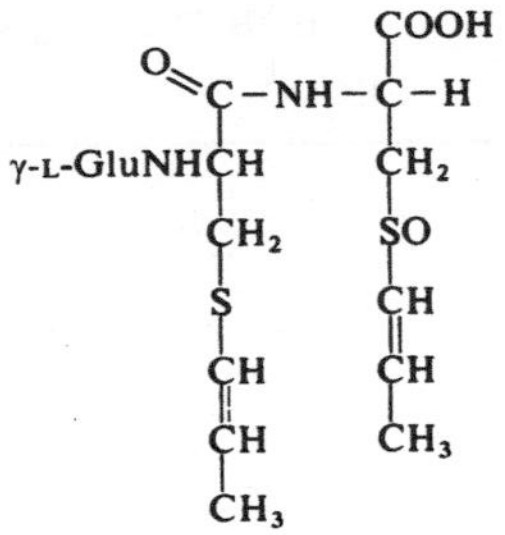 $\gamma$-L-GluNHCH, with O=C—NH—C—H, COOH, CH$_2$, SO, S—CH, CH=CH, CH$_3$ | *Allium schoenoprasum,* seeds | Alliaceae | | L | | L(R) No configuration known for chiral center at S | (215) | |
| γ-Glutamyl-S-(2-carboxypropyl)-cysteinylglycine (S-(2-carboxy-propyl)glutathione) | 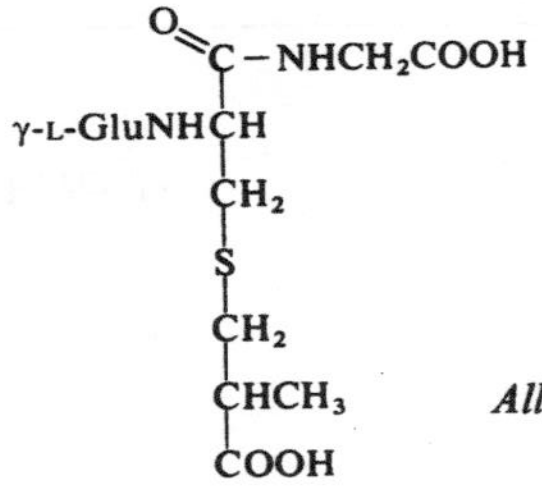 O=C—NHCH$_2$COOH, γ-L-GluNHCH, CH$_2$, S—CH$_2$, CHCH$_3$, COOH | *Allium cepa* | Alliaceae | Cryst. | L No configuration known for side chain | | | (354) | See Section V.11. (354) |
| | | *Allium sativum* | Alliaceae | | | | | (319) | |

Table 5 (*continued*)

| Compound | Formula | Plant species | Plant family | Character of isolate, comments | Configuration of: Glutamic acid moiety | The second amino acid | The third amino acid | Ref. | Chemical synthesis of γ-glutamyl derivative Method | Ref. |
|---|---|---|---|---|---|---|---|---|---|---|
| Ethyl ester of γ-Glutamyl-S-(2-carboxypropyl)-cysteinglycine (number and position of ethyl groups not indicated) | | *Allium cepa* | Alliaceae | Cryst., possibly artefact | L | L (R) | | (356) | | |
| γ-Glutamyl-γ-glutamylmethionine | 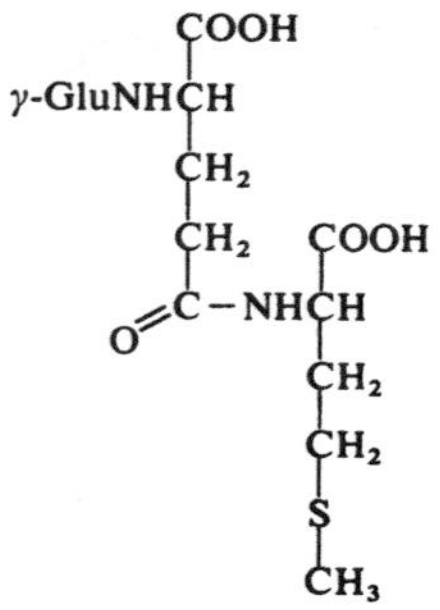 γ-GluNHCH (structure) | *Phaseolus radiatus* var. *typicus*, seeds | Leguminosae | Evap. residue | | | | (138) | 1 | (138) |

Table 6. γ-Glutamyl Derivatives of Amines Identified in Plants

| Compound | Formula | Plant species | Plant family | Character of isolate, Comments | Configuration of Gluta-mic acid moiety | The amine | Ref. | Occur-rence of amine | Chemical synthesis of γ-glutamyl derivative |  |
|---|---|---|---|---|---|---|---|---|---|---|
|  |  |  |  |  |  |  |  |  | Method | Ref. |
| N⁵-Methylglu-tamine (γ-Glu-tamyl methyl-amide) | γ-GluNHCH₃ | *Thea sinensis*, leaves | Theaceae | Evap. residue |  |  | (166) | Com-mon consti-tuent in plants | Reaction of N-ben-zyloxy-carbonyl-L-glu-tamic acid γ-methyl-ester with methyl-amine followed by hydro-genation | (89) |

Table 6 *(continued)*

| Compound | Formula | Plant species | Plant family | Character of isolate, comments | Configuration of Glutamic acid moiety | The amine moiety | Ref. | Occurrence of amine | Chemical synthesis of γ-glutamyl derivative | Method Ref. |
|---|---|---|---|---|---|---|---|---|---|---|
| Theanine (N⁵-Ethyl-L-glutamine, γ-L-Glutamyl-ethylamide) | γ-L-GluNHCH₂CH₃ | *Thea sinensis,* leaves | Theaceae | Cryst. | L | | (283) | Common constituent in Plants | Reaction of N-benzyloxy-carbonyl-L-glutamic acid γ-ethyl ester with ethyl-amine followed by hydrogenation | (284) |
| | | *Boletus badius* | Boletaceae | Cryst. | | | (27) | | L-glutamic acid γ-methyl ester + ethyl-amine | (68) |

| Compound | Structure | Source | Family | Form | Config. | Notes | Ref. | Distribution | | Ref. |
|---|---|---|---|---|---|---|---|---|---|---|
| N⁵-Isopropyl-L-glutamine (γ-L-Glutamyl-isopropylamide) | γ-L-GluNHCH(CH₃)₂ | *Lunaria annua,* seeds | Cruciferae | Cryst. | L | | (187) | See (187) | 1 | (187) |
| N⁵-(2-Hydroxyethyl)-L-glutamine (γ-L-Glutamyl-hydroxyethylamide) | γ-L-GluNHCH₂CH₂OH | *Lunaria annua,* seeds | Cruciferae | Cryst. | L | | (188) | Ubiquitous | 5 | (188) |
| N⁵-(1-Methyl-2-oxopentyl)-L-glutamine (γ-L-Glutamyl-2-amino-3-hexanone) | $\underset{\displaystyle \overset{\|}{O}}{\text{γ-L-GluNHCHCH}}\overset{\text{CH}_3}{\phantom{|}}\text{CH}_2\text{CH}_2\text{CH}_3$ | *Russula ochroleuca* | Russulaceae | Cryst. | L | Configuration not known for chiral center in amine | (376) | Unstable | | |
| N⁵-(2-Cyanoethyl)-L-glutamine (γ-L-Glutamyl-β-aminopropionitrile) | γ-L-GluNHCH₂CH₂C≡N | *Lathyrus odoratus,* seeds | Leguminosae | Cryst. | L | | (203, 299) | *Lathyrus* sp. including *L. odoratus* (12) | 4 | (299) |
| | | *Lathyrus pusillus.* seeds | Leguminosae | Cryst. | | | (44) | | | |
| | | *Lathyrus,* seeds | Leguminosae | Paper chromatography | | | (12) | | | |

14*

Table 6 (continued)

| Compound | Formula | Plant species | Plant family | Character of isolate, comments | Configuration of Glutamic acid — The amine moiety | Ref. | Occurrence of amine | Chemical synthesis of γ-glutamyl derivative — Method | Ref. |
|---|---|---|---|---|---|---|---|---|---|
| Coprine (N⁵-(1-hydroxy-cyclopropyl)-L-glutamine) | 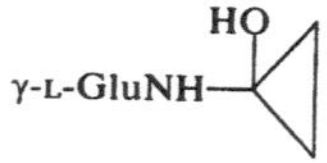 HO, γ-L-GluNH–⊲ | Coprinus atramentarius | Coprinaceae | Evap. residue | | (94) | Unstable | 4, See Section VI.3 | (201) |
| | | Coprinus atramentarius | Coprinaceae | Cryst. | L | (201) | | | |
| N⁵-(4-Hydroxyphenyl)-L-glutamine (γ-Glutaminyl-4-hydroxybenzene) | 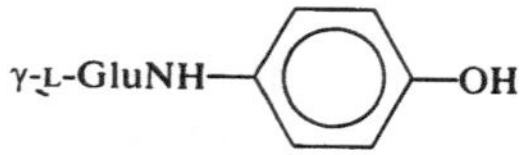 γ-L-GluNH–⬡–OH | Agaricus hortensis | Agaricaceae | Cryst. | L | (115) | Not recorded from plants | 5-oxo-proline + p-hydroxyaniline | (115) |
| | | Agaricus hortensis | Agaricaceae | Cryst. | L | (370, 371) | | 1 | (371) |
| Agaridoxin (N⁵-(3,4-Dihydro-xyphenyl)-L-glutamine) | 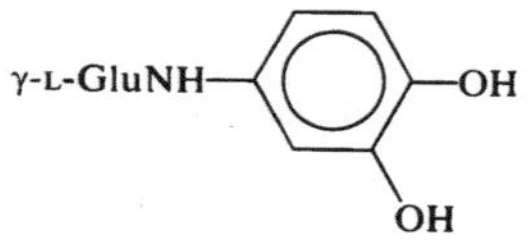 γ-L-GluNH–⬡(–OH)(–OH) | Agaricus campestris | Agaricaceae | Cryst. | L | (323) | Not recorded from plants | 4, See Section VI. 5 | (323) |

| Compound | Structure | Source | Family | Method | Config. | Refs. | Notes |
|---|---|---|---|---|---|---|---|
| N⁵-(3,4-Dioxo-cyclohexa-1,5-dien-1-yl)-L-glutamine (γ-Glutaminyl-3,4-benzoquinone) | 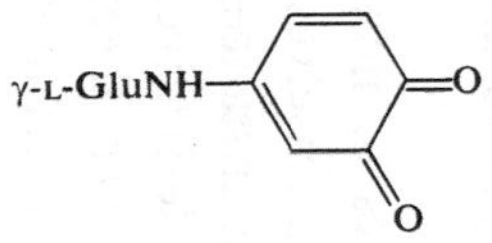 γ-L-GluNH— | Agaricus bisporus | Agaricaceae | Evap. residue | L | (369, 370) | Not recorded from plants |
| Agaritine (β-N-(γ-L-Glutamyl)-4-hydroxymethyl-phenylhydrazine, N⁵-[(4-Hydroxymethylphenyl)-amino]-L-glutamine) | γ-L-GluNHNH— | Agaricus bisporus | Agaricaceae | Cryst. | L | (147, 193) | Un-stable 2, (147) See Section VI.6 |
|  | 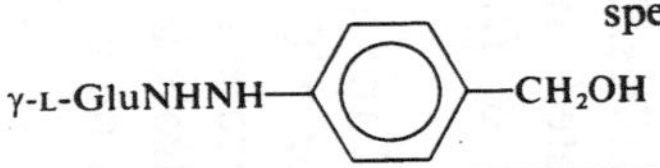 γ-L-GluNHNH—⟨⟩—CH₂OH | Agaricus species | Agaricaceae | Paper chromatography |  | (193) |  |
| N⁵-[(4-Hydroxy-phenyl)methyl]-L-glutamine (N⁵-(4'-Hydroxy-benzyl)-L-glutamine) | 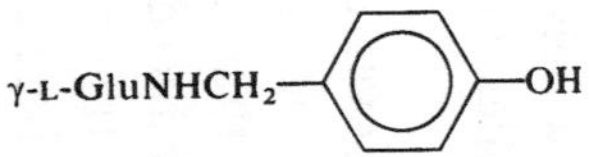 γ-L-GluNHCH₂—⟨⟩—OH | Fagopyrum esculentum, seeds | Polygonaceae | Cryst. | L | (173) | Fago-pyrum esculen-tum (172) |
| N⁵-[2-(5-oxo-2-(5H)-isoxazolyl)-ethyl]glutamine (2-(2-γ-Glutamyl-amino)ethyliso-xazolin-5-one) | 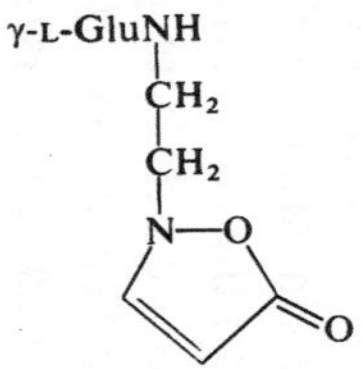 γ-L-GluNH—CH₂—CH₂— | Lathyrus odoratus, seedlings | Leguminosae |  |  | (185) | Lathyrus odoratus seedlings (185) |
|  |  | Lathyrus odoratus, roots | Leguminosae |  |  | (180) |  |

## 4. γ-Glutamyl Derivatives of Amines

The glutamyl derivatives of amines so far identified in plants are listed in Table 6. Again, information on the occurrence of the free amine in plants together with information on chemical syntheses is included. Unlike amino acids, the amines corresponding to the γ-glutamyl derivatives have in most cases not been identified in plants.

## 5. Distribution Patterns for the γ-Glutamyl Derivatives

From Tables 1—6 it is apparent that most of the compounds have been found in the families Leguminosae and Alliaceae, and in mushrooms. There is no doubt that the occurrence of sulphur-containing amino acids and γ-glutamyl derivatives is characteristic for Alliaceae; the enzymatic transformations and biological significance of these compounds are discussed in detail in Section VII. 6.2. The occurrence of γ-glutamyl derivatives in mushrooms is most likely also characteristic. As far as the numerous identifications of γ-glutamyl derivatives in Leguminosae are concerned it must be remembered that this family has proven to be an especially rich source of non-protein amino acids. Therefore, and because of both the economic importance of many leguminous species and of their toxicity (towards man and/or animals), the Leguminosae have been investigated much more thoroughly than other families of higher plants. Tables 1—6 also note the presence of γ-glutamyl derivatives in 18 other families. It seems likely that γ-glutamyl derivatives are very widely distributed, and that they will be identified in many more families as studies of free amino acids in plants are extended

As mentioned above, nearly all the γ-glutamyl derivatives of non-protein amino acids occur together with the amino acids themselves. With the reservation that the non-protein amino acids in some instances may be artefacts produced by hydrolysis of the γ-glutamyl derivatives (see Section IV. 10), this may indicate a general trend. However, there are numerous cases where a non-protein acid and its γ-glutamyl derivative occur together in one or a few species, whereas the free amino acid occurs alone in additional related species. In many cases the γ-glutamyl derivatives predominate in the fraction of free amino acids and other amphoteric plant constituents which give positive ninhydrin reactions. In other cases the concentrations are similar to those of the free amino acids or even lower.

*References, pp. 267—285*

There are not only examples of the simultaneous occurrence of γ-glutamyl derivatives of protein- and non-protein amino acids, but also of the occurrence of γ-glutamyl derivatives of protein amino acids alone or non-protein amino acids alone. A few γ-glutamyl derivatives of β- and γ-amino acids have been identified, sometimes together with derivatives of α-amino acids, sometimes only together with derivatives of amines. Rather few of the γ-glutamyl derivatives of amines occur together with derivatives of amino acids. However, the crucifer *Lunaria annua* contains three derivatives of amines and three of β- and γ-amino acids (*186, 187, 188*).

The identification of compounds depends, of course, on the efficiency of the analytical methods employed, and γ-glutamyl derivatives present in low concentrations are thus not always detected. In a number of the species listed in the tables only one or a few γ-glutamyl derivatives have been found, although others may be present. The results of an investigation on seeds of *Fagus silvatica* demonstrate the wide range of concentrations which can occur. In these seeds 10 γ-glutamyl derivatives were found, with γ-glutamylphenylalanine as the major constituent. γ-Glutamylalanine was present in the lowest concentration, *viz.* 1/6000 of the concentration of γ-glutamylphenylalanine (*177*). However, there are no grounds for believing that the application of appropriate analytical methods will reveal the presence of all possible γ-glutamyl derivatives in all species. Indeed, a number of detailed investigations of various plants have failed to · detect any γ-glutamyl derivatives whatsoever. Thus, whereas γ-glutamyl derivatives were found in a number of *Fagus* species, none were found in species from the related genera *Quercus* and *Castanea* (*129*).

Furthermore, in a very detailed investigation of the amino acid content of *Beta vulgaris* which permitted the identification of compounds present in extremely low concentrations only one γ-glutamyl amino acid, γ-glutamyl-γ-aminobutyric acid, was found (*54*).

There must, of course, be a connection between the free amino acids and the γ-glutamyl derivatives in plants, but there is no simple correlation between the concentration of an amino acid and that of the corresponding γ-glutamyl derivative. A simple correlation between concentrations would also require only one pool of amino acids in the plants, a condition which is known not to be fulfilled. The amino acids present in highest concentration are not always those whose γ-glutamyl derivatives occur in high concentration, and γ-glutamyl derivatives of the former may even be completely absent. Comparison of the concentrations of γ-glutamyl peptides and of free amino acids in various species of beans shows that the pattern for γ-glutamyl peptides bears no simple relation to that for the free amino acids [see for example (*126, 127*) for *Glycine max*, (*141*) for *Phaseolus radiatus*, (*390*) for *Phaseolus vulgaris*, and (*143, 144*) for *Vicia faba*]. The ability of various amino acids to act as acceptors of the γ-glutamyl residue in the reaction catalyzed by the γ-glutamyl-transferase from *Phaseolus vulgaris* (*333*) shows no correlation with the pattern of occurrence of γ-glutamyl derivatives (*390*) (compare Section VII. 3).

As is apparent from the tables, most of the γ-glutamyl derivatives have been found in seeds, although some have been identified in fresh plant parts. This is due partly to the fact that seeds have been more thoroughly investigated because of their availability and economic importance, but it also reflects the general tendency for non-protein amino acids and γ-glutamyl derivatives to occur in high concentrations in seeds but in low concentrations or not at all in fresh plant parts (cf. Section VII. 5).

γ-Glutamyl peptides have on occasion been utilized as index compounds in chemotaxonomy. The genus *Lathyrus* was divided into groups of species characterized by combinations of ninhydrin-reacting compounds in their seeds; γ-glutamyl-β-aminopropionitrile was used as one of the important index compounds (*12*). The genus *Vicia* has also been divided into groups using γ-glutamyl-β-cyanoalanine as a marker (*13*).

*Vigna mungo* and *V. radiata* can be clearly distinguished from each other on the basis of the fact that seeds of *V. mungo* contain a high concentration of γ-glutamyl-S-methylcysteine and the corresponding sulphoxide, whereas seeds of *V. radiata* are particularly rich in γ-glutamyl-methionine and the corresponding sulphoxide (*256*). The occurrence of γ-glutamylphenylalanine and of free amino acids in seeds from species of *Macrotyloma, Dolichos* and *Pseudovigna* seems to be characteristic (*39*). Field crops belonging to the family Leguminosae were classified into three groups (*143*): (1) Crops containing high concentrations of γ-glutamylphenylalanine and γ-glutamyltyrosine in the seeds, (2) Crops containing very low- or zero concentrations of γ-glutamylphenylalanine and γ-glutamyltyrosine, but containing high concentrations of other γ-glutamyl peptides in the seeds, and (3) Crops containing very low- or zero concentrations of γ-glutamyl peptides in the seeds. *Glycine max, Trifolium repens, Medicago sativa* and other crops utilized as oil plants and/or fodder plants belong to group (1). *Arachis hypogea* was shown recently to belong to this group, as predicted previously (*317*). A number of beans belonging to the genus *Phaseolus* and which are used for food belong to group (2) (e. g. *Phaseolus radiatus, P. angularis, P. vulgaris* and *P. lunatus*). *Vicia faba* is the only crop species belonging to group (3) which has so far been found.

# IV. Chemical, Spectroscopic, and Analytical Properties of γ-Glutamyl Derivatives

## 1. pK-Values

The behaviour of γ-glutamyl derivatives during most analytical and isolational procedures is dependent on their protolytic properties; hence some knowledge of these properties is a prerequisite for the interpretation of the data presented in the following sections. Although very few pK-values for γ-glutamyl derivatives are reported in the literature, it is possible to make reliable estimates from the known pK-values of related compounds. Table 7 gives the estimated values of $pK_1$ and $pK_2$ for glutamic acid-containing dipeptides. It is assumed that $pK_1$ for a γ-glutamyl amino acid is similar to those for glutamic acid and glutamine, that $pK_2$ for a γ-glutamyl-α-amino acid is similar to that for an acylated α-amino acid, that $pK_2$ for a γ-glutamyl-β-amino acid is similar to that for an acylated β-amino acid, that $pK_1$ for an α-glutamyl-α-amino acid and for a dipeptide with glutamic acid as C-terminal are similar to that for a normal dipeptide and that the $pK_2$-values for the latter two com-

Table 7. *pK-Values at 25° C For Amino Acids and Derivatives Related to γ-Glutamyl Derivatives and Estimated pK-Values For γ-Glutamyl Amino Acids, α-Glutamyl Amino Acids and Dipeptides With Glutamic Acid as C-Terminal and an α-Amino Acid as N-Terminal*

| Compound | $pK_1$ (Shift in charge from +1 to 0) | $pK_2$ (Shift in charge from 0 to −1) | Reference |
|---|---|---|---|
| Glutamic acid | 2.13 | 4.31 | *(257)* |
| Aspartic acid | 1.99 | 3.90 | *(257)* |
| Glutamine | 2.17 | 9.13 | *(257)* |
| Isoglutamine | 3.81 | 7.88 | *(257)* |
| β-Alanine | 3.55 | | *(257)* |
| γ-Aminobutyric acid | 4.03 | | *(257)* |
| Alanylalanine | 3.30 | | *(257)* |
| N-Acetylalanine | | 3.72 ($pK_1$) | *(169)* |
| N-Acetyl-β-alanine | | 4.45 ($pK_1$) | *(169)* |
| Glutaminylglutamic acid | 3.14 | 4.38 | *(257)* |
| Reduced glutathione | 2.12 | 3.53 | *(261)* |
| $N^5$-(2-Cyanoethyl)glutamine | 2.2 | 9.14 | *(203)* |
| γ-Glutamyl-α-amino acid | ~2.1 | ~3.7 | |
| γ-Glutamyl-β-amino acid | ~2.1 | ~4.4 | |
| α-Glutamyl-α-amino acid | ~3.0 | ~4.3 | |
| Dipeptide with glutamic acid as C-terminal and an α-amino acid as N-terminal | ~3.0 | ~4.3 | |

pounds are similar to that for glutamic acid. Also, the pK-values for the $NH_3^+$-groups in α- and γ-glutamyl derivatives will be widely different, as seen for the $pK_2$-values for glutamine and isoglutamine (Table 7).

## 2. Optical Activity

Rotations at the sodium D-line for the majority of the γ-glutamyl derivatives in either water or hydrochloric acid have been reported. Although the γ-glutamyl linkage is unstable towards acid (see Section IV. 9), it is possible to obtain reliable rotations for hydrochloric acid solutions if the rotation is measured within 5 min of the preparation of the solution; furthermore, one can check that no measurable change in rotation takes place in the first 15 min. The rotation of glutamine itself has previously been determined under these conditions (218).

For the γ-glutamyl derivatives of chiral amino acids and amines no simple relationship between the rotation of the glutamyl derivatives and the rotation of the individual constituents has been established, although some relationships have been tentatively proposed, based on rotation values of a restricted number of glutamic acid containing peptides (280). For the derivatives of achiral amines and amino acids it might be assumed that molecular rotations are close to those for glutamine itself and this holds true in some cases. Thus, L-glutamine has $[M]_D +9.2°$ in $H_2O$ and $+46.5°$ in 1 N HCl [(78), p. 1930]. $N^5$-Ethyl-L-glutamine has $[M]_D +12.3°$ in $H_2O$ (283), $N^5$-isopropyl-L-glutamine has $[M]_D +11.3°$ in $H_2O$ (187), $N^5$-(2-hydroxyethyl)-L-glutamine has $[M]_D +11.0°$ in $H_2O$ (188), γ-L-glutamyl-β-alanine has $[M]_D +12.0°$ in $H_2O$ and $+23.0°$ in 0.1 N HCl (186), γ-L-glutamyl-γ-aminobutyric acid has $[M]_D +12.0°$ in $H_2O$ (187), and coprine [$N^5$-(1-hydroxycyclopropyl)-L-glutamine] has $[M]_D +15.4°$ in $H_2O$ (201). However, even this simple relationship breaks down in many cases. Thus, $N^5$-(2-cyanoethyl)-L-glutamine has $[M]_D +35.8°$ in $H_2O$ (299), $N^5$-(4-hydroxyphenyl)-L-glutamine has $[M]_D +71.4°$ in 1 N HCl (115), $N^5$-[(4-hydroxyphenyl)-methyl]-L-glutamine has $[M]_D +40.1°$ in $H_2O$ (173), and $N^5$-[(4-hydroxymethylphenyl)amino]-L-glutamine has $[M]_D +18.7°$ in $H_2O$ (147).

No CD- or ORD-measurements have been reported for γ-glutamyl derivatives, but L-glutamine has a positive CD at 200 nm in $H_2O$ and at 208 nm in hydrochloric acid, similar to that of other simple α-L-amino acids (58).

## 3. Infrared Spectra

The infrared spectra of ten pairs of α- and γ-glutamyl dipeptides and of several related compounds have been measured in KBr discs (134, 137). The carbonyl stretching vibration of the peptide bond (Amide I) was invariably observed at higher wave number for the α-glutamyl peptides than for the γ-glutamyl peptides (1665—1685 cm$^{-1}$ versus 1635—1660 cm$^{-1}$). This difference was attributed to the influence of the charged amino group on the peptide bond in the α-glutamyl peptides. The vibration was also found at a high wave number for dipeptides with glutamic acid as the C-terminal. Thus for α-glutamylvaline, γ-glutamylvaline, and valylglutamic acid, $\nu$ (C=O) was found at 1675, 1645 and 1680 cm$^{-1}$, respectively. The Amide II i.r. absorption band could not be observed because of overlap with the deformation band of the $NH_3^+$-group and the stretching band of the $COO^-$-group. The N−H stretching vibration band was observed at higher wave number for the γ-glutamyl peptides than for both the α-glutamylpeptides and for dipeptides with glutamic acid as C-terminal. Thus for α-glutamyl-valine, γ-glutamylvaline, and valylglutamic acid $\nu$ (N−H) was found at 3260, 3350, and 3250 cm$^{-1}$, respectively.

## 4. NMR-Spectra

The PMR-spectra of γ-glutamyl derivatives have been studied primarily for the purpose of differenting between γ-glutamyl peptides, α-glutamyl peptides, and dipeptides with glutamic acid as C-terminal. Sequence determination on peptides, based on chemical shift changes of α-protons on change in ionization is well-known (298). However,

Table 8. *Changes in the Surroundings of α-Protons in Glutamic Acid Containing Dipeptides Due to Changes in Groups at α-Carbon*

|  | α-Proton in glutamic acid moiety | | | α-Proton in second amino acid | | |
|---|---|---|---|---|---|---|
|  | Titration step | | | Titration step | | |
|  | 1 | 2 | 3 | 1 | 2 | 3 |
| γ-Glutamyl-α-amino acid | + | − | + | − | + | − |
| α-Glutamyl-α-amino acid | − | − | + | + | − | − |
| Dipeptide with glutamic acid as C-terminal | + | − | − | − | − | + |

+ indicates charge change, − indicates no change.

glutamic acid-containing dipeptides present special problems and possibilities. Chemical shift changes of the α-protons will only take place as a result of changes in the charge of the amino group and carboxyl group at the α-carbon atom. Table 8 summarizes the changes. Measurements of NMR-spectra at ionization stages with net charge $+1$, zero, and $-2$ can be made by using $DCl + D_2O$, $D_2O$, and $NaOD + D_2O$. Measurement of spectra at the ionization stage with net charge $-1$ does not necessitate precise titration or pH-measurement in $D_2O$ but can easily be accomplished by the use of $D_2O$ and an excess of secondary phosphate.

These considerations have been corroborated by experimental studies which show the expected different titration shifts for the different isomers (*132, 137, 175*). Some typical results are collected in Table 9.

Table 9. δ-*Values for α-Protons in Glutamic Acid Containing Dipeptides*

|  | $DCl + D_2O$ | $D_2O$ | $D_2O + Na_2HPO_4$ | $D_2O + NaOD$ |
|---|---|---|---|---|
| α-Proton in glutamic acid moiety in: |  |  |  |  |
| γ-Gluala | 4.15 | 3.82 | 3.77 | 3.25 |
| γ-Glutyr | 4.01 | 3.72 | 3.69 | 3.13 |
| α-Gluala | 4.15 | 4.10 | 4.05 | 3.37 |
| α-Glutyr | 4.07 | 4.00 | 3.97 | 3.35 |
| Alaglu | 4.53 | 4.27 | 4.19 | 4.15 |
| Tyrglu | 4.45* | 4.25 | 4.23 | 4.15 |
| α-Proton in second amino acid moiety in: |  |  |  |  |
| γ-Gluala | 4.29 | 4.31 | 4.15 | 4.15 |
| γ-Glutyr | 4.65 | 4.60 | 4.48 | 4.40 |
| α-Gluala | 4.48 | 4.28 | 4.22 | 4.17 |
| α-Glutyr | 4.70 | 4.48 | 4.46 | 4.49 |
| Alaglu | 4.20 | 4.15 | 4.17 | 3.52 |
| Tyrglu | 4.30 | 4.20 | 4.18* | 3.60 |

Values from (*175*)

* Indicates that exact δ-values are difficult to obtain because of superposition of signals.

The chemical shifts for the α-protons can also be used to differentiate between isomers of derivatives. Thus for ten pairs of γ- and α-glutamyl dipeptides it was shown that the α-protons in the glutamic acid moieties in $D_2O$ had δ-values from 3.93 to 4.16 ppm in the α-isomers and δ-values 3.71 to 3.82 ppm in the γ-isomers (*131, 137*).

It has been observed that the two methyl groups in α-glutamyl-α-aminoisobutyric acid are magnetically non-equivalent in aqueous and basic solution but magnetically equivalent in acid solution, whereas the methyl groups in γ-glutamyl-α-aminoisobutyric acid, $N^5$-isopropylglutamine and $N^1$-isopropylisoglutamine, N-glutaryl-α-aminoisobutyric acid and N-glutarylisopropylamine were magnetically equivalent at all ionization stages (*135, 137*). Similarly, the methylene protons in α-glutamylglycine were magnetically non-equivalent in aqueous and basic solution but magnetically equivalent in acid solution (*133, 137*). The non-equivalence cannot be explained by *cis-trans* isomerism of the peptide bond, but is presumably due to interaction with the α-amino group and the ionized γ-carboxyl group (*135*).

A conformational study of glutathione based on PMR-spectroscopy in the whole pH-region from strongly acid to strongly basic solutions has recently been described (*65*).

In addition to these general studies, PMR-spectroscopy has been used in a number of cases in the elucidation of the structure of individual γ-glutamyl derivatives; see for example references (*72, 94, 100, 129, 147, 173, 176, 185, 188, 202, 323, 371, 376*).

No systematic $^{13}$C-NMR spectroscopy studies of γ-glutamyl derivatives have been reported, although chemical shift values are well-known for glutamic acid and glutamine (*45*) and for the reduced and oxidized forms of glutathione (*122*). Changes in chemical shifts taking place on titration have also been determined for glutamine (*121*) and glutamic acid (*268*). $^{13}$C-NMR-spectroscopy has also been used in the elucidation of the structure of individual γ-glutamyl derivatives; see for example references (*72, 94, 100, 202, 376*).

## 5. Gas Chromatography, Mass Spectroscopy, and Combined Gas Chromatography — Mass Spectroscopy

Separation and quantitative determination of γ-glutamyl-amino acids has been performed by reaction with a mixture of 2,2,3,3,3-pentafluoro-1-propanol and pentafluoropropionic anhydride followed by gas chromatography (*379*).

In sequence determination studies by mass spectroscopy a number of peptides containing α- or γ-glutamyl linkages were included. N-butyl and N-decanoyl derivatives of methyl esters were used and it was possible from the degradation patterns to distinguish between α- and γ-linkages (*152, 343*). In a later study the problem of distinguishing between α- and γ-linkages was approached by the use of decanoyl and benzyloxycarbonyl derivatives of methyl esters; in this manner structures of glutamic acid-containing di-, tri- and tetrapeptides, including the four isomeric dipeptides containing glutamic acid and lysine, could be determined (249). Preliminary studies indicate that trimethylsilyl derivatives may be used for combined gas chromatography-mass spectroscopy of γ-glutamyl derivatives (32).

Mass spectroscopy has been used in special cases for structure determinations; see for example Ref. (72, 94, 100, 138, 173, 371). Gas chromatography-mass spectroscopy has been employed in the elucidation of the structure of agaridoxin (323).

## 6. Paper Chromatography

It has been stated in the literature that γ-glutamyl derivatives should, in accordance with the differences in acid strength, have higher $R_f$-values on paper chromatograms run in phenol:water:ammonia than the corresponding α-glutamyl derivatives [(78), p. 1482], but published $R_f$-values do not reveal a simple pattern (190, 191, 279). α-Glutamyl- and γ-glutamyl-γ-aminobutyric acid have the same $R_f$-value in this solvent (50). On the other hand, it is well established that α-glutamyl derivatives have higher $R_f$-values than the corresponding γ-glutamyl derivatives in paper chromatography in the following solvents: propanol:acetic acid:water (8:1:1) (83), butanol:acetic acid:water (4:1:1), and pyridine:acetic acid:ethyl acetate:water (5:1:5:3) (199). $R_f$-values for the individual γ-glutamyl derivatives in various standard solvent systems are in most cases available in the references describing their isolation and identification. As a rule of thumb it can be mentioned that in various paper chromatographic systems γ-glutamyl derivatives have $R_f$-values intermediate between those of glutamic acid and the second constituent (and γ-glutamyltripeptides on two-dimensional chromatograms occupy intermediate positions in relation to those of the three constituent amino acids).

## 7. Paper Electrophoresis

In keeping with the low isoelectric point for γ-glutamyl-α-amino acids relative to those for α-glutamyl-α-amino acids, and dipeptides with glutamic acid as C-terminal and α-amino acid as N-terminal (compare Table 7), it has been shown that γ-glutamyl derivatives have higher mobility than the corresponding α-glutamyl derivatives in paper electrophoresis at pH 4 (199, 363). High-voltage paper electrophoresis has been employed extensively in the analysis of γ-glutamyl derivatives [see for example refs. (57, 324)]; the results are consistent with those expected from examination of Table 7. It must, however, be remembered that the derivatives of β- and γ-amino acids have different pK-values and therefore also show different behaviour in electrophoresis.

## 8. Ion-Exchange Chromatography

The normal procedure for automatic ion-exchange amino acid analysis involves adsorption to, and elution from, a strongly acid resin. The elution behaviour under these conditions is determined mainly by the pK-value corresponding to the change from net charge $+1$ to net charge zero (although adsorption also plays a role, especially for compounds containing aromatic rings). The relevant pK in the case of neutral amino acids, monoaminodicarboxylic acids, and dipeptides containing glutamic acid and a neutral amino acids is $pK_1$. Correspondingly, it has been shown for 10 pairs of α- and γ-glutamyl derivatives of amino acids that the γ-glutamyl derivatives of α-amino acids are eluted from the amino acid analyzer close to aspartic acid and glutamic acid and before the second amino acid in the derivative, whereas the corresponding α-glutamyl derivatives are eluted after the second amino acid. Furthermore, γ-glutamyl-β-alanine was found to be eluted before α-glutamyl-β-alanine on the analyzer, but in this case the second amino acid, β-alanine, was eluted after both the γ- and α-derivatives (131, 137). These general rules are consistent with the pK-values estimated for glutamic acid-containing peptides (Table 7) and with the elution behaviour reported for 12 γ-glutamyl derivatives (392), 17 γ-glutamyl derivatives and 13 α-glutamyl derivatives (137), and 12 α-glutamyl derivatives (4).

Under certain conditions γ-glutamylaspartic is eluted from the amino acid analyzer in two bands. Whether one or two peaks occur is determined by the volume of buffer in which the peptide is applied to the resin; the phenomenon therefore probably reflects differences in elution behaviour of the species at different stages of ionization. None of the other α- and γ-glutamyl derivatives which have been investigated showed this type of behaviour (136, 137), although similar problems have been encountered with γ-glutamyl-β-cyanoalanine (275).

Acidic amino acids and peptides can also be separated by ion-exchange chromatography on basic resins in the acetate (or formate) form, using elution with aqueous acetic (or formic) acid. The separation under these conditions is a function of the pK-value corresponding to change from net charge zero to net charge $-1$. For monoaminodicarboxylic acids, and dipeptides containing glutamic acid and a neutral amino acid, this is the $pK_2$-value. In keeping with the $pK_2$-values given in Table 7, γ-glutamyl-α-amino acids are eluted from the ion-exchange resin after glutamic acid but close to aspartic acid [see for example (177)]. For γ-glutamyl derivatives of β-amino acids the order of elution is reversed, glutamic acid appearing after such derivatives [see for example (186)].

Ion-exchange chromatography of amino acids on acid resins is a very efficient analytical procedure but is difficult to adapt to preparative separations. On the other hand, ion-exchange chromatography of acidic amino acids and peptides on basic resins is an efficient technique for preparative separations and is used mainly for that purpose. Thus, nearly all isolations of acid γ-glutamyl derivatives from natural material have employed this technique.

## 9. Degradation and General Chemical Properties of γ-Glutamyl Derivatives

The amide bond in glutamine is much more labile towards treatment with strong acid than normal peptide bonds or the amide bond in asparagine [(78), p. 1933, (328)]. When the α-amino group in glutamine is protected, as for example in a peptide bond, the rate of acid hydrolysis is much lower, indicating participation of the α-amino group in the reaction (29, 222, 328). Similarly, the γ-linkage in glutathione is very acid-labile (103, 149), and the same is true for other γ-glutamyl derivatives. The lability toward acid has therefore often been used to distinguish between α- and γ-glutamyl linkages. As an exception, γ-glutamylphenyl-alanine was found to be stable for 7 days at room temperature in 2 N HCl (177). The rate constant for acid hydrolysis of the γ-glutamyl bond has been determined to be about four times larger than the rate constant for acid hydrolysis of α-glutamyl bonds (247).

In a detailed study of $N^6$-(γ-glutamyl)lysine (1) and $N^6$-(α-glutamyl)-lysine (2) it was again shown that the γ-derivative was more easily hydrolyzed. It was also demonstrated that both the α- and the γ-isomer gave the cyclic $N^6$-(α,γ-glutamyl)lysine (3) as an intermediate during the acid hydrolysis, and probably interconverted *via* this intermediate. The proportion of the cyclic intermediate and of the γ-isomer which could be formed from the α-isomer through partial hydrolysis was much larger than the proportion of the cyclic intermediate and the α-isomer formed from the γ-isomer under the same conditions. This was ascribed to the more facile complete hydrolysis of the γ-isomer. Ring-opening of the pure cyclic imide with acid was found to favour the α-derivative 2:1 over the γ-glutamyl derivative. This ratio was reversed when the cyclic imide was treated with base. α-Glutamylglycine and α-glutamylalanine when treated with strong acid yielded small amounts of the corresponding γ-glutamyl peptides, whereas the γ-glutamyl peptides of glycine and alanine were rapidly cleaved under identical conditions and no glutamyl rearrangement could be observed (*168*).

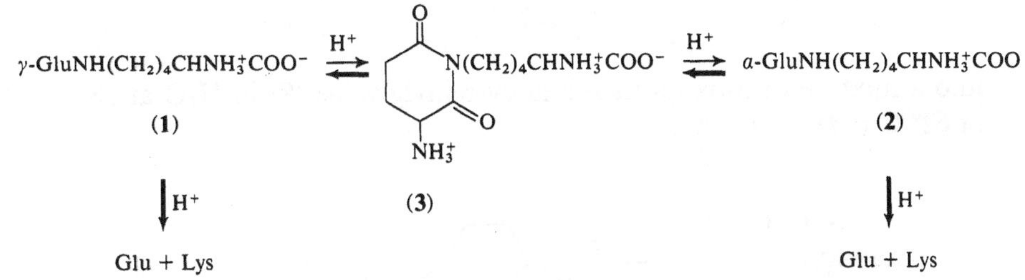

$$\gamma\text{-GluNH(CH}_2)_4\text{CHNH}_3^+\text{COO}^- \underset{}{\overset{H^+}{\rightleftarrows}} \quad\quad \underset{}{\overset{H^+}{\rightleftarrows}} \quad \alpha\text{-GluNH(CH}_2)_4\text{CHNH}_3^+\text{COO}$$

N(CH₂)₄CHNH₃⁺COO⁻

(1)                                    (3)                                    (2)

$\downarrow H^+$                                                                $\downarrow H^+$

Glu + Lys                                                        Glu + Lys

In strong base glutamine is hydrolyzed easily to give ammonia (*300*), but base hydrolysis has not been studied as thoroughly as acid hydrolysis [cf. (*78*), p. 1933].

Of greater interest, however, is the observation that glutamine is easily decomposed by heating in weak acid, neutral or alkaline solutions to give ammonia and 5-oxoproline (pyroglutamic acid, 2-pyrrolidone-5-carboxylic acid) (**4**) (*5, 22, 29*), [(*78*), p. 1933], (*82, 174, 250, 346, 380*). In strongly alkaline solution the equilibrium between glutamine and 5-oxoproline is shifted in favour of glutamate (*380*).

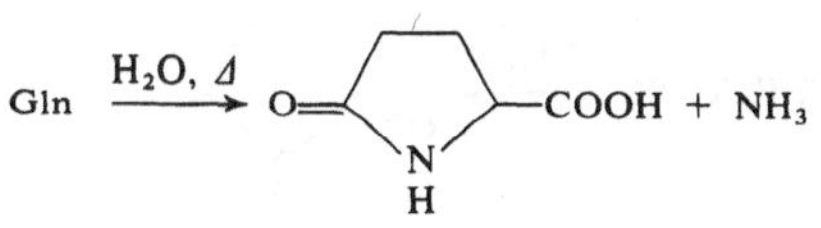

$$\text{Gln} \xrightarrow{\text{H}_2\text{O, } \Delta} \text{O}=\!\!\!\left\langle \begin{array}{c} \\ \text{N} \\ \text{H} \end{array}\right\rangle\!\!\!-\text{COOH} + \text{NH}_3$$

(**4**)

Again, the presence of the non-acylated α-amino group is necessary for the reaction, as demonstrated by the stability of $N^2$-chloroacetylglutamine (*70*). The reaction is catalyzed by both acid and base (*70*) and has been shown to follow pseudo-first-order kinetics between pH 3 and 6 (*1*).

Glutathione readily undergoes a similar degradation. Thus it has been shown that oxidized glutathione (**5**) is transformed on boiling with water to 5-oxoproline and the diketopiperazine "anhydrodiglycylcysteine" (**6**) (*103*), and that reduced glutathione (**7**) is transformed

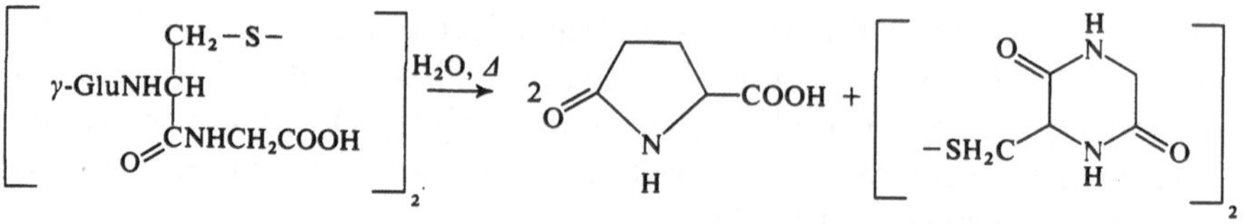

$$\left[\begin{array}{c} \text{CH}_2-\text{S}- \\ \gamma\text{-GluNHCH} \\ \underset{O}{\overset{}{\text{CNHCH}_2\text{COOH}}} \end{array}\right]_2 \xrightarrow{\text{H}_2\text{O},\,\varDelta} 2\; \underset{O}{\overset{}{\diagdown}}\!\!\diagup\!\!-\text{COOH} \;+\; \left[\begin{array}{c} \text{(diketopiperazine)} \\ -\text{SH}_2\text{C} \end{array}\right]_2$$

(5)                          (4)                          (6)

into a mixture of 5-oxoproline and cysteinylglycine (8) in $H_2O$ at 38° C or 62° C (*148, 149, 209*).

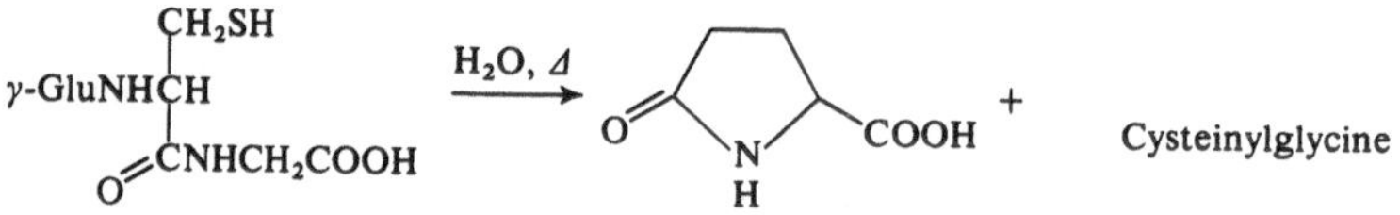

$$\begin{array}{c} \text{CH}_2\text{SH} \\ \gamma\text{-GluNHCH} \\ \underset{O}{\overset{}{\text{CNHCH}_2\text{COOH}}} \end{array} \xrightarrow{\text{H}_2\text{O},\,\varDelta} \underset{O}{\overset{}{\diagdown}}\!\!\diagup\!\!-\text{COOH} \;+\; \text{Cysteinylglycine}$$

(7)                          (4)                          (8)

Similarly it has been shown that γ-glutamylglycine, γ-glutamylglutamic acid and γ-glutamylaspartic acid can be transformed on heating in aqueous solution into 5-oxoproline and the free second amino acid from the peptide. On the other hand, α-glutamylglutamic acid and α-glutamyltyrosine were found to be transformed into the corresponding 5-oxoproline derivatives on prolonged heating in aqueous solutions.

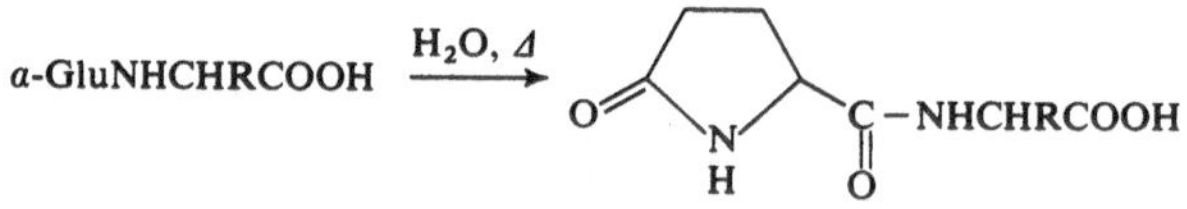

$$a\text{-GluNHCHRCOOH} \xrightarrow{\text{H}_2\text{O},\,\varDelta} \underset{O}{\overset{}{\diagdown}}\!\!\diagup\!\!\overset{}{\underset{O}{\text{C}}}-\text{NHCHRCOOH}$$

α-Glutamylglycine gave a mixture of the 5-oxoproline derivative, glycine, and 5-oxoproline, whereas only aspartic acid and 5-oxoproline could be identified in the reaction mixture obtained starting from γ-glutamyl-aspartic acid (*192*).

$N^6$-(γ-Glutamyl)lysine likewise yields 5-oxoproline and lysine when heated in $H_2O$ or glacial acetic acid, whereas $N^6$-(α-glutamyl)lysine yielded the corresponding 5-oxoproline derivative. By treatment with acid the latter could be opened to give $N^6$-(α-glutamyl)lysine again (*168*).

It can thus be concluded that all γ-glutamyl derivatives are readily cleaved in strong acid to give glutamic acid and the second constituent, and in neutral solution to give 2-pyrrolidone-5-carboxylic acid and the second constituent. γ-Glutamyl dipeptides can be cleaved enzymatically in neutral solution into the constituent amino acids. It has been shown that the equilibrium constants for these reactions in the direction of hydrolysis are much smaller than those for the reactions of normal dipeptides (87).

When treated with $HNO_2$, glutamine gives 2 molar equivalents of $N_2$ (29). The same is true for the reaction of $N^5$-methylglutamine, $N^5$-ethylglutamine (197), glutathione (103, 106), γ-glutamylalanine, γ-glutamylglutamic acid, and S-acetylglutathione (282). γ-Glutamyl-γ-glutamylglutamic acid and γ-glutamylglutamine give 3 molar equivalents of $N_2$ with $HNO_2$. The reaction is specific for γ-glutamyl derivatives and probably involves formation of the lactone of 2-hydroxyglutaric acid (282).

During the quantitative reaction with ninhydrin (344) one molar equivalent of $CO_2$ is produced by glutathione (235, 344), glutamine (82) and other γ-glutamyl peptides (281). α-Glutamyl-α-amino acids and glutamic acid-containing dipeptides with glutamic acid as C-terminal give no $CO_2$ with ninhydrin (281). The quantitative reaction with ninhydrin has been used in a number of cases to establish the presence of the γ-linkage.

A number of γ-glutamylpeptides have been subjected to dinitrophenylation with subsequent hydrolysis. This procedure gives dinitrophenylglutamic acid and thus indicates that glutamic acid is N-terminal. However, distinction between α- and γ-glutamyl linkages is not possible by this means. On the other hand, the behaviour of γ-glutamyl peptides towards Edman degradation is different from that of α-glutamyl peptides. Normally, the C-terminal amino acid or peptide is liberated in the cyclization step, together with the phenylthiohydantoin, by treatment with HCl at room temperature. However, with γ-glutamyl derivatives it is necessary to treat the phenylthiocarbamyl derivative with HCl at 100° C (66, 294).

The C-terminal of a peptide can be labelled with tritium by exchange. When peptides containing glutamic acid with γ-linkage are subjected to this treatment, not only the C-terminal but also glutamic acid becomes labelled. This labelling can be used as a sensitive indicator of the establishment of γ-glutamyl linkages (270), and has been employed in the determination of the structure of γ-glutamyl derivatives (66).

## 10. Isolation Procedures and Stability of γ-Glutamyl Derivatives

Nearly all isolations of γ-glutamyl derivatives have involved extraction of the naturally-occurring material with water, aqueous methanol or aqueous ethanol, followed by isolation of an "amino acid" fraction by adsorption of the amino acids and peptides to a strongly acid exchange resin and subsequent elution with aqueous ammonia or pyridine. For the acid γ-glutamyl derivatives, i. e. γ-glutamyl di- and tripeptides in which the other amino acids are neutral, the second step in the isolation is normally fractionation by ion-exchange chromatography on a strongly basic ion-exchange resin in the acetate form, using aqueous acetic acid as eluent (compare Section IV. 8). In both the ion-exchange procedure and the concentration steps involved there is some risk of degrading the γ-glutamyl derivatives. It has, however, been shown that glutamine can be recovered quantitatively after passage through a column of strongly acid ion-exchange resin provided that all operations are performed at 0—6° C over a period of no more than 4—6 hours (331).

During fractionation on a basic ion-exchange resin using aqueous acetic acid, transformation of γ-glutamyl derivatives into 5-oxoproline may occur and once again it is advisable to work at low temperature. It is also known that glutamic acid can be transformed to 5-oxo-proline by elution from a strongly acid ion-exchange resin using ammonia (382). When glutamic acid in aqueous ethanol is applied to the resin transformation to the γ-ethyl ester can also occur (262), and the γ-ethyl ester can presumably be transformed further into 5-oxoproline. When heated with amines, 5-oxoproline can give the corresponding $N^5$-substi-tuted glutamines (197, 198); the γ-ethyl ester of glutamic acid can also be transformed to glutamine with ammonia and to $N^5$-substituted glutamines with amines. However, treatment of the γ-ethyl ester of glutamic acid with ammonia or amines under varying conditions gives mainly 5-oxoproline (11). The possibility has to be considered that γ-glutamyl derivatives of amines and amino acids are artefacts produced from either 5-oxoproline or the γ-ethyl ester of glutamic acid during isolation. However, in experiments in which [14]C-labelled glutamic acid, glutamine, 5-oxoproline or γ-glutamylalanine (the last mentioned being labelled in the glutamic acid moiety or the alanine moiety in different experiments) in a standard mixture of amino acids or in a plant extract were subjected to the ion-exchange procedure, it was shown that no γ-glutamyl derivatives were formed. The experiments showed that about 5% of glutamine was cyclized to 5-oxoproline when the amino acids were applied to the acid ion-exchange resin either in water or in aqueous ethanol. γ-Glutamylalanine was not degraded. When glutamic acid was applied in aqueous ethanol solution about 9% was converted into 5-oxo-proline, but no loss of glutamic acid took place when it was applied in water (128). It was also shown in a number of experiments that γ-glutamyl di- and tripeptides could not be formed by a chemical transpeptidation occurring during the isolation procedure (176). On the other hand, formation of γ-glutamyl peptides by enzymatic transpeptidation reactions, also in fermentation broths, is possible (cf. Section VII. 3).

## 11. Establishment of the γ-Linkage

In the preceding sections the problem of distinguishing between γ-glutamyl derivatives, α-glutamyl derivatives and peptides with glutamic acid as C-terminal has been discussed at length, and a number of methods for doing so have been described. In conclusion it may be mentioned that unambiguous chemical distinction can be achieved by means of quantitative reaction with ninhydrin or $HNO_2$, or by synthesis (see Section VI). Titration shifts in PMR-spectroscopy and mass spectroscopy, as well as the tritium-exchange method described in Section IV. 9, also provide clear results.

A number of the methods which depend on pK-values, including paper electrophoresis and ion-exchange chromatography, provide clear results for the derivatives of neutral α-amino acids. For derivatives of β- and γ-amino acids and amines this is not necessarily the case. The lability towards acid is also indicative, but it is not easy to exploit it in quantitative measurements.

It can be said generally that differentiation is straightforward when both the α- and the γ-isomers are available. When only one isomer is available, definite proof of structure can be provided only by one of the unambiguous methods mentioned above. A review concerned with identification of γ-glutamyl derivatives has been published (247).

## V. Isolation and Structure Determination of Individual Compounds

Structure determination of most of the γ-glutamyl derivatives listed in Tables 1—6 has been trivial. It involved establishment of the γ-glutamyl linkage and hydrolysis followed by paper chromatographic identification, by isolation of the constituents and, in numerous cases, unambiguous synthesis (see Section VI). With some of the γ-glutamyl derivatives of non-protein amino acids elucidation of the structure presents special problems. When the non-protein amino acids also occur free, their structures have normally been elucidated independently of the structure determinations of the γ-glutamyl derivatives. These structure determinations of non-protein amino acids will not be dealt with here as they are not considered to be relevant to a discussion specifically concerned with γ-glutamyl derivatives.

In the case of a few γ-glutamyl derivatives of non-protein amino acids the parent amino acid has not been found in natural material. Furthermore, with a number of the γ-glutamyl derivatives of amines specific problems have been encountered. In these cases a detailed individual discussion is of interest.

## 1. γ-L-Glutamyl-β-cyano-L-alanine

2 g of γ-L-glutamyl-β-L-cyanoalanine were isolated from 540 g of seeds of *Vicia sativa* by ion-exchange chromatography. The structure was established by elementary analysis, by acid hydrolysis to give equimolar amounts of ammonia, aspartic acid and glutamic acid, by treatment of the dicyclohexylammonium salt with sodium in ammonia containing methanol followed by hydrolysis to give glutamic acid and 2,4-diamino-butanoic acid, and by end-group determination using dinitrofluoro-benzene. Confirmation of the structure and establishment of the L-configurations was achieved by synthesis (*275, 276*). The same structure, although without determination of configuration, was proposed inde-pendently on the basis of similar evidence (*339*).

## 2. Linatine (1-[(N-γ-L-Glutamyl)amino]-D-proline)

Linatine (**9**), the $B_6$ antagonist from flaxseed (see Section VIII. 2), was isolated by standard ion-exchange fractionations and was subjected to a final purification on a silica gel column. 20 mg of amorphous material, but giving the correct elemental analysis, were obtained from 2.5 kg of seeds. Structure elucidation was based on hydrolysis to give L-glutamic acid and 1-aminoproline (**10**), hydrogenolysis to give proline and glutamine, end-group determination with dinitrofluorobenzene, quantitative reaction with ninhydrin to establish the γ-linkage, and reaction of 1-aminoproline with $HNO_2$ to give $CO_2$ and NO and with

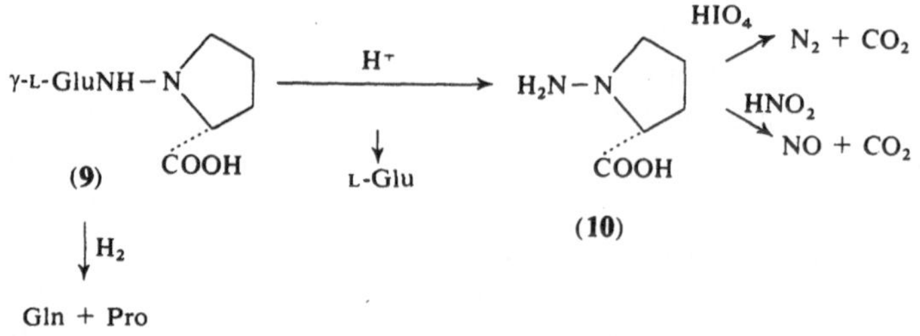

periodic acid to give $CO_2$ and $N_2$. The D-configuration in the 1-amino-proline moiety was established by synthesis (see Section VI. 2) (*158*).

## 3. γ-Glutamylvalylglutamic Acid

This tripeptide was obtained by standard ion-exchange operations and chromatography on cellulose columns, 160 mg being obtained from 5 kg of fresh plant material. Mild acid hydrolysis gave 1 molar equivalent of glutamic acid and a dipeptide which, when again subjected to vigorous hydrolysis, gave equivalent amounts of glutamic acid and valine. End-group determination on the dipeptide using dinitrofluorobenzene showed valine to be N-terminal. The structure was further supported by molecular weight determination and by N-analysis of both the tripeptide and dipeptide (350).

## 4. γ-L-Glutamyl-L-phenylalanyl-L-willardiine

35 mg of the crystalline tripeptide was obtained from 2.3 kg of seeds of *Fagus silvatica,* in which it occurred together with numerous γ-glutamylamino acids. The structure was established by UV- and PMR-spectroscopy, by partial hydrolysis to give γ-glutamylphenylalanine and phenylalanylwillardiine, and by end-group determination with dinitro-fluorobenzene (176). The L-configurations were established by synthesis (309).

## 5. γ-L-Glutamyl-γ-L-glutamyl-L-phenylalanine

35 mg of the crystalline tripeptide was obtained from 330 g of seeds of *Fagus silvatica* var. *purpurea.* The structure was established by PMR-spectroscopy, by partial hydrolysis to give γ-glutamylglutamic acid and γ-glutamylphenylalanine, and by synthesis (129).

## 6. γ-Glutamylmarasmine (γ-L-Glutamyl-3-(methylthiomethylsulphinyl)-L-alanine)

Crystalline γ-glutamylmarasmine (11) was isolated by standard ion-exchange operations from dried carpophores of *Marasmius* species. The yield was about 1% of the weight of the starting material. Acid hydro-lysis gave glutamic acid which was identified by paper and thin-layer chromatography. Reductive cleavage with Raney-Ni gave γ-L-glutamyl-L-alanine, identified by comparison of its specific rotation and IR spectrum with those of authentic material. A slightly acidic solution of the dipeptide slowly decomposed to give γ-glutamyl-2-aminopropenoic acid (12) and presumably the sulphenic acid $CH_3-S-CH_2-SOH$ (13). The complete structure was established by PMR-spectroscopy at 270 MHz and by $^{13}$C-NMR spectroscopy at 20 MHz. The two methylene protons on the C-atom between the two sulphur atoms had δ-values differing by about 0.2 ppm. This difference must be due mainly to the proximity of the chiral sulphoxide group. Mass spectroscopy (field desorption) confirmed the molecular weight. The structure proposed was also in agreement with the results of enzymatic degradations (see Section VII. 6.3.) (72).

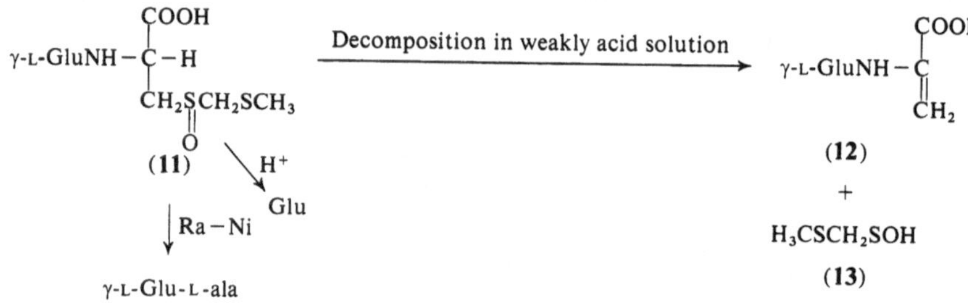

## 7. Lentinic Acid (2-γ-L-Glutamylamino-4,6,8,10,10-pentaoxo-4,6,8,10-tetrathiaundecanoic Acid)

Lentinic acid (14) was originally isolated from *Lentinus edodes*, and was established to have the elemental composition $C_{12}H_{22}N_2O_{10}S_4$ and to be a derivative of γ-L-glutamyl-L-cysteine sulphoxide. Elucidation of the structure of the $C_4H_9S_3O_4$-fragment by PMR spectroscopy was difficult, since the protons in the latter exchanged rapidly with deuterium (384). Lentinic acid was subsequently isolated from fruit bodies of *Micromphale perforans* and *Collybia hariolorum* by the same ion-exchange operations as used for γ-glutamylmarasmic acid. The yield was about 1% of the dry-weight of the starting material. Acid hydrolysis produced L-glutamic acid and reductive cleavage with Raney-Ni gave γ-L-glutamyl-L-alanine. The structural assignment was based on $^{13}$C-NMR- and PMR-spectroscopy. Treatment of lentinic acid with hexafluoroacetone resulted in transformation at the γ-glutamyl residue to give a 5-oxazolidinone derivative (15) which could be used for measurement of the PMR-spectrum in deuterated dimethylsulphoxide at 270 MHz. Again, the protons in each of the methylene groups proved to be non-equivalent. Mass spectroscopy (field desorption) confirmed the molecular weight. No information is available on the configuration at the chiral sulphoxide groups. The PMR-spectra of both the isolates from *Micromphale per-*

*forans* and *Collybia hariolurum* and the original isolate from *Lentinus edodes* were consistent with the presence of only one stereoisomer. However, a new isolate from *Lentinus edodes* contained 30% of a dia-stereoisomer. The structure proposed was also consistent with the results of enzymatic degradations (see Section VII. 6.3) (*100*).

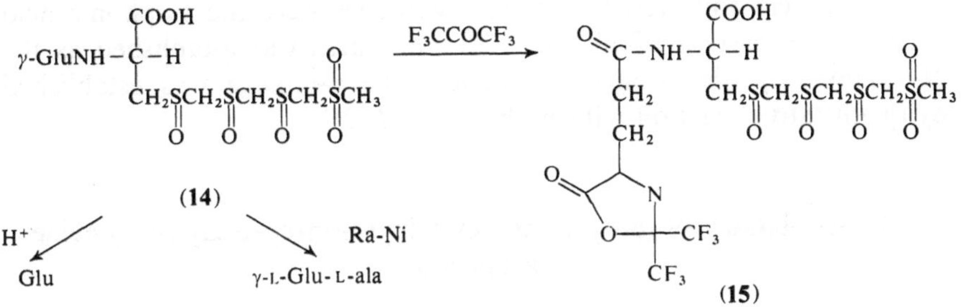

8. N,N-bis-(γ-Glutamyl)-3,3′-(1-methylethylene-1,2-dithio)-dialanine

590 mg of N,N-bis-(γ-glutamyl)-3,3′-(1-methylethylene-1,2-dithio)-di-alanine (16) was isolated from 1 kg of seeds of *Allium schoenoprasum*, using standard ion-exchange procedures and final purification on a cellu-lose column. Acid hydrolysis gave L-glutamic acid and 3,3′-(1-methyl-ethylene-1,2-dithio)-dialanine (17). The latter amino acid gave alanine on

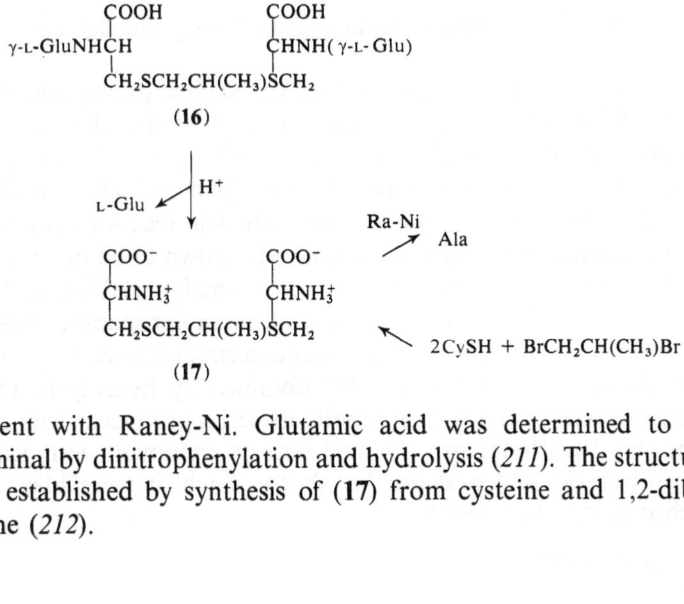

treatment with Raney-Ni. Glutamic acid was determined to be the N-terminal by dinitrophenylation and hydrolysis (*211*). The structure was finally established by synthesis of (17) from cysteine and 1,2-dibromo-propane (*212*).

## 9. γ-L-Glutamyl-L-cysteinyl-β-alanine

γ-L-Glutamyl-L-cysteinyl-β-alanine was isolated from seedlings of *Phaseolus aureus* by standard ion-exchange procedures, 229 mg of the pure crystalline disulphide being obtained from 61 kg of seedlings. Acid hydrolysis gave glutamic acid, cysteine, and β-alanine. Oxidation with performic acid followed by partial hydrolysis gave the sulphonic acid derivative of cysteinyl-β-alanine. Glutamic acid was established as the N-terminal by dinitrophenylation, and the γ-linkage was established by quantitative reaction with ninhydrin (*24, 25*).

## 10. γ-L-Glutamyl-S-(prop-1-enyl)-cysteinyl-S-(prop-1-enyl)-L-cysteine sulphoxide

This compound was isolated from seeds of *Allium schoenoprasum.* The structure was established by enzymatic hydrolysis to give L-glutamic acid, S-(prop-1-enyl)-cysteine and S-(prop-1-enyl)-cysteine sulphoxide, determination of the N-terminal by dinitrophenylation and determination of the C-terminal by enzymatic treatment with carboxypeptidase A, which only hydrolyzes normal peptide bonds and therefore gave rise to S-(prop-1-enyl)-cysteine sulphoxide and γ-glutamyl-S-(prop-1-enyl)-cysteine. The L-configuration of the S-(prop-1-enyl)-cysteine sulphoxide was established by enzymatic analysis (*215*).

## 11. γ-L-Glutamyl-S-(2-carboxypropyl)-L-cysteinylglycine

251 mg of crystalline γ-glutamyl-S-(2-carboxypropyl)-cysteinylglycine (**18**) was isolated from 2 kg of *Allium cepa* by ion-exchange followed by final purification on a cellulose column. Acid hydrolysis gave glutamic acid, S-(2-carboxypropyl)-cysteine (**19**) and glycine, which were identified by paper chromatography. Quantitative ninhydrin reaction produced one molar equivalent of $CO_2$. Glutamic acid was shown to be the N-terminal by dinitrophenylation. The structure was finally established by synthesis from glutathione and 3-bromo-2-methylpropanoic acid. The configuration of the cysteine moiety was determined to be L by measurement of the specific rotation of (**19**) obtained by hydrolysis. (**19**) was identified by comparison with synthetic material produced from cysteine and 3-bromo-2-methylpropanoic acid and by treatment with Raney-Ni to give isobutyric acid and alanine, both of which were identified by paper chromatography (*354*).

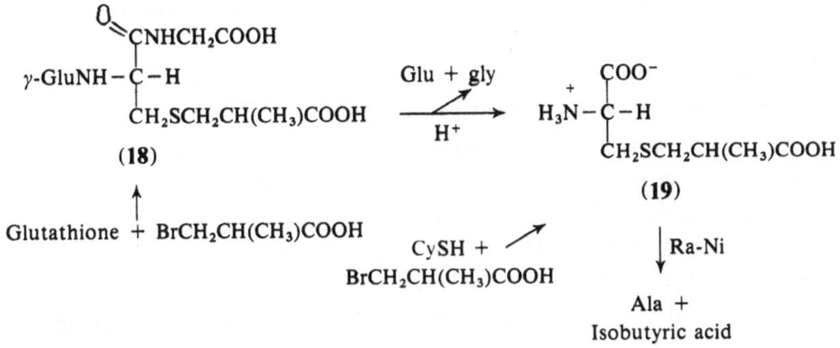

$$
\begin{array}{c}
\underset{\displaystyle \text{(18)}}{
\underset{\displaystyle \text{CH}_2\text{SCH}_2\text{CH(CH}_3)\text{COOH}}{
\gamma\text{-GluNH}-\overset{\displaystyle \overset{O}{\diagdown}\text{CNHCH}_2\text{COOH}}{\underset{|}{\text{C}}}-\text{H}}}
\quad \xrightarrow[\;H^+\;]{\text{Glu} + \text{gly}} \quad
\underset{\displaystyle \text{(19)}}{
\underset{\displaystyle \text{CH}_2\text{SCH}_2\text{CH(CH}_3)\text{COOH}}{
\underset{|}{\overset{+}{\text{H}_3\text{N}}}-\overset{\displaystyle \text{COO}^-}{\underset{|}{\text{C}}}-\text{H}}}
\end{array}
$$

Glutathione + BrCH$_2$CH(CH$_3$)COOH

CySH + $\nearrow$
BrCH$_2$CH(CH$_3$)COOH

Ra-Ni

Ala +
Isobutyric acid

## 12. γ-Glutamyl-γ-glutamylmethionine

γ-Glutamyl-γ-glutamylmethionine was obtained from seeds of *Phaseolus radiatus* var. *typicus* by ion-exchange and final purification on a cellulose column. Acid hydrolysis produced glutamic acid and methionine in the ratio 2 : 1. Mild hydrolysis gave methionine, γ-glutamylmethionine, and glutamic acid. The structure was finally established by mass spectroscopy of the triethylester and by comparison of the mass spectrum and the PMR-spectrum with those of synthetic material (*138*).

## 13. N⁵-(1-Methyl-2-oxopentyl)-L-glutamine

118 mg of N⁵-(1-methyl-2-oxopentyl)-L-glutamine was isolated from 2.6 kg of *Russula ochroleuca* by ion-exchange fractionation and chromatography on a carbon column. The structure was established by elementary analysis, ¹³C- and PMR-spectroscopy and hydrolysis to give glutamic acid (no rotation given) and the unstable ketoamine. CD data were interpreted in support of the L-configuration of the glutamic acid. The presence of the γ-linkage is consistent with the titration shifts observed in the PMR-spectra (*376*).

## 14. N⁵-(2-Cyanoethyl)-L-glutamine

N⁵-(2-Cyanoethyl)-L-glutamine was isolated from seeds of *Lathyrus odoratus* (*203*). The structure was established by elemental analysis, acid hydrolysis with the subsequent isolation of L-glutamic acid and β-alanine and by synthesis. The γ-linkage was inferred from the acid lability (*299*).

## 15. Coprine (N⁵-(1-Hydroxycyclopropyl)-L-glutamine)

The isolation of coprine (**20**) and its structure determination was reported recently by two different groups (*94, 201, 202*). In one large-scale isolation 7.3 g of crystalline coprine was obtained from 62.5 kg of *Coprinus atramentarius* by ion-exchange chromatography (*202*). The second isolation described gave 12 mg of coprine as an amorphous solid from 10 g of the mushrooms (*94*). The structure was established by elementary analysis, PMR-spectroscopy at 100 and 220 MHz, $^{13}$C-NMR-spectroscopy and mass spectroscopy, and by various degradation reactions.

Acid hydrolysis gave L-glutamic acid and the 1-hydroxycyclopropyl-ammonium ion. The later ion is relatively stable, whereas the corresponding free amine is unstable. Propionic acid could also be identified in the acid hydrolysis mixture. Treatment with carbonate buffer at pH 10 gave 5-oxo-L-proline and propionamide. Catalytic hydrogenation gave N⁵-isopropyl-L-glutamine, L-glutamine and acetone.

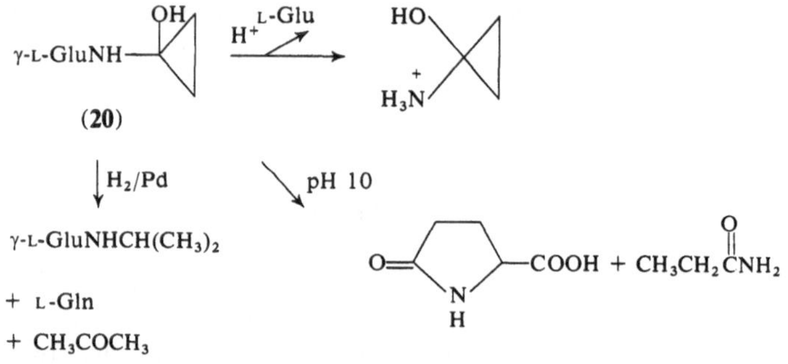

The γ-linkage was established by the formation of glutamine and N⁵-isopropylglutamine, and the structure was confirmed by synthesis (*202*).

## 16. N⁵-(3,4-Dioxocyclohexa-1,5-dien-1-yl)-L-glutamine

This compound is unstable and has not been isolated in a pure state directly from naturally-occurring material. The structure determination was carried out on material produced by enzymatic oxidation of N⁵-(4-hydroxyphenyl)-L-glutamine and was based on elementary analysis, UV-spectroscopy, IR-spectroscopy of two acetylated derivatives, and acid hydrolysis to give glutamic acid (*369*).

## 17. Agaritine (N$^5$-[(4-Hydroxymethylphenyl)amino]-L-glutamine)

Two different procedures for isolation of agaritine (21) have been described (*34, 147, 193*), the first giving a yield of 350 mg of crystalline material from 4.5 kg of mushrooms. The structure was established by elementary analysis, UV- and IR-spectroscopy, hydrolysis to give L-glutamic acid (L-configuration determined enzymatically), and enzymatic hydrolysis followed by oxidation of the liberated hydrazine with SeO$_2$ to give a diazonium ion which could itself be transformed into p-hydroxybenzyl alcohol (*193*). The structure was finally confirmed by synthesis. Agaritine could be degraded to glutamic acid, nitrogen, and benzyl alcohol using FeCl$_3$ (*147*).

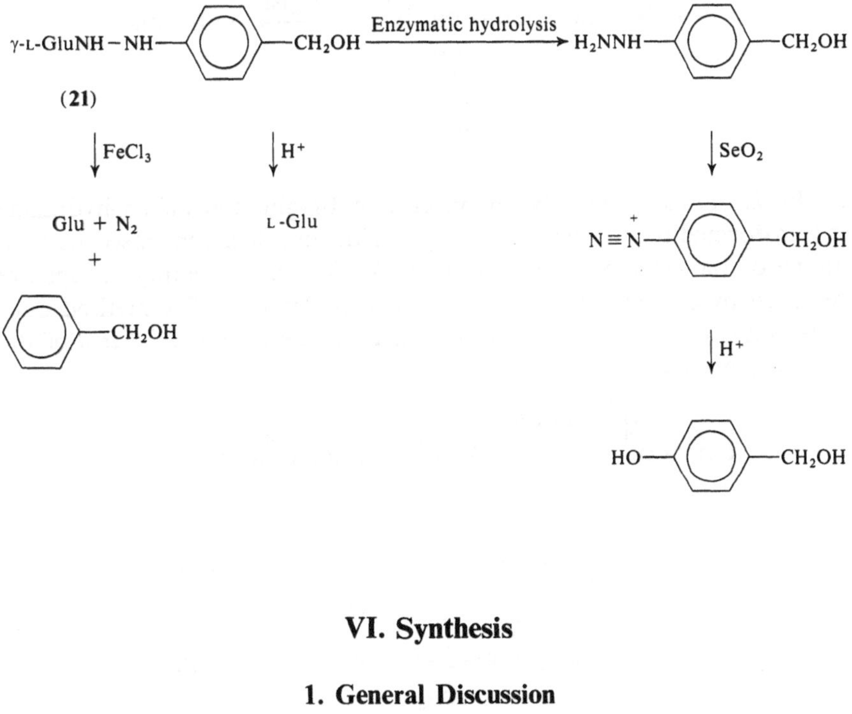

## VI. Synthesis

### 1. General Discussion

Several reviews are available on the synthesis of peptides containing γ-glutamyl linkages [see for example Ref. (*78*), p. 1093, (*117, 313, 383*)]. Whereas considerable ingenuity has been used in developing new syntheses of γ-glutamyl derivatives, nearly all syntheses of the compounds described in the present review employed one of five standard methods, all of which have proven dependable even in the hands of workers who are not specially trained in peptide synthesis.

In *the first method* benzyloxycarbonyl-L-glutamic acid α-benzyl ester (or another ester) is condensed with an amine or the ester of an amino acid, e.g. by use of ethyl chloroformate. The protecting groups are subsequently removed by catalytic hydrogenation (and in the case of e.g. an ethyl ester, by hydrolysis).

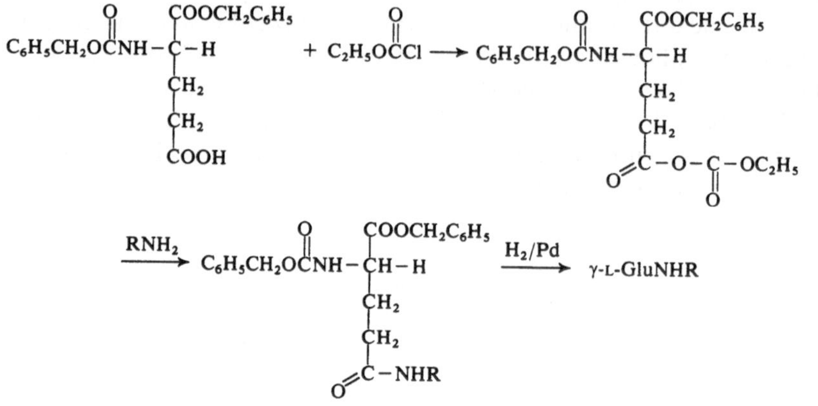

In *the second method* benzyloxycarbonyl-L-glutamic acid γ-hydrazide is transformed to the corresponding γ-azide and then condensed with an amine or the ester of an amino acid. Again, the protecting groups are removed by catalytic hydrogenation (and hydrolysis). The method does not always give exclusively the γ-isomer but sometimes mixtures of the α- and γ-isomers.

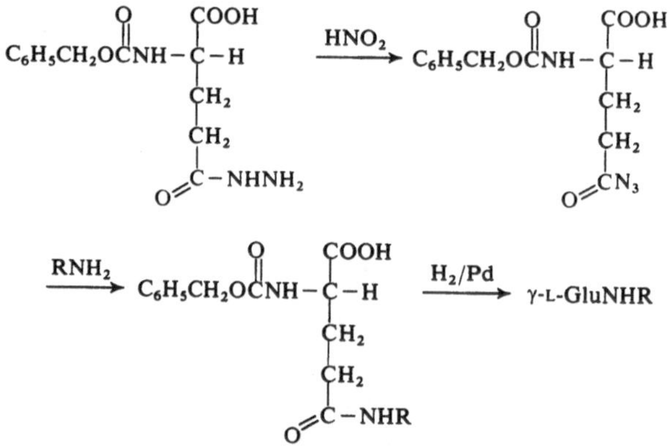

In *the third method* benzyloxycarbonyl-L-glutamic anhydride is condensed with an amine, amino acid or amino acid ester with subsequent removal of the benzyloxycarbonyl group by hydrogenation. This method gives a mixture of the α- and γ-isomers, which are easily separated by ion-exchange chromatography. Since so many γ-glutamyl derivatives of rare non-protein amino acids are known, and since these amino acids are often available only in small amounts, the method is advantageous since the second amino acid can be used without derivatization.

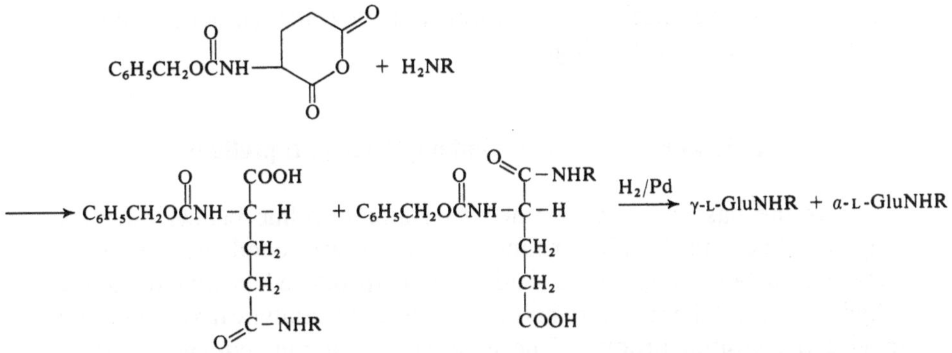

In *the fourth method* phthaloyl-L-glutamic anhydride is condensed with an amine, an amino acid or an amino acid ester with subsequent removal of the phthaloyl group by hydrazinolysis. This method gives exclusively — or mainly the γ-derivative (but compare Section VI. 4.).

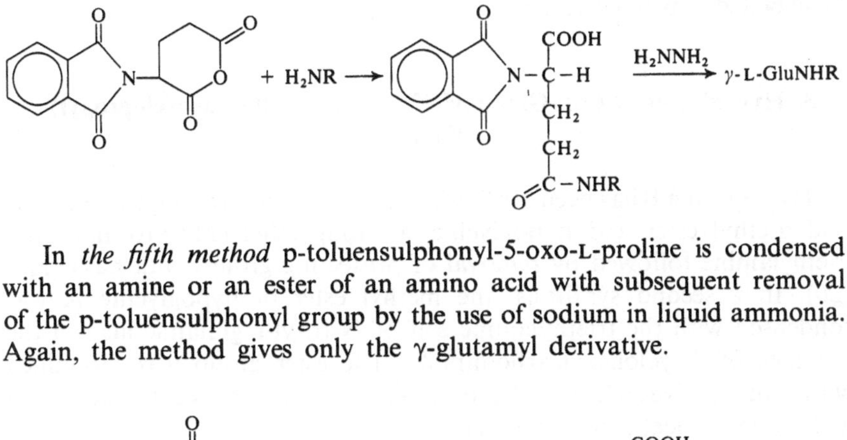

In *the fifth method* p-toluensulphonyl-5-oxo-L-proline is condensed with an amine or an ester of an amino acid with subsequent removal of the p-toluensulphonyl group by the use of sodium in liquid ammonia. Again, the method gives only the γ-glutamyl derivative.

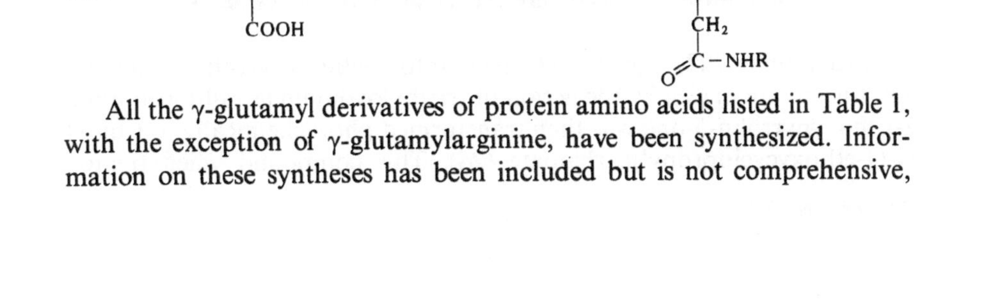

All the γ-glutamyl derivatives of protein amino acids listed in Table 1, with the exception of γ-glutamylarginine, have been synthesized. Information on these syntheses has been included but is not comprehensive, references being given to only one or two efficient syntheses for each compound. For the remaining γ-glutamyl derivatives, information on chemical syntheses is included in Tables 2—6. In a few cases not all the syntheses of a particular compound are included. The Tables also give information as which of the five standard methods listed above have been used in the synthesis; therefore only a few specific synthetic problems are discussed in the following.

## 2. Linatine (1-[(N-γ-L-Glutamyl)amino]-D-proline)

D-Proline has been transformed to 1-aminoproline benzyl ester *via* 1-nitroproline and 1-aminoproline. The ester was condensed with N-benzyloxycarbonyl-L-glutamic acid γ-azide (produced *in situ* from the γ-hydrazide) and the reaction mixture subjected to catalytic hydrogenation to give the desired product. The analogous reaction sequence starting with L-proline was also carried out (*158*). Another synthesis has also been reported, based on condensation of N-trifluoroacetyl-L-glutamic acid α-ethyl ester and D-proline benzyl ester with dicyclohexylcarbodiimide followed by removal of protecting groups. The method was also used for synthesis of the corresponding derivatives of L-proline and of [14]C-labelled D-proline (*245*).

## 3. Hypoglycine B (γ-L-Glutamyl-(2 S,1′R)-3-methylenecyclopropyl)-alanine

Hypoglycine B has been synthesized from N-trifluoroacetyl-L-glutamic acid α-ethyl ester and hypoglycine A, using ethyl chloroformate and triethylamine followed by removal of protecting groups with base (*119, 120*). In a second synthesis, the methyl ester of hypoglycine A was condensed with the triethylamine salt of N-trityl-L-glutamic acid in the presence of dicyclohexylcarbodiimide. The ester group was saponified with aqueous NaOH, and the trityl group was removed by treatment with aqueous acetic acid (*90, 91*).

## 4. Coprine (N⁵-(1-hydroxycyclopropyl)-L-glutamine)

Succinimide (**22**) was transformed into 2-ethoxypyrrolin-5-one (O-ethylsuccinimide) (**23**) by the action of triethyloxonium tetrafluoroborate. 2-Ethoxypyrrolin-5-one was then transformed photochemically to t-butyl N-(1-ethoxycyclopropyl)carbamate (**24**). This compound when treated with aqueous hydrochloric acid gave the crystalline 1-hydroxycyclo-propylammonium hydrochloride (**25**). On condensation with N-phthaloyl-L-glutamic anhydride in tetrahydrofuran in the presence of triethyl-amine, followed by hydrazinolysis, the hydrochloride gave coprine (**26**). The amount of isocoprine ($N^1$-(1-hydroxycyclopropyl)-L-isoglutamine) formed was normally negligible. However, in a large-scale synthesis of coprine (250 g) a lesser amount of isocoprine (13 g) was isolated. Syntheses of a number of structurally related compounds have also been described (*202*).

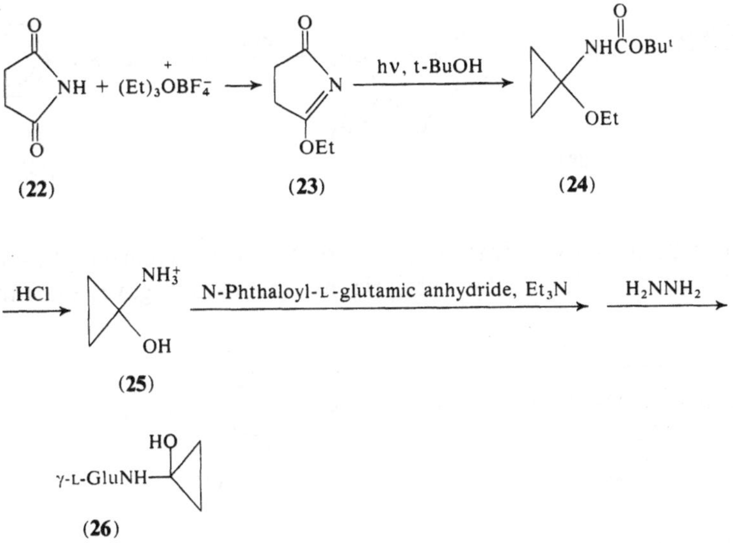

$$\text{(22)} \quad \text{NH} + (\text{Et})_3\text{OBF}_4^- \longrightarrow \text{(23)} \xrightarrow{\ h\nu,\ t\text{-BuOH}\ } \text{(24)}$$

(**22**)                (**23**)                (**24**)

$$\xrightarrow{\text{HCl}} \text{(25)} \xrightarrow{\text{N-Phthaloyl-L-glutamic anhydride, Et}_3\text{N}} \xrightarrow{\text{H}_2\text{NNH}_2}$$

(**25**)

γ-L-GluNH—

(**26**)

## 5. Agaridoxin ($N^5$-(3,4-Dihydroxyphenyl)-L-glutamine)

Isopropylidene-3,4-dioxyaniline (**27**) was condensed with N-phthaloyl-L-glutamic anhydride to give N-phthaloyl-(γ-L-glutamyl)-3,4-(isopropyl-idenedioxy)-anilide (**28**). Hydrazine treatment gave (γ-L-glutamyl-3,4-(isopropylidenedioxy)-anilide (**29**), which again on treatment with $BCl_3$ in $CH_2Cl_2$ gave agaridoxin (**30**) (*323*).

## 6. Agaritine ($N^5$-[(4-Hydroxymethylphenyl)amino]-L-glutamine)

Because of the instability of p-hydroxymethylphenylhydrazine (**31**) all steps in the synthesis had to be carried out under neutral con-ditions. Condensation of the hydrazine, produced by $LiAlH_4$-reduction

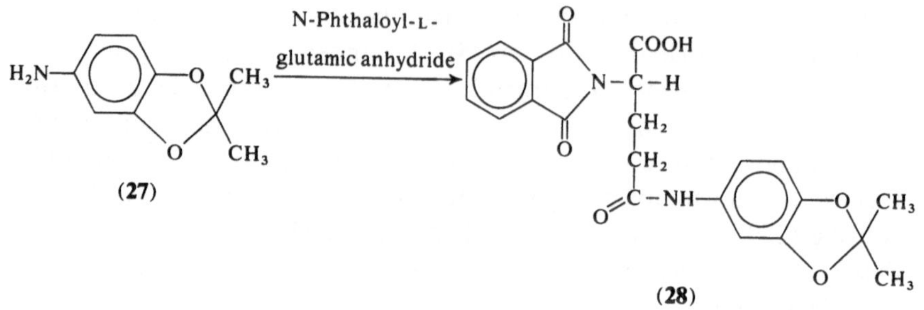

(27)

(28)

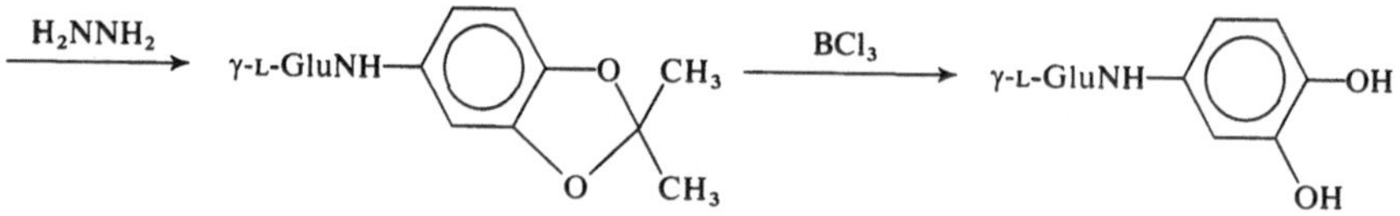

(29)

(30)

of p-carboxyphenylhydrazine (32), with N-benzyloxycarbonyl-L-glutamic acid γ-azide followed by catalytic hydrogenation gave agaritine (33) (*147*).

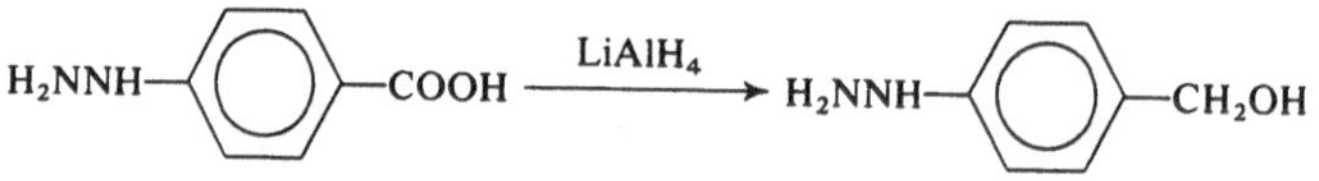

(32)

(31)

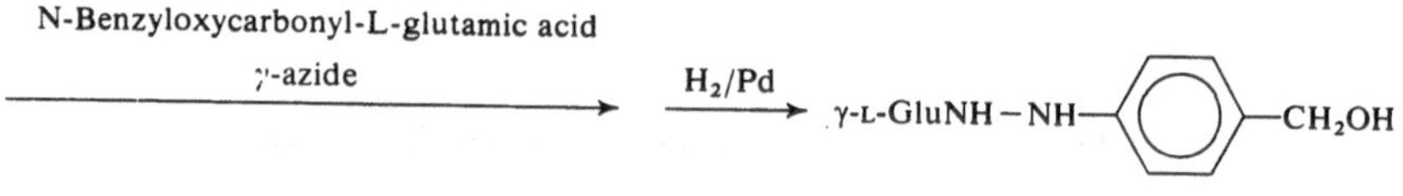

N-Benzyloxycarbonyl-L-glutamic acid

γ-azide

$\xrightarrow{\quad H_2/Pd \quad}$

(33)

# VII. Biosynthesis and Transformations of γ-Glutamyl Derivatives in Plants

γ-Glutamyl derivatives are found widely distributed in higher plants and mushrooms but their occurrence is not haphazard (*c. f.* Section III, 5); the biosynthesis and transformation of γ-glutamyl derivatives must therefore be enzymatically controlled. The occurrence of a particular free amino acid is by no means always indicative of the occurrence of the corresponding γ-glutamyl derivative, and in many cases there seems to be no simple correlation between the concentrations of free amino acids and those of the γ-glutamyl derivatives. Free amino acids in plants are presumably present in various metabolic pools. It is therefore conceivable that the occurrence of an amino acid in a specific pool results in the occurrence of the corresponding γ-glutamyl derivative. However, this would require the existence of some very specific pools and therefore specific systems for handling individual amino acids. Specific and controlled systems must therefore be responsible for the production of the γ-glutamyl derivatives.

On the other hand, the large number — and wide distribution — of γ-glutamyl derivatives may indicate that some widely distributed general biosynthetic systems are operative, and γ-glutamyl derivatives must therefore be considered in relation to the biosynthesis of glutathione and glutamine. Furthermore, the role of transpeptidases in the biosynthesis and degradation of γ-glutamyl derivatives, as well as the enzymatic hydrolysis of the γ-glutamyl linkage must be considered. In connection with the discussion of these general processes it is relevant to mention the proposed role of γ-glutamyl amino acids in the transport of amino acids across cell membranes in animals and microorganisms and to consider the possibilities for similar processes in plants.

In a few instances detailed information on the biosynthesis and transformation of individual γ-glutamyl derivatives is available. This material is discussed separately after the discussion of processes of a more general nature.

## 1. Glutathione in Plants and Its Relationship to Other Plant γ-Glutamyl Derivatives

### 1.1. Occurrence and Isolation of Glutathione

Glutathione is generally believed to be a constituent of all living organisms including plants (220, 221, 364). It has only seldom been recorded as a plant constituent, possibly because of its occurrence in low concentrations and difficulties in isolation (glutathione is eluted very late with acetic acid from a strongly basic ion-exchange resin in the acetate form (177)). The presence of glutathione in *Helianthus annuus, Alocasia odora* (arum), and *Zea mays* has been reported, although on weak evidence (23). Glutathione has been isolated from potato tubers, and it has been shown that a general increase in the content of glutathione in the tissues of the tuber is associated with sprouting (80, 259). Glutathione has been isolated from wheat germ (316), from seeds of *Phaseolus vulgaris* (104), from broad bean and vegetable marrow (105), from peanuts (272), from orange juice (116), from latex of *Hevea brasiliensis* (204) and from seeds of *Fagus silvaticus* (177). Glutathione is present in the cells and the medium of *Nicotiana tabacum* grown in tissue culture (*14*). It is also present in large amounts in wheat germ (*179*); recent Japanese patents describe the large scale isolation of glutathione from embryo buds of wheat, rice, and corn (*98, 99*).

A number of studies on glutathione reductase indicate the occurrence of glutathione in various plants (*3, 81, 96, 206*). Only scant information is available on the concentration levels in plants, but it has been shown that there is a rapid increase in the level of reduced glutathione in *Phaseolus vulgaris* during germination (*104*). This may, however, partly be due to conversion of the oxidized form to the reduced form, rather than to an active synthesis of the peptide (*311*). Similar marked concentration increases have been reported during the early stages of development of barley seedlings (*389*) and, as mentioned above, during the sprouting of potato tubers (*80, 259*). The seasonal variation of glutathione concentration in needles of *Picea abies* has also been studied (*48*). In the few instances where the determination or isolation of glutathione together with other γ-glutamic derivatives has been reported the levels of glutathione were relatively low. Mung bean seedlings, for example, contained much larger amounts of γ-glutamylcysteinyl-β-alanine (homoglutathione) than of glutathione (*24*), and only 5 mg of the latter were obtained together with 3 g of γ-glutamylphenylalanine from 2.3 kg of *Fagus silvatica* seeds (*177*).

## 1.2. Biosynthesis of γ-Glutamylcysteine and Glutathione

As described in a number of recent reviews (*219, 220, 221*) the biosynthesis of glutathione in animals and microorganisms takes place by the following two reactions:

$$\text{L-Glutamate} + \text{L-cysteine} + \text{ATP} \overset{M^{2+}}{\rightleftarrows} \text{γ-L-Glutamyl-L-cysteine} + \text{ADP} + P_i$$

$$\text{γ-L-Glutamyl-L-cysteine} + \text{ATP} + \text{glycine} \overset{M^{2+}}{\rightleftarrows} \text{Glutathione} + \text{ADP} + P_i$$

γ-Glutamylcysteine synthetase has been isolated from various sources, the most purified preparations described being obtained from rat kidney. The enzyme is specific for L-glutamate, although some activity was observed with some substituted glutamic acids. L-Cysteine and α-L-aminobutyric acid are about equally active as acceptor amino acids; substantial activity was also obtained with S-methyl-L-cysteine, DL-2-aminopent-4-enoic acid, β-chloro-L-alanine, 2-aminopentanoic acid, DL-homocysteine, L-threonine, and L-alanine when the reaction was carried out in the presence of magnesium ions. Substitution of manganese ions for magnesium ions leads to lower specificity for the acceptor amino acids (*220, 254, 303*). Similar specificity is shown by the enzyme isolated from bovine lens (*271*).

Partial purification of γ-glutamylcysteine synthetase from bean seedlings (*372, 373*) and wheat germ (*374*) has been achieved. The general properties of these preparations are similar to those of the enzyme from animal sources, but no detailed specificity studies have been carried out. Ammonia or hydroxylamine could be used as substrates instead of cysteine but this may be caused by the presence of glutamine synthetase in the preparation.

Like γ-glutamylcysteine synthetase glutathione synthetase has been isolated from various sources, the most active preparation being obtained from yeast (*220, 225*). The enzyme from this source was capable of catalyzing the formation of γ-glutamyl-α-aminobutyrylglycine (opthalmic acid) and γ-glutamylalanylglycine (noropthalmic acid) from the corresponding γ-glutamyl amino acids; no activity towards the γ-glutamyl derivatives of glycine, leucine, and β-alanine as substrates instead of γ-glutamyl-α-aminobutyric acid or γ-glutamylalanine was observed and no activity was observed when β-alanine, L-aspartate, or L-alanine were used as substrates instead of glycine.

Glutathione synthetase is also present in *Phaseolus vulgaris* seedlings (*372, 374*) and the enzyme isolated from wheat germ has been partially purified (*375*). The enzyme from *Phaseolus* was not very specific for glycine which could be replaced by a number of other amino acids (*374*).

It can be concluded that the biosynthesis of glutathione in plants occurs by the same pathway as in all other living organisms. The specificity of the enzymes involved in plants is not well known. The possibility that some γ-glutamyl derivatives are produced by γ-glutamylcysteine synthetase and glutathione synthetase cannot be excluded. On the other hand, glutathione is an essential constituent of plants and probably plays one or more well-defined roles (see next section); therefore it does not seem likely that the biosynthetic system for such a compound is responsible for many of the γ-glutamyl derivatives occurring in high concentrations.

### 1.3. The Function of Glutathione in Plants

The roles so far ascribed to glutathione in plants are all dependent on its ability to undergo oxidation and reduction and therefore to serve as an electron carrier and as reducing agent for other SH-groups (*118*, p. 279). These properties are absent in nearly all the γ-glutamyl

derivatives found in plants and are therefore of little relevance to a
discussion of these compounds. It may, however, be mentioned, that
a number of studies of glutathione reductase from various plants sources
have been reported (3, 19, 48, 62, 81, 96, 206, 297, 381).

## 1.4. Degradation of Glutathione in Plants

The degradation of glutathione in animals and microorganisms
probably takes place mainly by transpeptidation:

Glutathione + L-amino acid $\rightleftarrows$ γ-L-Glutamyl-L-amino acid +
L-cysteinylglycine

followed by hydrolysis of cysteinylglycine:

L-Cysteinylglycine + $H_2O$ → L-Cysteine + glycine

and by the following reactions

γ-L-Glutamyl-L-amino acid → 5-oxo-L-proline + L-amino acid

5-oxo-L-proline + ATP + $2H_2O$ → L-Glutamate + ADP + $P_i$

(219, 220) (cf. Section VII.7). Transpeptidases catalyzing the first of these
reactions are known to occur in certain plants and are discussed in
Section VII. 3. Although simple hydrolytic cleavage of the two peptide
bonds can possibly also take place in plants there are no reports of
such reactions. Only limited information is available on the simple
hydrolytic cleavage of glutathione in animals and microorganisms (219,
220).

## 2. Biosynthesis of Glutamine in Plants

Glutamine is synthesized in plants, animals and microorganisms
through the following reaction:

$$M^{2+}$$
$$\text{L-Glutamate} + ATP + NH_3 \rightarrow \text{L-Glutamine} + ADP + P_i$$

In plants this reaction is the first step for the introduction of nitrogen
from ammonia into organic compounds and is therefore of major importance for nitrogen metabolism in general (223, 224). A partially
purified preparation of the enzyme catalyzing the reaction has been
isolated from a number of different plant sources, including peas (46,
267, 345), soybean root nodules (205), rice plant roots (124), pea leaves

(251) and chloroplasts from *Vicia faba* (95). The specificity of the enzyme has not been greatly studied, but the enzyme from pea leaves is known to be able to use both ammonia and $H_2NOH$ as substrates. The same enzyme can use α-methylglutamic acid or D-glutamic acid instead of L-glutamic acid, but with reduced efficiency (252).

Glutamine synthetase may be responsible for the biosynthesis of some γ-glutamyl derivatives of amines, especially those of the lower aliphatic amines. However, a specific enzyme differing from glutamine synthetase is presumably responsible for the synthesis of N-ethyl-L-glutamine (theanine) in tea leaves, in which this compound occurs in fairly high concentrations (293) (cf. Section VII. 5.1.). On the other hand, it was found that theanine could be synthesized from glutamate, ATP and ethylamine using an enzyme preparation from *Lunaria annua* seedlings (63). In this case the enzyme responsible may well be glutamine synthetase. Theanine is not present in *Lunaria annua*, but the plant contains γ-glutamyl derivatives of isopropylamine (187) and ethanolamine (188); enzymatic experiments were not performed with the latter two amines. It seems unlikely that glutamine synthetase is involved in the biosynthesis of any γ-glutamyl derivatives of aromatic amines (see Table 6).

### 3. γ-Glutamyl Transferases in Plants

γ-Glutamyl transferases (γ-glutamyl transpeptidases, γ-glutamyl-peptide: amino acid γ-glutamyltransferase, EC 2.3.2.2) catalyze the following reaction:

Glutathione + amino acid → γ-glutamylamino acid + cysteinylglycine.

They are widely distributed in animals and microorganisms and their properties have been investigated in detail (220, 221, 253). γ-Glutamyl transferases have also been identified in a number of plants (329), and an enzyme from *Phaseolus vulgaris* has been most intensively studied (73, 74, 333):

The specificity of the enzyme was investigated using γ-glutamylaniline instead of glutathione as a substrate. Numerous α-L-amino acids were found to function as good acceptors, whereas the D-amino acids investigated (D-methionine and D-asparagine) were poor acceptors. Strangely enough, L-cysteine, L-aspartic acid, and alanine were poor acceptors. The β- and γ-amino acids investigated (β-alanine, β-amino-butyric acid, β-aminoisobutyric acid, and γ-aminobutyric acid) could not, or only to a very small degree, act as acceptors, and L-proline (the only imino acid investigated) and $NH_3$ were also found not to take part in the reaction. β-Aminopropionitrile and 3-methylbutylamine both acted as acceptors.

The enzyme was also able to catalyze the following reaction:

2 γ-L-glutamylaniline → γ-L-glutamyl-γ-L-glutamylaniline + aniline

indicating that γ-glutamylaniline can act both as a donor and an acceptor of γ-glutamyl groups. Glutathione did not act as an acceptor. The enzyme also catalyzed hydrolysis of γ-glutamylaniline. The enzyme was found to have an optimum pH of 9.5, but in spite of this it was assumed to be responsible for the biosynthesis of γ-glutamyl dipeptides in many plants (333).

It was subsequently shown, using a more purified enzyme preparation, that the enzyme had no transpeptidase activity below pH 7.5, whereas the hydrolytic reaction showed optimum activity at both pH 9.5 and 6.5. The amino group of the acceptor molecule must presumably be in the non-protonated form during the transpeptidation reaction. From these data it was concluded that the normal function of the enzyme was to catalyze hydrolysis (73). Despite this, it has repeatedly been suggested that γ-glutamyltranspeptidase is responsible for the biosynthesis of many γ-glutamyl derivatives (330, 347, 364).

γ-Glutamyl transpeptidase activity has been found in many species of Leguminosae, but only very low activity was detected in Iris bulbs and none whatsoever in bulbs of Allium cepa and roots of Brassica rapa (333).

In studies on germinating seeds of Allium schoenoprasum no trans-peptidase activity was detected. On the other hand, an enzyme catalyzing the hydrolytic cleavage of γ-glutamyl bonds was found. The enzyme has an optimum pH around 8—8.3 and catalyzed hydrolysis of the γ-glutamyl bond not only in glutathione, γ-glutamylisoleucine, and γ-glutamyl-p-nitroaniline but also glutamine, although the hydrolysis of glutamine may be due to the presence of more than one enzyme in the preparation (213). Homogenates of sprouting bulbs of Allium cepa catalyze hydrolysis of γ-glutamyl peptides. The pH in the homogenates is around 6, but even at this low value the hydrolytic activity is sufficient to explain the hydrolysis of γ-glutamyl peptides during sprouting of onions (214).

Transpeptidase activity was further studied in extracts of root and sprouting bulbs of Allium cepa. The pH optimum for the hydrolytic activity was 9.0, and it was again concluded that hydrolytic activity was adequate to explain the hydrolysis of γ-glutamyl peptides during sprouting of onions. It was confirmed that no enzymatic activity was present in dormant bulbs (8). Subsequently a partially purified enzyme was obtained from sprouted onion bulbs and was shown to have both hydrolytic activity and transpeptidase activity with an optimum pH of 9, although the transpeptidation action was presumably the essential one. The possibility that the crude onion extract also contained a genuine γ-glutamyl hydrolase could, however, not be excluded. It was suggested that γ-glutamyl peptides disappeared from sprouting onions by transpeptidation instead of hydrolysis (*302*).

A partially purified γ-glutamyl transpeptidase with an optimum pH of 8.0 has also been obtained from shoots of *Asparagus officinalis* which like *Allium* species belongs to Liliaceae. Its specificity was investigated with γ-glutamylaniline as γ-glutamyl donor. A number of α-L-amino acids, including alanine and cysteine, proved to be good acceptors, whereas aspartic acid, glutamic acid and glutathione inhibited the reaction. Proline did not, or at least only to a very limited degree, act as acceptor. The enzyme was also able to hydrolyse γ-L-glutamylaniline. No suggestions were made for the role of the enzyme in *Asparagus*, in which no γ-glutamyl derivatives have so far been identified (*64*).

γ-Glutamyl transpeptidase activity has also been established in germinating seeds of *Glycine max*. The seeds contain γ-glutamyl-phenylalanine and γ-glutamyltyrosine. The amounts of these dipeptides scarcely change during the first 20 hours of germination but decrease rapidly in the subsequent 50 hours, and no dipeptides are present after 70 hours of germination (*126*). It was shown that the γ-glutamyl transpeptidase activity was low in the first 20 hours of germination, but reached a maximum between 20 and 48 hours and dropped abruptly after 72 hours, thereby indicating that this enzyme was responsible for the disappearance of the dipeptides (*130*).

In some cases γ-glutamyl transpeptidase has been found together with γ-glutamyl derivatives in Basidiomycetes. Thus an enzyme catalyzing the following reaction

Lentinic acid + L-amino acid → γ-glutamylamino acid +
+ desglutamyllentinic acid

has been found in fruit bodies of *Lentinus edodes*. The desglutamyllentinic acid is subsequently degraded by a C–S lyase (see Section VII. 6.3.) (*385*). The enzyme has been partially purified and its properties studied, and both hydrolytic and transpeptidase activity have been found. Hydrolysis of γ-glutamyl-p-nitroaniline gives rise to nitroaniline, glutamic acid, γ-glutamyl-γ-glutamyl-p-nitroaniline, γ-glutamyl-γ-glutamyl-γ-glutamyl-p-nitroaniline, γ-glutamyl-γ-glutamyl-γ-glutamyl-γ-glutamyl-p-nitroaniline, and γ-glutamylglutamic acid. Thus γ-glutamyl-p-nitroaniline can serve as both donor and acceptor of the γ-glutamyl group. The optimum pH with γ-glutamyl-p-nitroaniline as donor was about 7—8, both with no extra acceptor and with S-methylcysteine as acceptor. Glutathione and lentinic acid can also serve as γ-glutamyl donors, lentinic acid being the more efficient substrate. The enzyme preparation exhibited no glutaminase activity, nor could glutamine serve as γ-glutamyl donor. The specificity of the acceptor was also investigated. A number of L-amino acids, including cysteine, S-alkylcysteines, S-alkylcysteine sulphoxides, methionine, lysine, arginine, asparagine and glutamine proved to be acceptors (as judged from the difference in the amounts of p-nitroaniline liberated with and without added acceptor). γ-Aminobutyric acid, proline and 4-hydroxyproline did not act as acceptors, and aspartic and glutamic acid were inhibitory (111). The optimum pH for hydrolysis of lentinic acid was 6.5, whereas the optimum pH for transpeptidation with lentinic acid as γ-glutamyl donor was >9. When valine was used as acceptor, γ-glutamyllentinic acid, desglutamyllentinic acid, γ-glutamylvaline, and glutamic acid were found in the reaction mixture (112). The enzyme has been used to obtain pure desglutamyllentinic acid from lentinic acid (387). The anion activation of the enzyme has been studied thoroughly (388).

The enzyme is also present in *Micromphale perforans* and *Collybia hariolorum*, both of which are sources of lentinic acid (100), and a γ-glutamyl transpeptidase is present in *Marasmius* species containing γ-glutamylmarasmine (72).

A γ-glutamyltransferase has been isolated from *Agaricus bisporus,* a mushroom from which $N^5$-(4-hydroxyphenyl)-L-glutamine, $N^5$-(3,4-dioxocyclohexa-1,5-dien-1-yl)-L-glutamine, and agaritine ($N^5$-[(4-hydroxymethylphenyl)-amino]-L-glutamine) have been isolated (see Table 6). The enzyme catalyzes both hydrolysis and transfer of the γ-glutamyl residue and a number of γ-glutamyl derivatives of aromatic amines and hydrazines including agaritine, $N^5$-(4-hydroxyphenyl)-L-glutamine and γ-(O-benzyl)-L-glutamic acid can serve as γ-glutamyl donors. γ-L-Glutamyl-hydrazine, L-glutamine, and $N^5$-ethyl-L-glutamine can also serve as γ-glutamyl donors, although with low efficiency. On the other hand, glutathione, γ-glutamyl-β-aminoisobutyric acid, and γ-glutamylphenyl-alanine were not degraded by the enzyme. The γ-glutamyl group could be transferred to water (hydrolysis), but hydroxylamine, hydrazine, ammonia, aromatic hydrazines and aromatic amines also served as acceptors. The enzyme was thus able to establish the following equilibrium:

$$\text{Arylhydrazine} + N^5\text{-(4-hydroxyphenyl)-glutamine} \rightleftarrows$$
$$\rightleftarrows \gamma\text{-glutamylarylhydrazine} + \text{p-hydroxyaniline}$$

and might therefore be responsible for interchange between the various γ-glutamyl derivatives present in *Agaricus*. Amino acids did not serve as acceptors, and the optimum pH of the enzyme was found to be 7. The presence of the enzyme has been established in a number of species of the genus *Agaricus,* all of which contain agaritine (69, 195).

From the information above it is clear that γ-glutamyl transferases play a role in the metabolism of γ-glutamyl derivatives. However, it is also apparent that with the present state of knowledge it is impossible to conclude that the presence of such enzymes can in all cases explain the occurrence of γ-glutamyl derivatives.

In a number of species of Basidiomycetes enzymes are present which appear to be specific for the γ-glutamyl derivatives occurring in the same species. However, it seems more likely that these enzymes are involved in degradation rather than in biosynthesis.

In higher plants γ-glutamyltranspeptidase may be responsible for the production of some of the γ-glutamyl peptides which are present in the plants. However, the optimum pH for the transpeptidase activity is much higher than the pH normally found in plants, and at lower pH values hydrolytic activity will dominate. Furthermore, specificity studies indicate that the formation of γ-glutamyl derivatives of β- and γ-amino acids and imino acids is not catalyzed by such enzymes.

Transpeptidases may be responsible for the formation of tripeptides containing the γ-glutamyl-γ-glutamyl residue (γ-glutamyl-γ-glutamyl-phenylalanine and γ-glutamyl-γ-glutamylmethionine, see Tables 3 and 5) in some plants, since the corresponding dipeptides (γ-glutamylphenyl-alanine and γ-glutamylmethionine) are present in large amounts in the same plants (see Tables 1 and 4) and can presumably behave as acceptors in transpeptidation reactions. However, for some of the other tripeptides, *viz* γ-glutamylvalylglutamic acid and γ-glutamylphenylalanyl-willardiine (see Table 3), transpeptidation cannot be invoked to explain their biosynthesis. The remaining tripeptides are probably related to glutathione (see Table 5). Furthermore, transpeptidation reactions alone give no explanation for the specific distribution patterns found for γ-glutamyl peptides (see Section III. 5).

In connection with considerations on the biosynthesis of γ-glutamyl peptides by transpeptidation it is of interest to describe the production of these compounds during glutamic acid fermentation by *Corynebacterium glutamicum*. This organism produces not only large amounts of L-glutamic acid but also γ-L-glutamyl-L-glutamic acid, γ-L-glutamyl-L-glu-tamine, γ-L-glutamyl-L-valine, γ-L-glutamyl-L-leucine and γ-L-glutamyl-γ-L-glutamyl-L-glutamic acid in the fermentation broth (86). The γ-glutamyl peptides are produced enzymatically from glutamic acid with an enzyme catalyzing transpeptidation and hydrolysis of γ-glutamyl peptides. γ-Glutamylglutamic acid is produced directly, the driving force being the high concentration of glutamic acid, whereas the other four dipeptides are produced by transpeptidation. The equilibrium constants for hydrolysis of various γ-glutamylpeptides at pH 6 were determined and shown to be compatible with the formation of γ-glutamylglutamic acid at high glutamic acid concentrations (87). The optimum pH for transpeptidation was about 8.5—9.5, whereas the optimum pH for hydrolysis/synthesis was around 6. At pH 6 only 11% — and at pH 7 only 30% of the maximum transpeptidation activity was found. A large number of L-amino acids were found to act as acceptors in the transpeptidation reaction when γ-glutamylglutamic acid or glutathione were used as donors. D-Amino acids were not active as acceptors. A large number of γ-L-glutamyl-L-amino acids also acted as γ-glutamyl donors with L-threonine or L-glutamic acid as acceptor. γ-D-Glutamyl-L-valine and γ-D-glutamyl-L-methionine were active as donors but with only low activity. The enzyme has been found in a large number of species of bacteria (*88*).

## 4. Degradation Systems for γ-Glutamyl Derivatives in Plants

As described in Section VII. 3, the γ-glutamyl transferases from plants are all able to catalyze hydrolysis of the γ-glutamyl linkage and may therefore play a role in degradation reactions. In a number of specific cases degradative pathways have been established (see Section VII. 6). The variation in content of γ-glutamyl derivatives in different parts of plants and throughout the life cycle of the plants shows clearly that degradation reactions must be of importance under some conditions. No information is available on degradative enzymes other than transpeptidases in plants. Even for glutathione (see Section VII. 1.4) and glutamine enzymes responsible for degradation have not been identified in plants.

## 5. Occurrence of γ-Glutamyl Derivatives in Different Parts of Plants and During the Life Cycles of Plants

Whereas no systematic investigations of the concentration levels of γ-glutamyl derivatives throughout the life cycles of plants have been reported, a number of reports concerning isolation which contain information on concentrations and changes in concentrations have appeared. γ-Glutamyl derivatives present in seeds often disappear at germination, as shown for γ-glutamyl-S-methylcysteine, γ-glutamyl-S-methylcysteine sulphoxide and γ-glutamylleucine in *Phaseolus lunatus* (*278*), γ-glutamylleucine, γ-glutamylmethionine and γ-glutamylmethionine sulphoxide in *Phaseolus radiatus* L. var. *typicus* (*141, 142*), and γ-glutamyl-β-pyrazol-1-ylalanine in *Cucumis sativus* (*41*). γ-Glutamylleucine and γ-glutamyl-S-methylcysteine could be found only in germinating cotyledons and in ripening seeds and not in roots, stems or leaves of *Phaseolus vulgaris* (*366*). γ-Glutamylphenylalanine and the many γ-glutamyl derivatives of sulphur-containing amino acids in bulbs of *Allium cepa* are not present in other parts of the plant (*354, 355*). γ-Glutamylphenylalanine and γ-glutamyltyrosine present in seeds of *Glycine max* disappear at germination (*126*) and cannot be found in pods, leaves, and stems but only in ripening seeds (*139, 365*). The concentrations of γ-glutamyl-S-methylcysteine and γ-glutamyl-Se-methyl-selenocysteine which are present in ripening pods and seeds of *Astragalus bisulcatus* decrease during germination and are very low in leaves and flowers (*240*). None of the six γ-glutamyl derivatives present in seeds of *Lunaria annua* were present in fresh parts of the plant (*188*).

The presence of γ-glutamyl derivatives is not restricted to seeds or other storage organs (compare Tables 1—6). Thus in *Iris tingitana* γ-glutamylalanine, γ-glutamyl-β-alanine, and γ-glutamyl-β-aminoisobutyric acid were all present in bulbs, roots, and leaves. The two β-amino acid derivatives occurred in highest concentration in bulbs, whereas γ-glutamylalanine occurred in highest concentration in roots (*231*). In seeds of *Fagus silvatica* var. *purpurea* γ-glutamylphenylalanine, γ-glutamylwillardiine and other γ-glutamyl derivatives occur in large concentrations, whereas in seedlings γ-glutamylleucine and γ-glutamylisoleucine are present in fairly high concentration (*129*). In pea seedlings the amount of γ-L-glutamyl-D-alanine increases significantly during the first 8 days of germination and thereafter rapidly declines (*67*). Linatine has been reported to be present in all parts of *Linum usitatissimum*, although degradation of linatine in seeds takes place during germination (*245*).

## 6. Specific Systems for Biosynthesis and Degradation of Individual γ-Glutamyl Derivatives

A number of γ-glutamyl derivatives have attracted special interest, for example because of their toxicity or their role in liberation of volatile flavouring compounds, and detailed studies on their biosynthesis and degradation have been carried out.

### 6.1. γ-L-*Glutamyl*-β-*cyano*-L-*alanine*

β-Cyano-L-alanine and γ-L-glutamyl-β-cyano-L-alanine are present in seeds of various species of *Vicia* and *Lathyrus* (see Table 2), including *V. sativa* and *V. angustifolia*. The concentrations of the γ-glutamyl derivative is much higher in seedlings than in seeds of *V. sativa* (*274*), and β-cyanoalanine is nearly- or totally absent from the seedlings of *V.*

*sativa* and *V. angustifolia.* The amount of β-cyanoalanine bound as the γ-glutamyl derivative in 5 day old seedlings is the same as the total amount of free and bound β-cyanoalanine in seeds, indicating synthesis does not take place during germination (*274*). β-Cyanoalanine is an intermediate in the transformation of cysteine (or O-acetylserine) and cyanide into asparagine (*155*), and this pathway must be operative in seedlings of *Lathyrus sylvestris, L. odoratus* and *Vicia villosa,* where [14]C from both cyanide and β-cyanoalanine is incorporated into asparagine. However, in seedlings of *V. sativa* and *V. angustifolia* the transformations of β-cyanoalanine to asparagine appears to occur slowly if at all. On the other hand, [14]C from cyanide is incorporated to a high degree into the γ-glutamyl derivative, indicating that synthesis of β-cyanoalanine can take place if cyanide is provided. [14]C from serine is incorporated to only a very low extent into γ-glutamyl-β-cyanoalanine at this stage in the life cycle of the plants (*274*). During maturation of seeds of *Vicia sativa* [14]C from serine is incorporated into both free β-cyanoalanine and the γ-glutamyl derivative without the provision of cyanide, indicating that this is the normal point in the life cycle for synthesis of β-cyanoalanine (*236*). γ-Glutamyl-β-cyanoalanine thus appears to serve as a storage supply for β-cyanoalanine in *V. sativa,* in which species the latter compound is not transformed into asparagine. In other species in which β-cyanoalanine is easily transformed into asparagine the γ-glutamyl derivative is present in much lower concentrations if at all.

### 6.2. γ-Glutamyl Derivatives of Sulphur-Containing Amino Acids in Allium Species

A number of γ-glutamyl derivatives of S-substituted cysteines have been found in *Allium* species, generally together with the free amino acids (see Tables 4 and 5). *Allium* species contain a C–S-lyase which catalyzes cleavage of S-substituted cysteine sulphoxides but not S-substituted cysteines themselves. The products of the enzymatic reaction are ammonia, pyruvic acid and a volatile sulphur compound. S-prop-1-enylcysteine sulphoxide gives rise to thiopropanal S-oxide, which is the lachrymatoric substance of *Allium cepa* and other *Allium* species. Other volatile sulphur compounds derived from cysteine derivatives are also biologically active as lachrymators, flavouring substances and antibiotics (*62a, 181, 301, 347*). Although it has not always been specifically stated (see, however, *110*), γ-glutamyl derivatives are not substrates for the rather specific C–S lyase. On the other hand, the possibility of hydrolysis or transpeptidation (see Section VII. 3) coupled with the action of the C–S lyase indicates a degradative route for γ-glutamyl derivatives; they may therefore play a role as reservoirs of the cysteine derivatives.

It has been proposed that a number of S-substituted cysteine derivatives are produced by addition of the SH-group in cysteine to a suitable compound with a double bond, although there is no experimental evidence for such reactions (*347*). More specifically, it has been proposed that N,N-bis-(γ-glutamyl)-3,3′-(1-methylethylene-1,2-dithio)-dialanine, present in seeds of *Allium schoenoprasum,* is produced by addition of γ-glutamylcysteine to γ-glutamyl-S-(prop-1-enyl)-cysteine, both of which are present in the same seeds (although γ-glutamylcysteine was identified only as the disulphide) (*211*).

[14]C-Valine is incorporated into the S-substituents in S-(2-carboxypropyl)-glutathione and S-(2-carboxypropyl)-cysteine in *Allium sativum,* and on this basis it was proposed that valine was transformed into methacrylic acid to which was added glutathione or cysteine, respectively (*321*). In this connection studies on the biosynthesis of S-(prop-1-enyl)-cysteine sulphoxide in *Allium cepa* which support a pathway from valine *via* methacrylic acid, S-(2-carboxypropyl)-cysteine, and S-(prop-1-enyl)-cysteine are of interest (*75*).

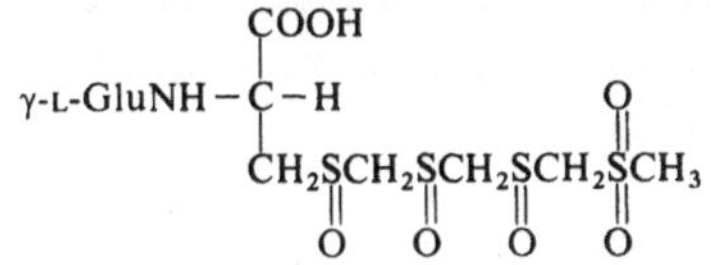

(34)

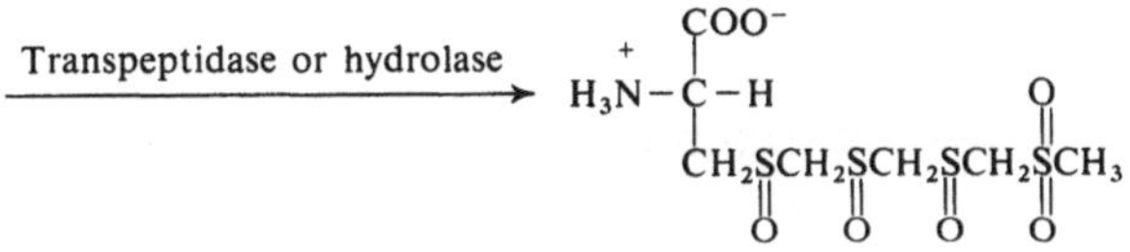

(35)

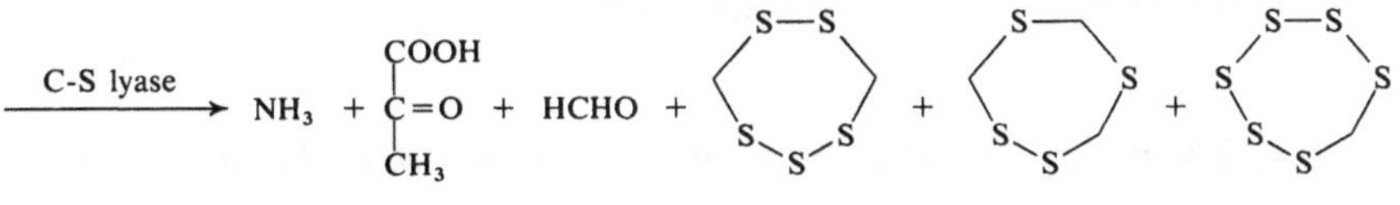

(36)   (37)   (38)

### 6.3. Lentinic Acid and γ-Glutamylmarasmine

Lentinic acid (**34**) in *Lentinus edodes* is degraded first by the action of a transpeptidase to desglutamyllentinic acid (**35**), as described in Section VII. 3. Desglutamyllentinic acid is then cleaved by action of a

C–S lyase to give ammonia, pyruvic acid, formaldehyde, and various volatile sulphur compounds including lenthionine (36) (385, 386). The C–S lyase is specific for S-substituted sulphoxides but can also use cystine as substrate (110). Lenthionine is the major flavouring substance obtained from *Lentinus edodes* and its structure has been established as 1,2,3,5,6-pentathiepane (36). Among other constituents produced from des-glutamyllentinic acid are 1,2,4,6-tetrathiepane (37), and 1,2,3,4,5,6-hexathiepane (38). Lenthionine can be synthesized from sodium poly-sulphide and methylene chloride (227, 228). The same degradation of lentinic acid takes place in *Micromphale perforans* and *Collybia hariolorum* (100).

γ-Glutamylmarasmine (39) in *Marasmius* species is similarly degraded first to marasmine (40) by the action of a transpeptidase as described in Section VII. 3. Marasmine is then cleaved by a C–S lyase to give ammonia, pyruvic acid and an unstable sulphur compound, probably methylthiomethylsulphenic acid (41), which again decomposes, probably *via* the sulphinic ester $CH_3SCH_2SSOCH_2SCH_3$ (42), to give various odorous sulphur compounds. The C–S lyase shows broad specificity, cleaving L-cysteine, S-alkyl- or aralkyl- and aryl-L-cysteines, L-djenkolic acid, and the corresponding sulphoxides (72).

$$\gamma\text{-}L\text{-}GluNH - \underset{\underset{CH_2SCH_2SCH_3}{|}}{\overset{\overset{COOH}{|}}{C}} - H \quad \xrightarrow{\text{Transpeptidase or hydrolase}} \quad H_3N^+ - \underset{\underset{CH_2SCH_2SCH_3}{|}}{\overset{\overset{COO^-}{|}}{C}} - H$$

(39)                                                                   (40)

C-S lyase
$\xrightarrow{\phantom{xxx}}$ [HOSCH$_2$SCH$_3$] $\longrightarrow$ [H$_3$CSCH$_2$SSCH$_2$SCH$_3$]
‖
O

(41)                          (42)

$\longrightarrow$ H$_3$CSCH$_2$SSCH$_2$SCH$_3$ + H$_3$CSCH$_2$S(O)SCH$_2$SCH$_3$

### 6.4. Theanine (N⁵-Ethyl-L-glutamine) and N⁵-Methyl-L-glutamine

Theanine was originally isolated from tea leaves *(Camellia sinensis)*, of which it constitutes up to 1.7% of the dry weight (283). It is of importance for the flavour of tea, both in itself and as a precursor of catechins.

Theanine is present in all parts of the plants except the fruits as the dominant free amino acid (305) although there is more glutamine than theanine in the sap of young tea plants (304). The largest amounts of theanine are found in the roots (325, 335).

[14]C-labelled theanine is produced in tea seedlings from labelled glutamic acid (*290*) or ethylamine (*153*). Theanine biosynthesis is presumed to take place mainly in the roots (*160, 161, 162, 377*) but some synthesis also takes place in young buds, stems, and leaves (*163*).

An enzyme catalyzing formation of theanine from glutamic acid, ethylamine, and ATP has been identified in tea seedlings and partially purified (*287, 289*). The enzyme is specific for L-glutamic acid and for lower aliphatic amines. The enzyme preparation also displayed glutamine synthetase activity but this proved to be due to the presence of a separate glutamine synthetase. Theanine synthetase does not use ammonia as substrate, as demonstrated by competition experiments and by the additivity of the two activities (*293*). The enzyme was compared with enzyme preparations from seeds of *Pisum sativum* and from pigeon liver as regards its ability to catalyze the synthesis of theanine. The enzymes from these sources appear, however, to be unspecific glutamine synthetases which catalyze the synthesis of both glutamine and theanine (*291, 292, 293*). The enzyme from seedlings of *Lunaria annua* which catalyzes the formation of theanine (*63*) is probably also an unspecific glutamine synthetase (compare Section VII. 2.), and the reported very low formation of theanine and $N^5$-methylglutamine from ethylamine and methylamine, respectively in *Oryza sativa* (*163*) may also be due to the action of an unspecific glutamine synthetase. Ethylamine in tea plants is derived from alanine, as shown by incorporation experiments on tea seedlings using [14]C-labelled alanine (*325*) and by the partial purification of a L-alanine decarboxylase from cotyledons and roots of tea seedlings (*326*).

Theanine is degraded in tea seedlings, especially in the presence of light (*165, 167*). No enzymes responsible for the degradation have been identified, but it has been shown that the ethyl group in theanine is incorporated mainly into the phloroglucinol nucleus in catechins. The transformation probably takes place *via* acetate. However, acetate cannot be substituted for theanine, indicating that a special pool is involved (*154*).

$N^5$-Methylglutamine is also present in tea leaves (*166*), [14]C-labelled methylamine is incorporated into $N^5$-methylglutamine in seedlings and excised shoots (*164, 166, 318*). The synthesis is believed to take place in the roots, and the $N^5$-methylglutamine is then supposed to be translocated to the shoots, in which most of the radioactivity is transferred into caffeine (*164*). However, $N^5$-methylglutamine can be degraded in excised shoots to methylamine, and may only be a storage or transport compound for methylamine (*318*).

A number of reviews on theanine in Japanese have been published (*159, 286, 288*).

*6.5. $N^5$-(4-Hydroxyphenyl)-L-glutamine, Agaridoxin ($N^5$-(3,4-Dihydroxy-phenyl)-L-glutamine), and $N^5$-(3,4-Dioxocyclohexa-1,5-dien-1-yl)-L-glutamine*

$N^5$-(4-Hydroxyphenyl)-L-glutamine (**43**) can be transformed enzymatically to $N^5$-(3,4-dioxocyclohexa-1,5-dien-1-yl)-L-glutamine (**44**). An enzyme exhibiting tyrosinase activity which catalyzes this transformation has been isolated from *Agaricus bisporus* and partially purified (*369*). Agaridoxin ($N^5$-(3,4-dihydroxyphenyl)-L-glutamine) (**45**) is believed to be an intermediate in this transformation and is easily autoxidized to give the quinone (*323*).

In the intact gill tissue of *Agaricus bisporus* the precursor and the enzyme are largely segregated; the quinone is formed only after the gill tissue is macerated (*371*).

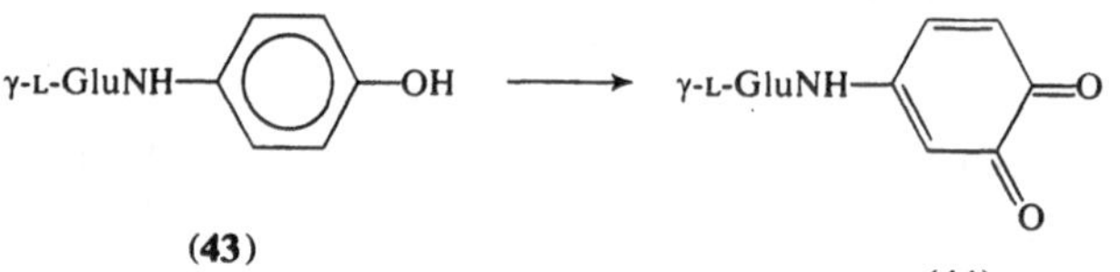

(43)                     (44)

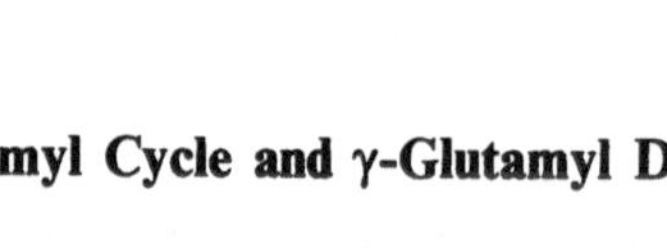

(45)

# 7. The γ-Glutamyl Cycle and γ-Glutamyl Derivatives in Plants

## 7.1. The Cycle in Animals and Microorganisms

In animals glutathione, γ-glutamyl amino acids and 5-oxoproline are intermediates in the γ-glutamyl cycle which has been proposed as responsible for the transport of amino acids across cell membranes. The cycle involves the action of γ-glutamylcysteine synthetase, glutathione synthetase, γ-glutamyl transpeptidase, γ-glutamyl cyclotransferase (catalyzing the transformation of a γ-glutamyl amino acid into free amino acid and 5-oxoproline), 5-oxoprolinase (catalyzing the opening of 5-oxoproline into glutamic acid with the concomitant cleavage of ATP into ADP and inorganic phosphate) and a peptidase cleaving cysteinylglycine (*220, 221*). The cycle may also be operative in some prokaryotes, although not in *Escherichia coli* (*221*). All the enzymes of the cycle are present in *Saccharomyces cerevisae*, thus indicating that the cycle may operate in this organism (*226*), and there is also evidence for its operation in phytoplankton (*182*).

## 7.2. Evidence for the Presence in Plants of the Enzymes Involved in the γ-Glutamyl Cycle

Plants contain γ-glutamylcysteine synthetase and glutathione synthetase (Section VII. 1.2.), γ-glutamyl transpeptidase (Section VII. 3.) and probably a peptidase cleaving cysteinylglycine (Section VII. 1.4.). There is no reported evidence for the presence of γ-glutamylcyclotransferase and 5-oxoprolinase. However, it has recently been shown that 5-oxoproline is not metabolically inactive in plants since it can be transformed to glutamic acid (*217*). Since only very limited information is available on the transport of amino acids across cell membranes in plants (see the following section) it is not possible to determine whether the γ-glutamyl cycle is operative in plants.

## 7.3. Transport of Amino Acids Across Cell Membranes in Plants

Only limited information is available regarding the transport of amino acids across cell membranes in plants (*243, 263*). Active carrier-mediated transport of amino acids into plant cells has been demonstrated in a number of different systems. In some cases competition experiments have indicated a single transport system for all amino acids, but in other cases there is evidence to indicate the existence of different transport systems for different groups of amino acids.

The active transport of amino acids and their accumulation in vacuoles has been demonstrated with carrot slices (*16, 17, 33*). Active transport of amino acids has been shown for leaf strips of barley (*2, 273, 308*) and for leaf slices of barley (*200*). In the latter investigation competition experiments indicated a single carrier system for all amino acids. However, experiments on the uptake of amino acids into barley roots indicated different transport systems for lysine + arginine, proline, and methionine, respectively (*310*).

A single carrier-mediated system has been found in leaves of *Egeria densa* (*260*) and a single uptake system for all α-amino acids has been proposed on the basis of experiments with seedling root tips of *Cucumis melo* and *Caesalpinia tinctoria (367)*. Experiments with cultured tobacco cells also indicate a single active uptake system for amino acids. The transport of cysteine was found to be inhibited by a wide range of amino acids, although neutral aliphatic amino acids were generally most effective. Since S-methylcysteine was found to be an effective inhibitor whereas cysteine methyl ester was relatively ineffective, it was suggested, in agreement with previous conclusions, that amino acids are transported as amphions. It was also pointed out that there is a close correlation between the ability of the tested amino acids to inhibit the transport system and their ability to accept the γ-glutamyl moiety in model studies with γ-glutamyl transpeptidase *(85)*. The transpeptidase used for reference was a preparation from rat kidney *(327)*; the results obtained with the enzyme from rat kidney correlate to some extent with corresponding results obtained with a γ-glutamyl transpeptidase from *Phaseolus vulgaris (333)*.

Separate active uptake mechanisms have been found for arginine and lysine in sugar cane cells *(207)*. Experiments on the uptake of arginine, glutamic acid, and alanine into soybean root cells indicated the existence of a number of uptake systems for basic, acidic and neutral amino acids, respectively *(151)*. Again, experiments on the uptake of amino acids into leaf fragments of *Pisum sativum* indicate a carrier-mediated transport system with a high specificity for neutral and positively charged L-amino acids but with low specificity for L-aspartic acid, L-glutamic acid and D-amino acids *(28)*.

Studies on the uptake of amino acids and oligopeptides into embryos of *Hordeum vulgare* indicate independent transport systems *(97)* whilst uptake studies with chloroplasts from *Pisum sativum* have indicated the presence of two carrier systems, one transporting glycine and aliphatic amino acids (L-alanine, L-leucine, L-isoleucine, and L-valine) and the other transporting L-serine, L-threonine and L-methionine. The carrier systems may not perform active transport but merely facilitated diffussion *(244)*. Other studies have also indicated the possibility of transport of amino acids into chloroplasts *(285)*.

## 8. Conclusions

The evidence presented in Section VII. 1—7 indicates that a single general role cannot be assigned to all the γ-glutamyl derivatives found in plants, although the possibility that some may have specific roles cannot be excluded. The γ-glutamyl amino acids may serve as storage compounds and in fact in some cases γ-glutamyl derivatives of protein amino acids occur in such high concentrations in seeds that they must be taken into account in determining the content of essential amino acids in the seeds (*cf. 39*). It has been suggested that γ-glutamyl peptides generally serve as special storage forms for non-protein amino acids. At the time of seed maturation there is an upward movement of soluble nitrogen in plants, and non-protein amino acids are translocated to the seeds together with other substances. The seeds, now faced with the problem of dealing with increasing quantities of non-protein amino acids which cannot be incorporated into proteins, convert them into γ-glutamyl peptides. This process may also be a detoxification mechanism (*237*), but even if this is the case the fact that the plant produces non-protein amino acids at all remains to be explained. Furthermore, no explanation is provided for the specific distribution patterns of γ-glutamyl derivatives (*cf.* Section III. 5.).

Some γ-glutamyl derivatives may be produced by the action of transpeptidase, although the evidence is conflicting (see Section VII. 3.). The production of γ-glutamyl peptides in bacterial fermentation broths by reversal of enzymatic hydrolysis has been demonstrated (see Section VII. 3.). Since the concentrations of free amino acids in ripening seeds may attain high values, similar processes may occur in plants.

Finally, γ-glutamyl peptides may be involved in the transport of amino acids across cell membranes in plants (*cf.* Section VII. 7.). However, this fact offers no explanation for their occurrence in high concentrations, since the operation of the γ-glutamyl cycle in animals does not lead to the accumulation of γ-glutamyl peptides.

## VIII. γ-Glutamyl Derivatives with Toxic or Other Specific Biological Properties

A number of γ-glutamyl derivatives from plants are toxic towards animals or have other specific biological actions. In most cases the action is due to the second amino acid or amine, but γ-glutamyl derivatives may be of importance if they contain the bulk of the toxic component in the plant material in question.

### 1. γ-L-Glutamyl-β-cyano-L-alanine and $N^5$-(2-Cyanoethyl)-L-glutamine

$N^5$-(2-Cyanoethyl)-L-glutamine was originally isolated from seeds of *Lathyrus odoratus* as the agent causing lathyrism in rats (*299*). γ-L-Glutamyl-β-cyano-L-alanine was likewise isolated from seeds of *Vicia sativa* and showed to be responsible for the toxicity of these seeds to-

wards rats, chickens, and humans (275). The γ-glutamyl derivatives as well
as β-cyano-L-alanine, and β-aminopropionitrile are toxic, causing neuro-
logical and skeletal abnormalities described as lathyrism. 2,4-Diamino-
butyric acid is also a lathyrogen (9, 55, 340).

## 2. Linatine (1-[(N-γ-L-Glutamyl)amino]-D-proline)

Linatine was originally isolated as an vitamin $B_6$ antagonist from
seeds of *Linum usitatissinum* (flax seeds). The compound is toxic towards
both chickens and *Azotobacter* species. 1-Amino-D-proline has the same
toxic effect (158). The toxicity is apparently due to formation of the
hydrazone of pyridoxal 5'-phosphate (296, 341). 1-Amino-L-proline and
1-amino-D-proline were found to be equally toxic towards chickens, but
the D-isomer was 50 times more toxic than the L-isomer towards
*Azotobacter vinelandii* (158). A review on naturally occurring $B_6$ antagonists has been published (157).

## 3. Hypoglycine B (γ-L-Glutamyl-(2S,1'R)-3-(methylenecyclopropyl)-alanine)

Hypoglycine B and hypoglycine A ((2S,1'R)-3-(methylenecyclopropyl)
alanine were originally isolated from unripe fruits of *Blighia sapida* as
the agents responsible for the toxicity and hypoglycaemic action of this
material (92). The extensive literature on the toxicity and pharmacolo-
gical properties of hypoglycine A and B has been collected in a symposium report (146).

## 4. γ-L-Glutamyl-Se-methylseleno-L-cysteine

This compound is the principal toxic component of seeds of *Astragalus bisulcatus,* in which it occurs together with the free selenium-
containing amino acid (237). The toxicity of selenium-containing amino
acids towards animals has been reviewed (208).

## 5. Coprine (N⁵-(1-Hydroxycyclopropyl)-L-glutamine)

Coprine was isolated as the agent responsible for the toxicity displayed
by the mushroom *Coprinus atramentarius* when ingested together with
ethanol (94, 202). Coprine has no effect on liver aldehyde dehydrogenase

*in vitro* (*94*), but the hydrochloride of 1-aminocyclopropanol was found to be a potent inhibitor of the enzyme both *in vivo* and *in vitro*. 1-Aminocyclopropanol is therefore probably responsible for the increase of *in vivo* acetaldehyde levels caused by coprine (*336*).

## 6. $N^5$-(3,4-Dioxocyclohexa-1,5-dien-1-yl)-L-glutamine

This compound was originally established as being the inhibitor of sulphydryl enzymes present in *Agaricus bisporus* (*369*), and was also shown to inhibit the dehydrogenase of the tricarboxylic acid cycle in rat liver mitochondria. Several other sulphydryl enzymes including xanthine dehydrogenase are also inhibited by the quinone, which probably acts by addition of the SH-group in cysteine residues (*368*). The compound is bactericidal and it is suggested that it, or its precursor, $N^5$-(4-hydroxyphenyl)-L-glutamic acid, initiates or maintains the dormant or cryptobiotic state of the fungus spore (*362*). The compound has a cytostatic effect on mammalian tumor cells (*361*).

## IX. Compounds Structurally Related to γ-Glutamyl Derivatives from Plants

γ-Glutamyl derivatives are a well-delimited group of natural products. However, a few other compounds isolated from plants are structurally related to γ-glutamyl derivatives and these are listed in Table 10.

The list includes three asparagine derivatives, all of which are found in species of Cucurbitaceae. Their biosynthesis in seedlings of *Ecballium* and *Bryonia* has been studied and the compounds have been shown to be produced by a transferase reaction between asparagine and the appropriate amine (*63*). The formation of ethylamine, present in $N^4$-ethylasparagine (Table 10), by decarboxylation of alanine has been demonstrated in *Ecballium elaterium* (*30*).

4-Hydroxyglutamine (see below) and three derivatives of (2*S*,4*S*)-4-hydroxyglutamic acid have been identified in plants. No information is available on their biosynthesis. (2*S*,4*S*)-4-Hydroxyglutamic acid has been found in *Phlox decussata* (Polemoniaceae) (*352*) (for configurational assignment, see *52*) and 4-hydroxyglutamic acid of unspecified configuration has been identified in *Solanum melongena* (Solanaceae) (*269*).

4-Methylene-L-glutamine has been found in various higher plants, occurring in all cases together with 4-methylene-L-glutamic acid, and has been suggested as playing an important role in the translocation

Table 10. *Compounds Structurally Related to γ-Glutamyl Derivatives From Plants*

| Compound | Formula | Plant species | Plant family | Character of isolate, comments | Configuration of amino acid | Ref. | Occurrence of second constituent | Chemical synthesis of derivative Ref. |
|---|---|---|---|---|---|---|---|---|
| $N^4$-Methyl-asparagine | | *Corallocarpus epigaeus*, seeds | Cucurbitaceae | Evap. residue | | (42) | | (42) |
| $N^4$-Ethyl-L-asparagine | | *Ecballium elaterium* | Cucurbitaceae | Cryst. | L | (76) | | (40, 76) |
| | | *Bryonia dioica*, seeds | Cucurbitaceae | Paper chromatography | | (42) | | |
| $N^4$-(2-Hydroxyethyl)-L-asparagine | | *Bryonia dioica*, seeds and shoots | Cucurbitaceae | Cryst. | L | (51) | | (51) |
| | | *Ecballium elaterium*, seeds | Cucurbitaceae | Paper chromatography | | (42) | | |
| 4-Hydroxy-glutamine | | *Phlox decussata* | Polemoniaceae | Cryst. | | (21) | | |
| Pinnatanine ((2S,4S)-4-Hydroxy-$N^5$-(2-hydroxy-methylbuta-1,3-dien-1-yl)-glutamine) | 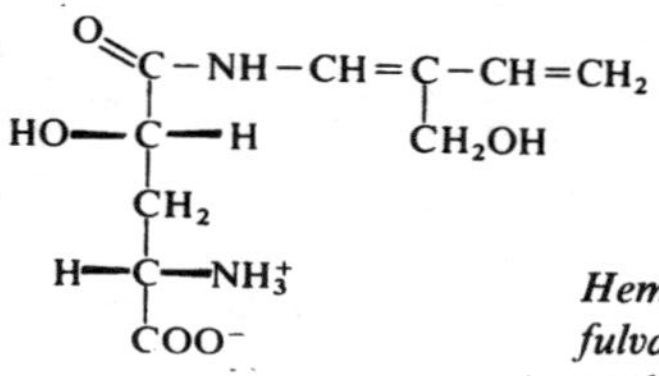 | *Staphylea pinnata*, seeds | Staphyleaceae | Cryst. | 2S,4S (L-allo) in glutamic acid moiety, not determined in amine | (79) | | |
| | | *Hemerocallis fulva*, seeds | Liliaceae | Paper chromatography | | (79) | | |

| Compound | Structure | Source | Family | Form | Configuration | Ref. | |
|---|---|---|---|---|---|---|---|
| Oxypinnatanine; (2S,4S,1'R)-4-Hydroxy-N⁵-(3-hydroxymethyl-2,5-dihydro-2-furyl)-glutamine | 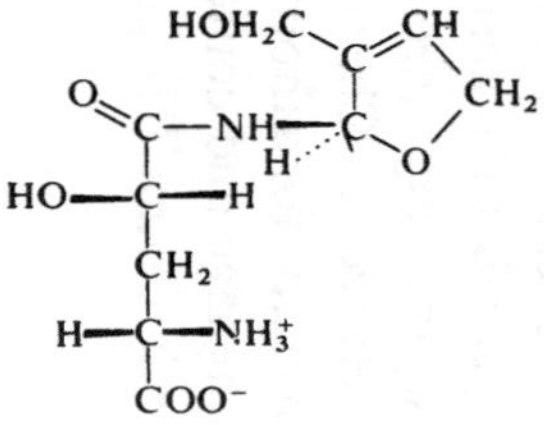 | *Staphylla pinnata*, seeds | Staphyleaceae | Cryst. | *2S,4S* (L-allo) in glutamic acid moiety, *1'R* in amine | *(79, 178)* | |
| | | *Hemerocallis fulva*, leaves | Liliaceae | Cryst. | | | |
| (2S,4S)-4-Hydroxy-N⁵-(2-hydroxy-benzyl)-glutamine (allo-L-4-Hydroxy-γ-glutamylsalicyl-amine) | 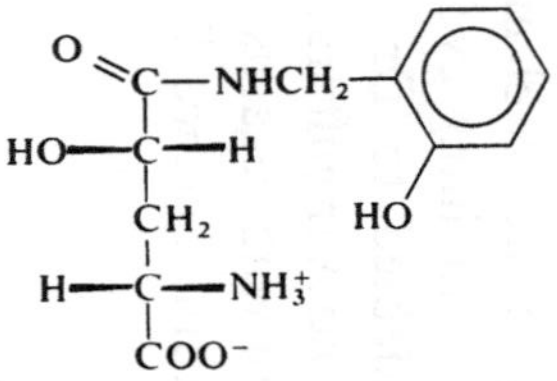 | *Fagopyrum esculentum*, seeds | Polygonaceae | Cryst. | *2S,4S* (L-allo) | *(173)* | *Fagopyrum esculentum (172)* |
| 4-Methylene-L-glutamine | | Various higher plants | | | L | *(52)* | |

of nitrogen in *Arachis hypogea* (Leguminosae), from which it was first isolated (*52*). Purified glutamine synthetase from peas catalyzes slow synthesis of 4-methyleneglutamine from 4-methyleneglutamic acid (*196*). 4-Hydroxyglutamine has been found in *Phlox decussata* (*21*).

Numerous other substituted glutamic acids are found in plants (*37*), whereas the structurally corresponding amides have not been detected. The glutamine synthetase isolated from peas was, however, able to catalyze the synthesis of 4-methylglutamine from 4-methylglutamic acid (*196*).

# X. Summary

A large number of γ-glutamyl derivatives of amines and amino acids have been isolated from plants. Many more presumably remain to be found, just as additional non-protein amino acids are being found in plants every year. The chemistry of the γ-glutamyl derivatives is simple and well-documented, but little is known about their significance, their production and degradation and about the role they play in the plant species in question. A general rationalisation regarding the latter aspects is unlikely to be forthcoming, but future-work will hopefully lead to a much-needed understanding of the biological significance of most of the compounds.

## Acknowledgements

We are indebted to Professor M. G. Ettlinger, University of Copenhagen, for calling our attention to the problems concerning the absolute configurations of amino acids containing cyclopropane rings (see Section III. 2.). We are also indebted to Dr. M. Hancock, Chemistry Department, Royal Veterinary and Agricultural University, for a thorough revision of the manuscript, both with regards to language and contents. Support from the Danish Natural Science Research Council is gratefully acknowledged.

## Addendum

In this section are mentioned articles which have appeared to us subsequent to completion of the main body of the review.

A second isolation of γ-L-glutamyl-3-(2-amino-4-pyrimidinyl)-L-alanine (γ-L-glutamyl-L-lathyrine) (Table 2) from seeds of *Lathyrus japonicus* has been described (*394*).

1-γ-L-Glutamyl-2-(2-carboxyphenyl)hydrazine (anthglutin) (45) has been isolated from the culture medium of *Penicillium oxalicum*. The compound is an inhibitor of γ-glutamyltransferase (*395, 396*).

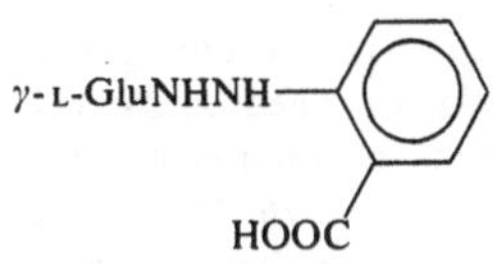

(45)

Epilentinic acid, a diastereoisomer of lentinic acid (Table 4, compare Section V. 7) has been isolated free from lentinic acid from *Micromphale cauvetii, M. foetidum* and *Collybia impudica* (*397*).

N⁵-(4-Hydroxyphenyl)-L-glutamine (Table 6) has been isolated from *Psalliota bisporus* (*398*).

A γ-glutamyl transferase from *Tricholoma shimeji* has been described (compare Section VII. 6.3.) (*399*).

During studies of the odorous substances from *Phallus impudicus* bis-γ-glutamylcystine has been isolated (compare Section VII. 6.3.) (*397*).

5-Oxoprolinase has been identified in several higher plants (compare Section VII. 7.2.) (*400*).

## References

*1.* ACREE, T. E., and C. Y. LEE: Kinetic study of the cyclization of L-glutamine to 2-pyrrolidone-5-carboxylic acid in a model system. J. Agric. Food Chem. **23**, 828 (1975).

*2.* AMAR, L., and L. REINHOLD: Loss of membrane transport ability in leaf cells and release of protein as a result of osmotic shock. Plant Physiol. **51**, 620 (1973).

*3.* ANDERSON, D. G., H. A. STAFFORD, E. E. CONN, and B. VENNESLAND: The distribution in higher plants of NADP⁺-linked enzyme systems capable of reducing glutathione. Plant Physiol. **27**, 675 (1953).

*4.* ARAI, S., M. YAMASHITA, M. NOGUGHI, and M. FUJIMAKI: Tastes of L-glutamyl oligopeptides in relation to their chromatographic properties. Agric. Biol. Chem. **37**, 151 (1973).

*5.* ARCHIBALD, R. M.: Chemical characteristics and physiological rôles of glutamine. Chem. Rev. **37**, 161 (1945).

6. Asen, S., J. F. Thompson, C. J. Morris, and F. Irreverre: Isolation of β-amino-isobutyric acid from bulbs of *Iris tingitana* var. *Wedgewood*. J. Biol. Chem. **234**, 343 (1959).

7. Augusti, K. T.: Chromatographic identification of certain sulfoxides of cysteine present in onion (*Allium cepa* Linn.) extract. Curr. Sci. **45**, 863 (1976).

8. Austin, S. J., and S. Schwimmer: L-γ-Glutamyl peptidase activity in sprouted onion. Enzymologia **40**, 273 (1971).

9. Barrow, M. V., C. F. Simpson, and E. J. Miller: Lathyrism. A review. Quart. Rev. Biol. **49**, 101 (1974).

10. Barsley, E. A.: Correlation of the configuration of some sulphoxides with (+)-S-methyl-L-cysteine S-oxide. Tetrahedron **24**, 3747 (1968).

11. Beecham, A. F.: The action of ammonia and other bases on γ-methyl and γ-ethyl L-glutamate. J. Amer. Chem. Soc. **76**, 4615 (1954).

12. Bell, E. A.: Associations of ninhydrin-reacting compounds in the seeds of 49 species of *Lathyrus*. Biochem. J. **83**, 225 (1962).

13. Bell, E. A., and A. S. L. Tirimanna: Associations of amino acids and related compounds in the seeds of forty-seven species of *Vicia*: their taxonomic and nutritional significance. Biochem. J. **97**, 104 (1965).

14. Bergmann, L., and H. Rennenberg: Efflux und Produktion von Glutathion in Suspensionskulturen von *Nicotiana tabacum*. Z. Pflanzenphysiol. **88**, 175 (1978).

15. Bielinska-Czarnecka, M.: Ninhydrin-reacting substances from apple spurs. J. Sci. Food Agr. **14**, 527 (1963).

16. Birt, L. M., and F. J. R. Hird: The uptake and metabolism of amino acids by slices of carrot. Biochem. J. **70**, 277 (1958).

17. — — Kinetic aspects of the uptake of amino acids by carrot tissue. Biochem. J. **70**, 286 (1958).

18. Black, D. K., and S. R. Landor: Allenes. Part XIX. Synthesis of (±)-hypoglycin A and configuration of the natural isomer. J. Chem. Soc. (C) **1968**, 228.

19. Boeck, D., and W. Grosch: Glutathion-Dehydrogenase (EC 1.8.5.1) aus Weizenmehl. Reindarstellung und Eigenschaften. Z. Lebensm.-Unters.-Forsch. **162**, 243 (1976).

20. Bonnet, R., P. Niviere et V. Labeyrie: Caractérisation d'amino-acides sulfoxidés dans la gaine et les limbes d'*Allium porum*. C. R. Acad. Sci., Ser. D **279**, 1919 (1974).

21. Brandner, G., and A. I. Virtanen: Isolierung und Synthese von γ-Hydroxy-L-Glutamin. Acta Chem. Scand. **17**, 2563 (1963).

22. Bray, H. H., S. P. James, I. M. Raffan, and W. V. Thorpe: The enzymic hydrolysis of glutamine and its spontaneous decomposition in buffer solutions. Biochem. J. **44**, 625 (1949).

23. Camp, W. H.: Glutathione in plants. Science **69**, 458 (1929).

24. Carnegie, P. R.: Isolation of a homologue of glutathione and other acidic peptides from seedlings of *Phaseolus aureus*. Biochem. J. **89**, 459 (1963).

25. — Structure and properties of a homologue of glutathione. Biochem. J. **89**, 471 (1963).

26. Carson, J. F., and F. F. Wong: Isolation of (+) S-methyl-L-cysteine sulfoxide and of (+) S-n-propyl-L-cysteine sulfoxide from onions as their N-2,4-dinitrophenyl derivatives. J. Org. Chem. **26**, 4997 (1961).

27. Casimir, J., J. Jadot et M. Renard: Séparation et caractérisation de la N-éthyl-γ-glutamine à partir de *Xerocomus badius*. Biochim. Biophys. Acta **39**, 462 (1960).

28. Cheung, Y. S., and P. S. Nobel: Amino acid uptake by pea leaf fragments. Specificity, energy sources, and mechanism. Plant Physiol. **52**, 633 (1973).

29. Chibnall, A. C., and R. G. Westall: The estimation of glutamine in the presence of asparagine. Biochem. J. **26**, 122 (1932).

30. Crocomo, O. J., and L. Fowden: Amino acid decarboxylases of higher plants. The formation of ethylamine. Phytochemistry **9**, 537 (1970).

31. DABROWSKA, T.: The isolation and identification of γ-L-glutamyl-L-glutamine from tillering nodes with roots of *Dactylis glomerata*. Bull. Acad. Pol. Sci., Ser. Sci. Biol. **19**, 95 (1971).

32. DALGAARD, L.: G.c.m.s. of γ-glutamyl derivatives, saccharopine, and aspergillomarasmines. Adv. Mass Spectroscopy **7**, 1644 (1978).

33. DALGARNO, L., and F. R. J. HIRD: Increase in the process of accumulation of amino acids in carrot slices with prolonged aerobic washing. Biochem. J. **76**, 209 (1960).

34. DANIELS, E. G., R. B. KELLY, and J. W. HINMAN: Agaritine: an improved isolation procedure and confirmation by synthesis. J. Amer. Chem. Soc. **83**, 3333 (1961).

35. DARDENNE, G. A.: Recherche, isolement et structure de nouveaux acides aminés libres dans des végétaux. Gembloux, pp. I—XII + 1—130 (1976).

36. DARDENNE, G., J. CASIMIR, and H. SØRENSEN: γ-L-Glutamyl-L-pipecolic acid in *Gleditsia caspica*. Phytochemistry **13**, 1515 (1974).

37. — — — 2(S),3(S)-3-hydroxy-4-methyleneglutamic acid from *Gleditsia caspica*. Phytochemistry **13**, 2195 (1974).

38. DARDENNE, G. A., and P. THONART: γ-Glutamylphenylalanine in *Dolichos* seeds. Phytochemistry **12**, 473 (1973).

39. DARDENNE, G. A., P. THONART, E. OTOUL et R. MARECHAL: Étude chimiotaxonomique dans les genres *Macrotyloma*, *Dolichos* et *Pseudovigna*. Phytochemistry **12**, 1983 (1973).

40. DINEEN, R. W., and D. O. GRAY: Improved synthesis and purification of N⁴-ethyl-L-asparagine. Org. Prep. Proced. Int. **9**, 39 (1977).

41. DUNNILL, P. M., and L. FOWDEN: γ-L-Glutamyl-β-pyrazol-1-yl-L-alanine, a peptide from Cucumber seeds. Biochem. J. **86**, 388 (1963).

42. — — The amino acids of seeds of the *Cucurbitaceae*. Phytochemistry **4**, 933 (1965).

43. — — The amino acids of the genus *Astragalus*. Phytochemistry **6**, 1659 (1967).

44. DUPUY, H. P., and J. G. LEE: The toxic component of the singletary pea *(Lathyrus pusillus)*. J. Amer. Pharm. Ass. **45**, 236 (1956).

45. DWEK, R. A.: Nuclear magnetic resonance (N.M.R.) in biochemistry, p. 154. Oxford: Clarendon Press. 1973.

46. ELLIOTT, W. H.: Isolation of Glutamine synthetase and glutamyl-transferase from green peas. J. Biol. Chem. **201**, 661 (1953).

47. ELOFF, J. N., and L. FOWDEN: The isolation of hypoglycin A and related compounds from *Billia hippocastanum*. Phytochemistry **9**, 2423 (1970).

48. ESTERBAUER, H., and D. GRILL: Seasonal variation of glutathione and glutathione reductase in needles of *Picea abies*. Plant Physiol. **61**, 119 (1978).

49. FÖLSCH, G.: Synthesis of phosphopeptides. V. Further dipeptides, tripeptides and O-phosphorylated derivatives of L-serine. Acta Chem. Scand. **20**, 459 (1966).

50. FOSKER, A. P., and H. D. LAW: L-Glutamyl-γ-aminobutyric acid and related compounds. J. Chem. Soc. **1965**, 7305.

51. FOWDEN, L.: A new asparagine derivative, N⁴-(2-hydroxyethyl)-L-asparagine from Bryony *(Bryonia dioica)*. Biochem. J. **81**, 154 (1961).

52. — The chemistry and metabolism of recently isolated amino acids. Annu. Rev. Biochem. **33**, 173 (1964).

53. — The acidic amino acids of tulip. Isolation of γ-ethylideneglutamic acid. Biochem. J. **98**, 57 (1966).

54. — Amino acid complement of plants. Phytochemistry **11**, 2271 (1972).

55. FOWDEN, L., D. LEWIS, and H. TRISTRAM: Toxic amino acids. Their action as antimetabolites. Advan. Enzymol. **29**, 89 (1967).

56. FOWDEN, L., and H. M. PRATT: Cyclopropylamino acids of the genus *Acer*. Distribution and biosynthesis. Phytochemistry **12**, 1677 (1973).

57. FOWDEN, L., H. M. PRATT, and A. SMITH: Nitrogenous constituents of *Billia hippocastanum* and *Acer pseudoplatanus*. Phytochemistry **11**, 3521 (1972).

58. Fowden, L., P. M. Scopes, and R. N. Thomas: Optical rotatory dispersion and circular dichroism. Part LXX. The circular dichroism of some less common amino acids. J. Chem. Soc. **1971** (C), 833.

59. Fowden, L., and A. Smith: Newly characterized amino acids from *Aesculus californica*. Phytochemistry **7**, 809 (1968).

60. — — Peptides from *Blighia sapida* seed. Phytochemistry **8**, 1043 (1969).

61. Fowden, L., A. Smith, D. S. Millington, and R. C. Sheppard: Cyclopropane amino acids from *Aesculus* and *Blighia*. Phytochemistry **8**, 437 (1969).

62. Foyer, C. H., and B. Halliwell: The presence of glutathione and glutathione reductase in chloroplasts: a proposed role in ascorbic acid metabolism. Planta **133**, 21 (1977).

62a. Freeman, G. G., and R. J. Whenham: Thiopropanol S-oxide, alkenylthiosulfinates and thiosulfonates: simulation of flavor components of *Allium* species. Phytochemistry **15**, 187 (1976).

63. Frisch, D. M., P. M. Dunnill, A. Smith, and L. Fowden: The specificity of amino acid biosynthesis in the Cucurbitaceae. Phytochemistry **6**, 921 (1967).

64. Fujii, J., and M. Izawa: γ-Glutamyl transpeptidase in green asparagus *(Asparagus officinalis)*. Agric. Biol. Chem. **31**, 767 (1967).

65. Fujiwara, S., G. Formicka-Kozlowska, and H. Kozlowski: Conformational study of glutathione by NMR. Bull. Chem. Soc. Jap. **50**, 3131 (1977).

66. Fukuda, M., T. Ogawa, and K. Sasaoka: Optical configuration of γ-glutamylalanine in pea seedlings. Biochim. Biophys. Acta **304**, 363 (1973).

67. Fukuda, M., A. Tokumura, T. Ogawa, and K. Sasaoka: D-Alanine in germinating *Pisum sativum* seedlings. Phytochemistry **12**, 2593 (1973).

68. Furuyama, T., T. Yamashita, and S. Senoh: The synthesis of L-theanine. Bull. Chem. Soc. Jap. **37**, 1078 (1964).

69. Gigliotti, H. J., and B. Levenberg: γ-Glutamyl transferase of *Agaricus bisporus*. J. Biol. Chem. **239**, 2274 (1964).

70. Gilbert, J. B., V. E. Price, and J. P. Greenstein: Effect of anions on the non-enzymatic desamidation of glutamine. J. Biol. Chem. **180**, 209 (1949).

71. Gmelin, R., and P. K. Hietala: S-[β-Carboxy-isopropyl]-L-Cystein, eine neue Aminosäure aus den Samen von *Acacia millefolia* und *Acacia willardiana* (Mimosaceae). Hoppe-Seyler's Z. physiol. Chem. **322**, 278 (1960).

72. Gmelin, R., H. Luxa, K. Roth, and G. Höfle: Dipeptide precursor of odour in *Marasmius* species. Phytochemistry **15**, 1717 (1976).

73. Goore, M. Y., and J. F. Thompson: γ-Glutamyl transpeptidase from kidney bean fruit. I. Purification and mechanism of action. Biochim. Biophys. Acta **132**, 15 (1967).

74. — — γ-Glutamyl transpeptidase from kidney bean fruit. II. Studies on the activating effect of sodium citrate. Biochim. Biophys. Acta **132**, 27 (1967).

75. Granroth, B.: Biosynthesis and decomposition of cysteine derivatives in onion and other *Allium* species. Ann. Acad. Fenn., Ser. A2 (1970), No. 154.

76. Gray, D. O., and L. Fowden: N-Ethyl-L-asparagine: a new amino-acid amide from *Ecballium*. Nature **189**, 401 (1961).

77. — — α-(Methylenecyclopropyl)glycine from *Litchi* seeds. Biochem. J. **82**, 385 (1962).

78. Greenstein, J. P., and M. Winitz: Chemistry of the Amino Acids. New York: Wiley. 1961.

79. Grove, M. D., D. Weisleder, and M. E. Daxenbichler: Pinnatanine and oxypinnatanine, novel amino acid amides from *Staphylea pinnata*. Tetrahedron **29**, 2715 (1973).

80. Guthrie, J. D.: Isolation of glutathione from potato tubers treated with ethylene chlorohydrin. J. Amer. Chem. Soc. **54**, 2566 (1932).

81. HALLIWELL, B., and C. H. FOYER: Properties and physiological function of a glutathione reductase purified from spinach leaves by affinity chromatography. Planta 139, 9 (1978).

82. HAMILTON, P. B., and R. R. TARR: Gasometric determination of glutamine amino acid carboxyl N in plasma and tissue filtrates by the ninhydrin-$CO_2$ method. J. Biol. Chem. 158, 375 (1945).

83. HANES, C. S., F. J. R. HIRD, and F. A. ISHERWOOD: Synthesis of peptides in enzymic reactions involving glutathione. Nature 166, 288 (1950).

84. HARINGTON, C. R., and T. H. MEAD: The synthesis of glutathione. Biochem. J. 29, 1602 (1935).

85. HARRINGTON, H. M., and I. K. SMITH: Cysteine transport into cultured tobacco cells. Plant Physiol. 60, 807 (1977).

86. HASEGAWA, M., N. FUKUDA, H. HIGUCHI, S. NOGUCHI, and I. MATSUBARA: The studies on the γ-glutamylpeptides in L-glutamic acid fermentation broths. Part I. Agric. Biol. Chem. 41, 49 (1977).

87. HASEGAWA, M., and I. MATSUBARA: γ-Glutamylpeptide formative activity of Corynebacterium glutamicum by the reverse reaction of the γ-glutamylpeptide hydrolytic enzyme. Agric. Biol. Chem. 42, 371 (1978).

88. — — The mechanism of the formation of γ-glutamylpeptides during L-glutamic acid fermentation contributed solely by γ-glutamyltranspeptidase. Agric. Biol. Chem. 42, 383 (1978).

89. HASHIZUME, T.: Amino acids in tea. II. Synthesis of related compounds of theanine. Nippon Nogei Kagaku Kaishi 25, 129 (1951). (In Japanese with English summary.)

90. HASSALL, C. H., and D. I. JOHN: Constitution of hypoglycin B. Tetrahedron Lett. 1959, 7.

91. — — Amino acids and peptides. III. Constitution of hypoglycin B. J. Chem. Soc. 1960, 4112.

92. HASSALL, C. H., and K. REYLE: Hypoglycin A and B, two biologically active polypeptides from Blighia sapida. Biochem. J. 60, 334 (1955).

93. HATANAKA, S., Y. NIIMURA, and K. TANIGUCHI: L-2-Aminohex-4-ynoic acid: A new amino acid from Tricholomopsis rutilans. Phytochemistry 11, 3327 (1972).

94. HATFIELD, G. M., and J. P. SCHAUMBERG: Isolation and structural studies of coprine, the disulfiram-like constituent of Coprinus atramentarius. Lloydia 38, 489 (1975).

95. HAYSTEAD, A.: Glutamine synthetase in the chloroplasts of Vicia faba. Planta 111, 271 (1973).

96. HENDLEY, D. D., and E. E. CONN: Enzymic reduction and oxidation of glutathione by illuminated chloroplasts. Arch. Biochem. Biophys. 46, 454 (1953).

97. HIGGINS, C. F., and J. W. PAYNE: Peptide transport by germinating barley embryos: Evidence for a single common carrier for di- and oligopeptides. Planta 138, 217 (1978).

98. HODA, Y., T. WATANABE, O. OKA, and T. SAEKI: Glutathione-containing liquids from embryo buds. Japan Kokai 76, 141, 899 (1976) (in Japanese).

99. — — — — Preparation method of glutathione-copper salt from embryo buds. Japan Kokai 76, 141, 900 (1976) (in Japanese).

100. HÖFLE, G., R. GMELIN, H. LUXA, M. N'GALAMULUME-TREVES, and S. HATANAKA: Struktur der Lentinsäure: 2-(γ-Glutamylamino)-4,6,8,10,10-pentaoxo-4,6,8,10-tetrathiaundecansäure. Tetrahedron Lett. 1976, 3129.

101. HÖRHAMMER, L., H. WAGNER, M. SEITZ, und Z. J. VEJDELEK: Zur Wertbestimmung von Knoblauchpräparaten. 1. Mitteilung: Chromatographische Untersuchungen über die genuinen Inhaltstoffe von Allium sativum L. Pharmazie 23, 462 (1968).

102. HOLT, C. V., und W. LEPPLA: Die Konstitution von Hypoglycin A und B. Hoppe-Seyler's Z. physiol. Chem. 313, 276 (1958).

*103.* HOPKINS, F. G.: On glutathione. A reinvestigation. J. Biol. Chem. **84**, 269 (1929).

*104.* HOPKINS, F. G., and E. J. MORGAN: Appearance of glutathione during the early states of germination of seeds. Nature **152**, 288 (1943).

*105.* — — Distribution of glyoxylase and glutathione. Biochem. J. **39**, 320 (1945).

*106.* HUNTER, G., and B. A. EAGLES: Glutathione — A critical study. J. Biol. Chem. **72**, 147 (1927).

*107.* ISHIKAWA, Y., S. HASEGAWA, T. KASAI, and Y. OBATA: Changes in amino acid composition during germination of soybean. IV. Identification of α- and γ-glutamyl-aspartic acid. Agric. Biol. Chem. **31**, 490 (1967).

*108.* ITO, K., and L. FOWDEN: New characterizations of amino acids and γ-glutamyl peptides from *Acacia georginae* seed. Phytochemistry **11**, 2541 (1972).

*109.* ITOH, M.: Peptides. I. Selective protection of α- or sidechain carboxyl groups of aspartic and glutamic acid. A facile synthesis of β-aspartyl and γ-glutamyl peptides. Chem. Pharm. Bull. **17**, 1679 (1969).

*110.* IWAMI, K., K. YASUMOTO, and H. MITSUDA: Enzymic cleavage of cysteine sulfoxide in *Lentinus edodes.* Agric. Biol. Chem. **39**, 1947 (1975).

*111.* IWAMI, K., K. YASUMOTO, K. NAKAMURA, and H. MITSUDA: Properties of γ-glutamyl-transferase from *Lentinus edodes.* Agric. Biol. Chem. **39**, 1933 (1975).

*112.* — — — — Reactivity of *Lentinus* γ-glutamyltransferase with lentinic acid as the principal endogenous substrate. Agric. Biol. Chem. **39**, 1941 (1975).

*113.* IZADDOST, M., B. G. HARRIS, and R. W. GRACY: Structure and toxicity of alkaloids and amino acids of *Sophora secundiflora.* J. Pharm. Sci. **65**, 352 (1976).

*114.* JADOT, J., J. CASIMIR et G. MAGHUIN: Identification de la L(+)cystathionine dans *Boletus erythropus.* Bull. Soc. Roy. Sci. Liège **40**, 355 (1971).

*115.* JADOT, J., J. CASIMIR et M. RENARD: Séparation et caractérisation du L(+)-γ-(p-hydroxy)anilide de l'acide glutamique à partir de *Agaricus hortensis.* Biochim. Biophys. Acta **43**, 322 (1960).

*116.* JANSEN, E. F., and R. JANG: Cysteine and glutathione in orange juice. Arch. Biochem. Biophys. **40**, 358 (1952).

*117.* JESCHKEIT, H., und G. LOSSE: Peptidsynthesen in der Aminodicarbonsäurereihe. Z. Chem. **5**, 81 (1965).

*118.* JOCELYN, P. C.: Biochemistry of the SH-group. London-New York: Academic Press. 1972.

*119.* JÖHL, A., und W. G. STOLL: Zur Konstitution von Hypoglycin B. Helv. Chim. Acta **42**, 156 (1959).

*120.* — — Synthese von γ-L-Glutamyl-hypoglycin A (Hypoglycin B). Helv. Chim. Acta **42**, 716 (1959).

*121.* JUNG, G., E. BREITMAIER, und W. VOELTER: Dissociationsgleichgewichte von Gluta-thion. Einer Fourier-Transform-[13]C-NMR spektroskopische Untersuchung der pH-Abhängigkeit der Ladungsverteilung. Eur. J. Biochem. **24**, 438 (1972).

*122.* JUNG, G., E. BREITMAIER, W. VOELTER, T. KELLER und C. TÄNZER: Fourier-Transform-[13]C-NMR-Spektroskopie biologisch aktiver Cysteinpeptide. Angew. Chem. **82**, 882 (1970).

*123.* KAKIMOTO, Y., A. KANAZAWA, T. NAKAJIMA, and I. SANO: Isolation of γ-L-glutamyl-L-β-aminoisobutyric acid from bovine brain. Biochem. Biophys. Acta **100**, 426 (1965).

*124.* KANAMORI, T., and H. MATSUMOTO: Glutamine synthetase from rice plant roots. Arch. Biochem. Biophys. **125**, 404 (1972).

*125.* KANAMARU, K., K. KATO, and M. NOGUCHI: Isolation and identification of γ-L-glutamyl-L-glutamic acid from tobacco cells in suspension culture. Agric. Biol. Chem. **38**, 2285 (1974).

*126.* KASAI, T., Y. ISHIKAWA, and Y. OBATA: Changes in amino acid composition during germination of soybean. II: Identification of two γ-glutamyl peptides and their change during germination. Agric. Biol. Chem. **30**, 979 (1966).

127. KASAI, T., Y. ISHIKAWA, Y. OBATA, and T. TSUKAMOTO: Changes in amino acid composition during germination of soybean. I. Changes in free amino acids, several nitrogen compounds and total amino acids. Agric. Biol. Chem. **30**, 973 (1966).

128. KASAI, T., and P. O. LARSEN: Acta Chem. Scand. **B 33**, 213 (1979).

129. KASAI, T., P. O. LARSEN, and H. SØRENSEN: Free amino acids and γ-glutamyl peptides in Fagaceae. Phytochemistry **17**, 1911 (1978).

130. KASAI, T., and Y. OBATA: Changes in amino acid compositions during germination of soybean. Part III. Changes in γ-glutamyltranspeptidase activity. Agric. Biol. Chem. **31**, 127 (1967).

131. KASAI, T., and S. SAKAMURA: Distinction between α- and γ-glutamyl dipeptides by means of NMR spectrometer and amino acid analyzer. Agric. Biol. Chem. **37**, 685 (1973).

132. — — Nuclear magnetic resonance spectra of glutamic acid containing dipeptides in relation to sequence determination. Agric. Biol. Chem. **37**, 2155 (1973).

133. — — Differences in NMR spectra between some α- and γ-glutamyl dipeptides. Agric. Biol. Chem. **38**, 1257 (1974).

134. — — Infrared spectra of α- and γ-glutamyl dipeptides. Nippon Nogei Kagaku Kaishi **48**, 521 (1974) (in Japanese with English summary).

135. — — NMR-spectra of α- and γ-L-glutamyl-α-amino-isobutyric acid and some related compounds. Agric. Biol. Chem. **39**, 239 (1975).

136. — — Elution behaviour of γ-L-glutamyl-L-aspartic acid during ion-exchange chromatography. J. Chromatogr. **103**, 189 (1975).

137. — — NMR and IR spectra and elution behaviors during ion exchange chromatography of glutamic acid-containing dipeptides in relation to sequence determination. J. Fac. Agric. Hokkaido Univ. **58**, 283 (1975).

138. KASAI, T., S. SAKAMURA, S. INAGAKI, and R. SAKAMOTO: Isolation and identification of γ-glutamyl-γ-glutamylmethione from green gram seed. Agric. Biol. Chem. **36**, 2621 (1972).

139. KASAI, T., S. SAKAMURA, S. OHASHI, and H. KUMAGAI: Amino acid composition of soybean. V. Changes in free amino acids, ethanolamine and two γ-glutamylpeptides content during the ripening period of soybean. Agric. Biol. Chem. **34**, 1848 (1970).

140. KASAI, T., S. SAKAMURA, and R. SAKAMOTO: Amino acid composition of green gram *(Phaseolus radiatus* var. *typicus)*. I. Isolation and identification of N-carboxymethyl-β-alanine and four γ-glutamyl peptides. Agric. Biol. Chem. **35**, 1603 (1971).

141. — — — Amino acid composition of green gram *(Phaseolus radiatus* L. var. *typicus* Prain).* Part II. Contents of free amino acids, γ-glutamyl peptides and protein amino acids in green gram seeds and seedlings. Agric. Biol. Chem. **35**, 1607 (1971).

142. — — — Isolation and identification of γ-L-glutamyl-L-methionine sulfoxide from green gram seed. Agric. Biol. Chem. **36**, 967 (1972).

143. KASAI, T., M. SANO, and S. SAKAMURA: Amino acid composition of broad bean *(Vicia faba)*. I. Pattern of acidic amino acid fractions. Nippon Nogei Kagaku Kaishi **49**, 313 (1975). (In Japanese with English summary.)

144. — — — Amino acid composition of broad bean (*Vicia faba* L.) Part II. $N^G$-Methylated arginines in broad bean seed. Agric. Biol. Chem. **40**, 2449 (1976).

145. KASAI, T., M. UEDA, S. SAKAMURA, and K. SAKATA: Amino acid composition of ladino clover *(Trifolium repens* L. var. *giganteum)*. (Studies on the components in acid amino acid fraction Part I.) Nippon Nogei Kagaku Kaishi **47**, 583 (1973). (In Japanese with English summary.)

146. KEAN, E. A. (Ed.): Hypoglycine. PAABS Symposium Series **3**, pp. 183. New York: Academic Press. 1976.

147. KELLY, R. B., E. G. DANIELS, and J. W. HINMAN: Agaritine: Isolation, degradation, and synthesis. J. Org. Chem. **27**, 3229 (1962).

148. Kendall, E. C., H. L. Mason, and B. F. McKenzie: A study of glutathione. III. The structure of glutathione. J. Biol. Chem. **87**, 55 (1930).

149. — — — A study of glutathione. IV. Determination of the structure of glutathione. J. Biol. Chem. **88**, 409 (1930).

150. Kerdel-Vegas, F., F. Wagner, P. B. Rusell, N. H. Grant, H. E. Alburn, D. E. Clark, and J. A. Miller: Selenocystathionine, a pharmacologically active factor in the seeds of *Lecythis ollaria*. Structure of the pharmacologically active factor in the seeds of *Lecythis ollaria*. Nature **205**, 1186 (1965).

151. King, J., and R. Hirji: Amino acid transport systems of cultured soybean root cells. Can. J. Bot. **53**, 2088 (1975).

152. Kiryushkin, A. A., A. I. Miroshinikov, Y. A. Ovchinnikov, B. V. Rosinov, and M. M. Shemyakin: Mass spectrometric determination of the type of amide bond in α- and γ-peptides of glutamic acid. Biochem. Biophys. Res. Commun. **24**, 943 (1966).

153. Kito, M., H. Inagaki, S. Konishi, and K. Sasaoka: Studies on the biosynthesis of theanine in tea seedlings. Incorporation of ethylamine-1-$^{14}$C into the ethylamine part of theanine. Mem. Res. Inst. Food Sci., Kyoto Univ. No. **25**, 34 (1963).

154. Kito, M., H. Kokura, J. Izaki, and K. Sasaoka: Theanine, a precursor of the phloroglucinol nucleus of cathecins in tea plants. Phytochemistry **7**, 599 (1968).

155. Kjaer, A., and P. O. Larsen: Non-protein amino acids, cyanogenic glycosides, and glucosinolates. Biosynthesis (Geissman. T. A.. ed.) (Specialist Periodical Reports). The Chemical Society, London **2**, 71 (1973).

156. — — Non-protein amino acids, cyanogenic glycosides, and glucosinolates. Biosynthesis (Bu'lock, J. D.. ed.) (Specialist Periodical Reports), The Chemical Society, London **4**, 179 (1976).

157. Klosterman, H. J.: Vitamin B$_6$ antagonists of natural origin. J. Agr. Food Chem. **22**, 13 (1974).

158. Klosterman. H. J.. G. L. Lamoreux. and J. L. Parsons: Isolation. characterization. and synthesis of linatine. A vitamin B$_6$ antagonist from flaxseed *(Linum usitatissimum)*. Biochemistry **6**, 170 (1967).

159. Konishi, S.: Physiological chemistry of two amides contained in tea tree. Chagyo Kenkyu Hokoku, Shiryo **22** (1970) (in Japanese).

160. Konishi, S., and Z. Kasai: Effect of shading on carbon dioxide-$^{14}$C assimilation by tea leaves and the metabolism and regulation of theanine and related compounds in tea plants. I. Nippon Dojo-Hiryogaku Zasshi **39**, 264 (1968) (in Japanese).

161. — — Metabolism and regulation of theanine and related compounds in tea plants. II. Synthesis of theanine from carbon-$^{14}$C dioxide in tea plants and sites of the synthesis. Nippon Dojo-Hiryogaku Zasshi **39**, 439 (1968) (in Japanese).

162. — — Metabolism and regulation of theanine and related compounds in tea plants. III. Translocation and metabolic changes of carbon-$^{14}$C dioxide assimilated products in tea plants during autumn. Nippon Dojo-Hiryogaku Zasshi **39**, 444 (1968) (in Japanese).

163. Konishi, S., T. Matsuda, and E. Takahashi: Synthesis of theanine and L-glutamic acid γ-methylamide in *Thea sinensis, Camellia sasanqua*, and *Oryza sativa*. V. Metabolism and regulation of theanine and related compounds. Nippon Dojo-Hiryogaku Zasshi **40**, 107 (1969) (in Japanese). (English summary in: Soil Science and Plant Nutrition **15**, 242 (1969).)

164. Konishi, S., M. Ozasa, and E. Takahashi: Metabolic conversion of N-methyl carbon of γ-glutamylmethylamide to caffeine in tea plants. Plant Cell Physiol. **13**, 365 (1972).

165. Konishi, S., and E. Takahashi: Degradation of theanine labelled with carbon-14 in tea seedlings. Nippon Dojo-Hiryogaku Zasshi **37**, 612 (1966) (in Japanese).

166. — — Existence and synthesis of L-glutamic acid γ-methylamide in tea plants. Plant Cell Physiol. **7**, 171 (1966).

167. — — Metabolism of theanine-N-ethyl-$^{14}$C and its metabolic redistribution in the tea plant. VI. Metabolism and regulation of theanine and related compounds. Nippon Dojo-Hiryogaku Zasshi **40**, 479 (1969) (in Japanese).

168. KORNGUTH, M. L., A. NEIDLE, and H. WAELSCH: The stability and rearrangement of ε-N-glutamyl-lysines. Biochemistry **2**, 740 (1963).

169. KORTÜM, G., W. VOGEL, and K. ANDRESSOV: Dissociation constants of organic acids in aqueous solution. London: Butterworth. 1961.

170. KOYAMA, M., and Y. OBATA: Isolation and structure of γ-L-glutamyl-L-β-phenyl-β-alanine, a new γ-glutamyl peptide from *Phaseolus angularis* W. F. Wight (Azuki bean). Agric. Biol. Chem. **30**, 472 (1966).

171. — — Synthesis of α- and γ-L-glutamyl dipeptides of L-β-phenyl-β-alanine. Agric. Biol. Chem. **31**, 738 (1967).

172. KOYAMA, M., Y. OBATA, and S. SAKAMURA: Identification of hydroxybenzylamines in buckwheat seeds (*Fagopyrum esculentum* Moench). Agric. Biol. Chem. **35**, 1870 (1971).

173. KOYAMA, M., Y. TSUJIZAKI, and S. SAKAMURA: New amides from buckwheat seeds (*Fagopyrum esculentum* Moench). Agric. Biol. Chem. **37**, 2749 (1973).

174. KREBS, H. A.: Metabolism of amino acids. IV. The synthesis of glutamine from glutamic acid and ammonia, and the enzymic hydrolysis of glutamine in animal tissues. Biochem. J. **29**, 1951 (1935).

175. KRISTENSEN, I., and P. O. LARSEN: Differentiation between α-glutamyl peptides, γ-glutamyl peptides, and α-aminoacylglutamic acids by PMR spectroscopy. Acta Chem. Scand. **27**, 3123 (1973).

176. — — γ-Glutamylwillardiine and γ-glutamylphenylalanylwillardiine from seeds of *Fagus silvatica*. Phytochemistry **13**, 2799 (1974).

177. KRISTENSEN, I., P. O. LARSEN, and H. SØRENSEN: Free amino acids and γ-glutamyl peptides in seeds of *Fagus silvatica*. Phytochemistry **13**, 2803 (1974).

178. KRUGER, G. J., L. M. DU PLESSIS, and N. GROBBELAAR: The structure of $N^5$-(3-hydroxymethyl-2,5-dihydro-2-furyl)-L-allo-γ-hydroxyglutamine, a new amino acid from *Hemerocallis fulva* L. (Day Lily). J. S. Afr. Chem. Inst. **29**, 24 (1976).

179. KUNINORI, T., and H. MATSUMOTO: Glutathione in wheat and wheat flour. Cereal Chem. **40**, 252 (1964).

180. KUO, Y. H., F. LAMBEIN, and R. VAN PARIJS: The presence of isoxazolin-5-one derivatives in root exudates of pea and sweet pea seedlings. Arch. Int. Physiol. Biochim. **84**, 169 (1976).

181. KUPIECKI, F. P., and A. I. VIRTANEN: Cleavage of allyl cysteine sulphoxides by an enzyme in onion (*Allium cepa*). Acta Chem. Scand. **14**, 1913 (1960).

182. KURELEC. B.. M. RIJAVEC. S. BRITVIC. W. E. G. MÜLLER. and R. K. ZAHN: Phytoplankton: presence of γ-glutamyl cycle. Comp. Biochem. Physiol. B **56**, 415 (1977).

183. KUTTAN, R., N. G. NAIR, A. N. RADHAKRISHNAN, T. F. SPANDE, H. J. YEH, and B. WITKOP: The isolation and characterization of γ-L-glutamyl-S-(*trans*-1-propenyl)-L-cysteine sulfoxide from sandal (*Santalum album* L.). An interesting occurrence of sulfoxide diastereoisomers in nature. Biochemistry **13**, 4394 (1974).

184. LAMBEIN, F., and R. VAN PARIJS: Isolation and characterization of 1-alanyl-uracil (willardiine) and 3-alanyl-uracil (isowillardiine) from *Pisum sativum*. Biochem. Biophys. Res. Comm. **32**, 474 (1968).

185. — — New isoxazolinone amino acids from *Lathyrus odoratus* seedlings. Biochem. Biophys. Res. Commun. **61**, 155 (1974).

186. LARSEN, P. O.: Amino acids and γ-glutamyl derivatives in seeds of *Lunaria annua* L. Acta Chem. Scand. **16**, 1511 (1962).

187. Larsen, P. O.: Amino acids and γ-glutamyl derivatives in seeds of *Lunaria annua* L. Part II. Acta Chem. Scand. **19**, 1071 (1965).

188. — Amino acids and γ-glutamyl derivatives in seeds of *Lunaria annua* L. Part III. Acta Chem. Scand. **21**, 1592 (1967).

189. Larsen, P. O., and H. Sørensen: γ-Glutamyl-phenylalanine and γ-L-glutamyl-L-tyrosine from seeds of *Aubrietia deltoidea* DC. Acta Chem. Scand. **21**, 2908 (1967).

190. Le Quesne, W. J., and G. T. Young: Amino acids and peptides. Part I. An examination of the use of carbobenzyloxy-L-glutamic anhydride in the synthesis of glutamyl-peptides. J. Chem. Soc. **1950**, 1954.

191. — — Amino acids and peptides. Part II. Synthesis of α- and γ-glutamyl-peptides by the azide route. J. Chem. Soc. **1950**, 1959.

192. — — Amino-acids and peptides. Part VII. The autohydrolysis of glutamyl-peptides. J. Chem. Soc. **1952**, 594.

193. Levenberg, B.: Isolation and structure of agaritine. a γ-glutamyl-substituted aryl-hydrazide derivative from Agaricaceae. J. Biol. Chem. **239**, 2267 (1964).

194. — Isolation and characterization of β-methylene-L-(+)-norvaline from *Lactarius helvus*. J. Biol. Chem. **243**, 6009 (1968).

195. — Agaritine and γ-glutamyltransferase (mushroom). Methods Enzymol. **17A**, 877 (1970).

196. Levintov, L., A. Meister, G. H. Hogeboom, and E. L. Kuff: The relation between the enzymic synthesis of glutamine and the glutamine-transfer reaction. J. Amer. Chem. Soc. **77**, 5304 (1955).

197. Lichtenstein, N.: Preparation of γ-alkylamides of glutamic acid. J. Amer. Chem. Soc. **64**, 1021 (1942).

198. Lichtenstein, N., and N. Grossowicz: Inhibition of the growth of *Staphylococcus aureus* by some derivatives of glutamic acid. J. Biol. Chem. **171**, 387 (1947).

199. Liefländer, M.: Über die Darstellung und das Zinkbindungsvermögen einiger Glutamylpeptide. Hoppe-Seyler's Z. physiol. Chem. **320**, 35 (1960).

200. Lien, R., and S. E. Rognes: Uptake of amino acids by barley leaf slices: kinetics, specificity, and energetics. Physiol. Plant. **41**, 175 (1977).

201. Lindberg, P., R. Bergman, and B. Wickberg: Isolation and structure of coprine, a novel physiologically active cyclopropanone derivative from *Coprinus atramentarius* and its synthesis via 1-aminocyclopropanol. Chem. Commun. **1975**, 946.

202. — — — Isolation and structure of coprine, the *in vivo* aldehyde dehydrogenase inhibitor in *Coprinus atramentarius*; synthesis of coprine and related cyclopropanone derivatives. J. Chem. Soc. Perkin I **1977**, 684.

203. MacKay, G. F., J. J. Lalich, E. D. Schilling, and F. M. Strong: A toxic factor from *Lathyrus odoratus*. Arch. Biochem. Biophys. **52**, 313 (1954).

204. McMullen, A. I.: Thiols of low molecular weight in *Hevea brasiliensis* latex. Biochim. Biophys. Acta **41**, 152 (1960).

205. McParland, R. H., J. G. Guevara, R. R. Becker, and H. J. Evans: The purification and properties of the glutamine synthetase from the cytosol of soybean root nodules. Biochem. J. **153**, 597 (1976).

206. Mapson, L. W., and F. A. Isherwood: Glutathione reductase from germinated peas. Biochem. J. **86**, 173 (1963).

207. Maretzki, A., and M. Them: Arginine and lysine transport in sugar-cane cell suspension cultures. Biochemistry **9**, 2731 (1970).

208. Martin, J. L.: Selenium assimilation in animals. In Organic selenium compounds: their chemistry and biology, p. 663. (Klayman, D. L., and W. H. H. Günther, eds.) New York: Wiley. 1973.

209. Mason, H. L.: Glutathione V. The spontaneous cleavage of glutathione in aqueous solution. J. Biol. Chem. **90**, 25 (1931).

210. MATIKKALA, E. J., and A. I. VIRTANEN: A new γ-glutamylpeptide, γ-L-glutamyl-S-(prop-1-enyl)-L-cysteine, in the seeds of Chives (Allium schoenoprasum). Acta Chem. Scand. 16, 2461 (1962).

211. — — New γ-Glutamylpeptides isolated from the seeds of chives: N,N'-bis-(γ-glutamyl)-cystine, N,N'-bis-(γ-glutamyl)-3,3'-(2-methylethylene-1,2-dithio)-dialanine, γ-glutamyl-S-propylcysteine. Acta Chem. Scand. 17, 1799 (1963).

212. — — Synthesis of 3,3'-(2-methylethylene-1,2-dithio)dialanine, an amino acid found as γ-glutamylpeptide in the seeds of chive (Allium schoenoprasum). Acta Chem. Scand. 18, 2009 (1964).

213. — — γ-Glutamylpeptidase (glutaminase) in germinating seeds of chive (Allium schoenoprasum). Acta Chem. Scand. 19, 1258 (1965).

214. — — γ-Glutamylpeptidase in sprouting onion bulbs. Acta Chem. Scand. 19, 1261 (1965).

215. — — A new type of γ-glutamyl tripeptide, γ-glutamyl-S-(prop-1-enyl)cysteinyl-S-(prop-1-enyl)-cysteine sulphoxide. Suom. Kemistilehti B 39, 201 (1966).

216. — — Isolation of γ-L-Glutamyl-L-arginine and γ-L-glutamyl-S-(2-carboxy-n-propyl)-L-cysteine from Allium cepa (onion). Suom. Kemistilehti B 43, 435 (1970).

217. MAZELIS, M., and H. M. PRATT: In vivo conversion of 5-oxoproline to glutamate by higher plants. Plant Physiol. 57, 85 (1976).

218. MEISTER, A.: The mechanism and specificity of the glutamine-α-keto acid transamination-deamidation. J. Biol. Chem. 210, 17 (1954).

219. — Glutathione synthesis. In "The Enzymes" (BOYER, P. D., ed.), 3rd. ed. Vol. 10, pp. 671—697. New York: Academic Press. 1974.

220. — Biochemistry of glutathione. In Metab. Pathways (GREENBERG, D. M., ed.), 3rd ed. `7, 101. New York: Academic Press. 1975.

221. MEISTER, A., and S. S. TATE: Glutathione and related γ-glutamyl compounds: biosynthesis and utilization. Ann. Rev. Biochem. 45, 559 (1976).

222. MELVILLE, J.: Labile glutamine peptides, and their bearing on the origin of the ammonia set free during the enzymic digestion of proteins. Biochem. J. 29, 179 (1935).

223. MIFLIN, B. J., and P. J. LEA: The pathway of nitrogen assimilation in plants. Phytochemistry 15, 873 (1976).

224. — — Amino acid metabolism. Ann. Rev. Plant Physiol. 28, 331 (1977).

225. MOOZ, E. D., and A. MEISTER: Tripeptide (glutathione) synthetase, purification, properties, and mechanism of action. Biochemistry 6, 1722 (1967).

226. MOOZ, E. D., and L. WIGGLESWORTH: Evidence for the γ-glutamyl cycle in yeast. Biochem. Biophys. Res. Commun. 1975, 1066.

227. MORITA, K., and S. KOBAYASHI: Isolation and synthesis of lenthionine, an odorous substance of shiitake, an edible mushroom. Tetrahedron Lett. 1966, 573.

228. — — Isolation, structure, and synthesis of lenthionine and its analogs. Chem. Pharm. Bull. 15, 988 (1967).

229. MORRIS, C. J., and J. F. THOMPSON: The identification of (+)-S-methyl-L-cysteine sulfoxide in plants. J. Amer. Chem. Soc. 78, 1605 (1956).

230. — — The isolation and characterization of γ-L-glutamyl-L-tyrosine and γ-L-glutamyl-L-phenylalanine from soybeans. Biochemistry 1, 706 (1962).

231. MORRIS, C. J., J. F. THOMPSON, and S. ASEN: The identification of γ-L-glutamyl-L-alanine and γ-L-glutamyl-L-valine from Iris leaf tissue. J. Biol. Chem. 239, 1833 (1964).

232. MORRIS, C. J., J. F. THOMPSON, S. ASEN, and F. IRREVERRE: The isolation of γ-L-glutamyl-β-aminoisobutyric acid from Iris bulbs. J. Biol. Chem. 236, 1181 (1961).

233. — — — — — The isolation of γ-L-glutamyl-β-alanine from Iris bulbs. J. Biol. Chem. 237, 2180 (1962).

234. Morris, C. J., J. F. Thompson, and R. M. Zacharius: The identification of γ-L-glutamyl-L-leucine and γ-L-glutamyl-L-methionine in kidney bean seeds *(Phaseolus vulgaris)*. J. Biol. Chem. **238,** 650 (1963).
235. Neuberger, A., and F. Sanger: The nitrogen of the potato. Biochem. J. **36,** 664 (1942).
236. Nigam. S. N.. and W. B. McConnell: Incorporation of serine-U-$^{14}$C into β-cyano-alanine and γ-glutamyl-β-cyanoalanine in *Vicia sativa*. Can. J. Biochem. **46,** 1327 (1968).
237. — — Seleno-amino compounds from *Astragalus bisulcatus*. Isolation and identification of γ-L-glutamyl-Se-methyl-seleno-L-cysteine and Se-methylseleno-L-cysteine. Biochim. Biophys. Acta **192,** 185 (1969).
238. — — Isolation and identification of L-cystathionine and L-selenocystathionine from the foliage of *Astragalus pectinatus*. Phytochemistry **11,** 377 (1972).
239. — — Isolation and identification of two isomeric glutamyl seleno cystathionines from the seeds of *Astragalus pectinatus*. Biochim. Biophys. Acta **437,** 116 (1976).
240. Nigam, S. N., J. Tu, and W. B. McConnell: Distribution of selenomethyl-seleno-cysteine and some other amino acids in species of *Astragalus,* with special reference to their distribution during the growth of *A. bisulcatus*. Phytochemistry **8,** 1161 (1969).
241. Niimura, Y., and S. Hatanaka: L-Threo- and L-erythro-2-amino-3-hydroxyhex-4-ynoic acids: New amino acids from *Tricholomopsis rutilans*. Phytochemistry **13,** 175 (1974).
242. — — Two γ-glutamylpeptides of acetylenic amino acids in *Tricholomopsis rutilans*. Phytochemistry **16,** 1435 (1977).
243. Nissen, P.: Uptake mechanism: inorganic and organic. Ann. Rev. Plant Physiol. **25,** 53 (1974).
244. Nobel, P. S., and Y. S. Cheung: Two amino acid carriers in pea chloroplasts. Nature. New Biol. **237,** 207 (1972).
245. Nugent. D. J.: Chemical synthesis and metabolism of linatine. Dissertation. North Dakota State University, 1970. xxiv + 87 pp. University Microfilms. Ann Arbor, Michigan, Publication No. 71—12, 504.
246. Obata, Y., and R. Kitasawa: Synthesis of γ-L-glutamyl-S-methyl-L-cysteine. Agric. Biol. Chem. **28,** 624 (1964).
247. Ogawa, T.: Identification of γ-glutamyl peptides. Kyoto Daigaku Shokuryo Kagaku Kenkyusho Hokoku **37,** 1 (1974).
248. Ohgishi, H.: Extraction of γ-L-glutamyl-L-lathyrine from plants. Japan Kokai 76 29. 213 (1976) (in Japanese).
249. Okada, K., and M. Kawase: Mass spectral differentiation of α- and γ-linkages in glutamyl oligopeptides and its application for structure elucidation of naturally occurring peptides. Chem. Pharm. Bull. **25,** 1497 (1977).
250. Olcott, H. S.: A method for the determination of glutamic acid in proteins. J. Biol. Chem. **153,** 71 (1944).
251. O'Neal, D., and K. W. Joy: Glutamine synthetase of pea leaves. Purification, stabilization, and pH optima. Arch. Biochem. Biophys. **159,** 113 (1973).
252. — — Glutamine synthetase of pea leaves. Divalent cation effects, substrate specificity, and other properties. Plant Physiol. **54,** 773 (1974).
253. Orlowski, M., and A. Meister: γ-Glutamyl transpeptidase (hog kidney). Methods Enzymol. **17 A,** 883 (1970).
254. — — Isolation of highly purified γ-glutamylcysteine synthetase from rat kidney. Biochemistry **10,** 372 (1971).
255. Otani, T. T., and A. Meister: ω-Amide and ω-amino acid derivatives of α-keto-glutaric acid and oxalacetic acid. J. Biol. Chem. **224,** 137 (1957).
256. Otoul, E., R. Marechal, G. Dardenne et F. Desmedt: Des dipeptides soufres differencient nettement *Vigna radiata* de *Vigna mungo*. Phytochemistry **14,** 173 (1975).

257. PERRIN, D. D.: Dissociation constants of organic bases in aqueous solution. Butterworth, London, 1965.
258. PETERSON, P. J., and P. J. ROBINSON: L-Cystathionine and its selenium analog in *Neptunia amplexicaulis*. Phytochemistry **11**, 1837 (1972).
259. PETT, L. B.: Changes in the ascorbic acid and glutathione contents of stored and sprouting potatoes. Biochem. J. **30**, 1228 (1936).
260. PETZOLD, U., A. BAUMERT und F. JACOB: Untersuchungen zur Aufnahme von α-Aminoisobuttersäure und L-Valin in Blätter von *Egeria densa* Planch. Wiss. Z. Humboldt. Berlin, Math.-Naturwiss. Reihe **25**, 137 (1976).
261. PIRIE, N. W., and K. G. PINHEY: The titration curve of glutathione. J. Biol. Chem. **84**, 321 (1929).
262. PLAISTED, P. H.: Clearing free amino acids solutions of plant extracts for paper chromatography. Contrib. Boyce Thompson Inst. **19**, 231 (1958).
263. POOLE, R. J.: Transport in cells of storage tissue. In Encyclopaedia of Plant Physiology (PIERSON, A., and M. H. ZIMMERMANN, eds.), New Series, Vol. 2A (LÜTTIGE, U., and M. G. PITMAN, eds.), Transport in Plants II. Part A. Cells, p. 229. Berlin-Heidelberg-New York: Springer. 1976.
264. PRZYBYLSKA, J., and B. CHWALEK: γ-L-Glutamyl-L-tyrosine and γ-L-glutamyl-L-phenylalanine in *Lotus corniculatus*. L. Bull. Acad. Pol. Sci., Ser. Sci. Biol. **18**, 249 (1970).
265. — — Free amino acids in different *Lotus* species. Acta Soc. Bot. Pol. **40**, 439 (1971).
266. PRZYBYLSKA, J., and W. KANIEWSKI: γ-L-Glutamyl-β-alanine and γ-L-glutamyl-γ-aminobutyric acid in young pods of *Vicia sativa* L. Bull. Acad. Pol. Sci., Ser. Sci. Biol. **16**, 615 (1968).
267. PUSHKIN, A. V., Z. G. EVSTIGNEEVA, and V. L. KRETOVICH: Purification and some characteristics of glutamine synthetase from pea seeds. Biokhimiya **39**, 533 (1974) (in Russian with English summary).
268. QUIRT, A. R., J. R. LYERLA, I. R. PEAT, J. S. COHEN, W. F. REYNOLDS, and M. H. FREEDMAN: Carbon-13 nuclear magnetic resonance titration shifts in amino acids. J. Amer. Chem. Soc. **96**, 570 (1974).
269. RAMASWAMY, S., and D. V. REGE: Occurrence of γ-hydroxyglutamic acid in brinjal. Curr. Sci. **41**, 681 (1972).
270. RAMPONI, G., G. CAPPUGI, and P. NASSI: Labelling of noncarboxy-terminal glutamic acid during C-terminal analysis by the tritiation method when the γ-glutamyl peptide linkage is present. Biochem. Biophys. Res. Commun. **41**, 642 (1970).
271. RATHBUN, W. B.: γ-Glutamyl-cysteine synthetase from bovine lens. II. Cysteine analog studies. Arch. Biochem. Biophys. **122**, 73 (1967).
272. REEVES, W. A., and J. D. GUTHRIE: Isolation of xanthine, guanine, adenine, proteose, oxalic acid, and glutathione from peanut kernels. Arch. Biochem. Biophys. **26**, 316 (1950).
273. REINHOLD, L., R. A. SHTARKSHALL, and D. GANOT: Transport of amino acids in barley leaf tissue. II. The kinetics of uptake of an unnatural analogue. J. Exp. Bot. **21**, 926 (1970).
274. RESSLER, C., Y. H. GIZA, and S. N. NIGAM: β-Cyanoalanine, product of cyanide fixation and intermediate in asparagine biosynthesis in certain species of *Lathyrus* and *Vicia*. J. Amer. Chem. Soc. **91**, 2766 (1969).
275. RESSLER, C., S. N. NIGAM, and Y. H. GIZA: Toxic principle in vetch. Isolation and identification of γ-L-glutamyl-L-β-cyanoalanine from common vetch seeds. Distribution in some legumes. J. Amer. Chem. Soc. **91**, 2758 (1969).
276. RESSLER, C., S. N. NIGAM, Y. GIZA, and J. NELSON: Isolation and identification from common vetch of γ-L-glutamyl-β-cyano-L-alanine, a bound form of the neurotoxin β-cyano-L-alanine. J. Amer. Chem. Soc. **85**, 3311 (1963).

277. Rinderknecht, H., D. Thomas und S. Aslin: γ-Glutamyl-S-methyl-cystein und andere Peptide in der Mondbohne (*Phaseolus lunatus* L.). Helv. Chim. Acta **41**, 1 (1958).
278. Roponen, I. E.: S-Methylcysteine and γ-glutamyl-S-methylcysteine in the fruit of the sea buckthorn *(Hippophae rhamnoides)*. Suom. Kemistilehti **B 44**, 163 (1971).
279. Rowlands, D. A., and G. T. Young: Amino acids and peptides. Part IX. γ-L-Glutamyl-L-alanine, -L-valine, and -L-leucine. J. Chem. Soc. **1952**, 3937.
280. Sachs, H.: The optical rotation and some reactions involving α- and γ-peptides of glutamic acid and alanine. Dissertation, Columbia University, 1954, iv + 55 pp. University Microfilms. Ann Arbor, Michigan, Publication No. 8818.
281. Sachs, H., and E. Brand: Optical rotation of peptides. VII. α- and γ-dipeptides of glutamic acid and alanine. J. Amer. Chem. Soc. **75**, 4608 (1953).
282. — — The reaction of nitrous acid with γ-glutamyl peptides. J. Amer. Chem. Soc. **76**, 3601 (1954).
283. Sakato, Y.: Studies on the chemical constituents of tea. Part III. On a new amide theanine. Nippon Nogei Kagaku Kaishi **23**, 262 (1950) (in Japanese with English summary).
284. Sakato, Y., T. Hashizume, and Y. Kishimoto: Studies on the chemical constituents of tea. Part V. Synthesis of theanine. Nippon Nogei Kagaku Kaishi **23**, 269 (1950) (in Japanese with English summary).
285. Santarius, K. A., and C. R. Stocking: Intracellular localization of enzyme in leaves and chloroplast membranes. Permeability to compounds involved in amino acid syntheses. Z. Naturforsch. **24b**, 1170 (1969).
286. Sasaoka, K.: Biosynthesis of theanine. Nippon Nogei Kagaku Kaishi **39** R1 (1965) (in Japanese).
287. Sasaoka, K., and M. Kito: Synthesis of theanine by tea seedling homogenate. Agric. Biol. Chem. **28**, 313 (1964).
288. — — Biochemistry of theanine. Chagyo Kenkyu Hokoku, Shiryo **2**, 12 (1970) (in Japanese).
289. Sasaoka, K., M. Kito, and H. Inagaki: Biosynthesis of theanine in tea seedlings. Synthesis of theanine by homogenate of tea seedlings. Agric. Biol. Chem. **27**, 467 (1963).
290. Sasaoka, K., M. Kito, S. Konishi, and H. Inagaki: Biosynthesis of theanine in tea seedlings. Incorporation of Glutamic acid-1-$C^{14}$ into theanine. Agric. Biol. Chem. **26**, 265 (1962).
291. Sasaòka, K., M. Kito, and Y. Onishi: Synthesis of theanine by pea seed acetone powder extract. Agric. Biol. Chem. **28**, 318 (1964).
292. — — — Synthesis of theanine by pigeon liver acetone powder extract. Agric. Biol. Chem. **28**, 325 (1964).
293. — — — Some properties of the theanine synthesizing enzyme in tea seedlings. Agric. Biol. Chem. **29**, 984 (1965).
294. Sasaoka, K., T. Ogawa, and M. Fukuda: Conjugated homoserine in pea seedlings. Mem. Res. Inst. Food Sci., Kyoto Univ. **32**, 16 (1971).
295. — — — Conjugated amino acids in the non-cationic fraction of pea seedlings. Agric. Biol. Chem. **36**, 383 (1972).
296. Sasaoka, K., T. Ogawa, K. Moritoki, and M. Kimoto: Antivitamin B-6 effect of 1-aminoproline on rats. Biochim. Biophys. Acta **428**, 396 (1976).
297. Schaedle, M., and J. A. Bassham: Chloroplast glutathione reductase. Plant Physiol. **59**, 1011 (1977).
298. Scheinblatt, M.: Determination of amino acid sequence in di- and tripeptides by nuclear magnetic resonance techniques. J. Amer. Chem. Soc. **88**, 2845 (1966).
299. Schilling, E. D., and F. M. Strong: Isolation, structure and synthesis of a lathyrus factor from *L. odoratus*. J. Amer. Chem. Soc. **77**, 2843 (1955).

*300.* SCHULZE, E., und E. BOSSHARD: Über das Glutamin. Ber. **16**, 312 (1883).

*301.* SCHWIMMER, S.: S-Allyl-L-cysteine sulfoxide lyase [*Allium cepa* (Onion)]. Methods Enzymol. **17 B**, 475 (1971).

*302.* SCHWIMMER, S., and S. J. AUSTIN: γ-Glutamyl transpeptidase of sprouted onion. J. Food Sci. **36**, 807 (1971).

*303.* SEKURA, R., and A. MEISTER: γ-Glutamylcysteine synthetase. Further purification, "half of the sites" reactivity, subunits, and specificity. J. Biol. Chem. **252**, 2599 (1977).

*304.* SELVENDRAN, R. R., and S. SELVENDRAN: Chemical changes in young tea plant *(Camellia sinensis)* tissues following application of fertilizer nitrogen. Ann. Bot. (London) **37**, 453 (1973).

*305.* — — Distribution of some nitrogenous constituents in the tea plant. J. Sci. Food Agr. **24**, 161 (1973).

*306.* SENEVIRATNE, A. S., and L. FOWDEN: The amino acids of the genus *Acacia*. Phytochemistry **7**, 1039 (1968).

*307.* SHEEHAN, J. C., and D. H. YANG: A new synthesis of cysteinyl peptides. J. Amer. Chem. Soc. **80**, 1158 (1958).

*308.* SHTARKSHALL, R. A., L. REINHOLD, and H. HAREL: Transport of amino acids in barley leaf tissue. I. Evidence for a specific uptake mechanism and the influence of "ageing" on accummulatory capacity. J. Exp. Bot. **21**, 915 (1970).

*309.* SHVACHKIN, Y. P., N. A. VOSKOVA, and G. A. KORSHUNOVA: Total synthesis of willardiine peptides isolated from *Fagus silvatica*. Zh. Obshch. Khim. **47**, 2631 (1977) (in Russian).

*310.* SOLDAL, T., and I. NISSEN: Multiphasic uptake of amino acids by barley roots. Physiol. Plant. **43**, 181 (1978).

*311.* SPRAGG, S. P., and E. W. YEMM: Glutathione and ascorbic acid in the metabolism of germinating peas. Biochem. J. **58**, Proc. Biochem. Soc. XI (1954).

*312.* SPÅRE, G., and A. I. VIRTANEN: On the lachrymatory factor in onion *(Allium cepa)* vapours and its precursor. Acta Chem. Scand. **17**, 641 (1963).

*313.* STELZEL, P.: Die Herstellung der Peptidbindung. Synthese homöomerer Peptide aus Aminosäuren. Asymmetrische Anhydride. In: Methoden der Organischen Chemie (HOUBEN-WEYL), 4. Auflage (Ed. E. MÜLLER), Vol. **15, 2**. Synthese von Peptiden. Teil II, p. 169. Stuttgart: G. Thieme. 1974.

*314.* SUGII, M., T. SUZUKI, and S. NAGASAWA: Isolation of (−)-(S)-propenyl-L-cysteine from garlic. Chem. Pharm. Bull. **11**, 548 (1963).

*315.* SUGII, M., T. SUZUKI, S. NAGASAWA, and K. KAWASHIMA: Isolation of γ-L-Glutamyl-S-allylmercapto-L-cysteine and S-allylmercapto-L-cysteine from garlic. Chem. Pharm. Bull. **12**, 1114 (1964).

*316.* SULLIVAN, B., and M. HOWE: The isolation of glutathione from wheat germ. J. Amer. Chem. Soc. **59**, 2742 (1937).

*317.* SUNDAR, R. S., K. HARIHARAN, and D. R. RAO: Major acidic peptides of *Arachis hypogea*. Lebensm.-wiss. Technol. **9**, 180 (1976).

*318.* SUZUKI, T.: Metabolism of methylamine in tea plant *(Thea sinensis)*. Biochem. J. **132**, 753 (1973).

*319.* SUZUKI, T., M. SUGII, and T. KAKIMOTO: New γ-Glutamyl peptides in garlic. Chem. Pharm. Bull. **9**, 77 (1961).

*320.* — — — γ-L-Glutamyl-S-allyl-L-cysteine, a new γ-glutamyl peptide in garlic. Chem. Pharm. Bull. **10**, 345 (1962).

*321.* — — — Metabolic incorporation of L-valine-$C^{14}$ into S-(2-carboxypropyl)glutathione and S-(2-carboxypropyl)cysteine in garlic. Chem. Pharm. Bull. **10**, 328 (1962).

*322.* SUZUKI, T., M. SUGII, T. KAKIMOTO, and N. TSUBOI: Isolation of (−)-S-allyl-L-cysteine from garlic. Chem. Pharm. Bull. **9**, 251 (1951).

323. SZENT-GYORGYI, A., R. H. CHUNG, M. J. BOYAJIAN, M. TISHLER, B. H. ARISON, E. F. SCHOENWALDT, and J. J. WITTICK: Agaridoxin, a mushroom metabolite. Isolation, structure and synthesis. J. Org. Chem. **41**, 1603 (1976).

324. SZEWSZUK, A., and G. E. CONNELL: Specificity of γ-glutamyl cyclotransferase. Can. J. Biochem. **53**, 706 (1975).

325. TAKEO, T.: L-Alanine as a precursor of ethylamine in *Camellia sinensis.* Phytochemistry **13**, 1401 (1974).

326. — L-Alanine decarboxylase in *Camellia sinensis.* Phytochemistry **17**, 313 (1978).

327. TATE, S. S., and A. MEISTER: Interaction of γ-glutamyl transpeptidase with amino acids, dipeptides, and derivatives and analogs of glutathione. J. Biol. Chem. **249**, 7593 (1974).

328. THIERFELDER, H., und E. VON CRAMM: Über glutaminhaltige Polypeptide und zur Frage ihres Vorkommens in Eiweiß. Hoppe-Seyler's Z. physiol. Chem. **105**, 58 (1919).

329. THOMPSON, J. F.: γ-Glutamyl transpeptidase (plant). Methods Enzymol. **17A**, 894 (1970).

330. THOMPSON, J. F., C. J. MORRIS, W. N. ARNOLD, and D. H. TURNER: γ-Glutamylpeptides in plants. Amino Acid Pools (ed. HOLDEN, J. T.). Amsterdam-London-New York: Elsevier. 54 (1962).

331. THOMPSON, J. F., C. J. MORRIS, and R. K. GERING: Purification of plant amino acids for paper chromatography. Anal. Chem. **31**, 1028 (1959).

332. THOMPSON, J. F., C. J. MORRIS, and R. M. ZACHARIUS: Isolation of (−)-S-methyl-L-cysteine from beans *(Phaseolus vulgaris).* Nature **178**, 593 (1956).

333. THOMPSON, J. F., L. K. TURNER, and R. K. GERING: γ-Glutamyl transpeptidase in plants. Phytochemistry **3**, 33 (1964).

334. TKACHUK, R., and V. J. MELLISH: γ-L-Glutamyl-L-cysteine: its isolation and identification from wheat germ. Can. J. Biochem. **55**, 295 (1977).

335. TORII, H., and I. OTA: Environmental variation of the chemical constituents of the tea leaf. IV. Distribution of nitrogeneous fractions in the tea seedling grown in dark. Nippon Nogei Kagaku Kaishi **33**, 125 (1959) (in Japanese with English summary).

336. TOTTMAR, O., and P. LINDBERG: Effects on rat liver acetaldehyde dehydrogenases *in vitro* and *in vivo* by coprine, the disulfiram-like constituent of *Coprinus atramentarius.* Acta Pharmacol. Toxicol. **40**, 476 (1977).

337. TOUZE-SOULET, J. M., et C. MONTANT: Étude de quelques formes combinées nouvelles de l'acide glutamique chez *Boletus edulis* Fr. ex. Bull. Bull. Soc. Chim. Biol. **44**, 451 (1962).

338. TRELEASE, S. F., A. A. D. SOMMA, and A. L. JACOBS: Selenoamino acid found in *Astragalus bisulcatus.* S-Methylcysteine and Se-methyl selenocysteine. Science **132**, 618 (1960).

339. TSCHIERSCH, B.: Occurrence of γ-glutamyl-β-cyanoalanines. Tetrahedron Lett. **1964**, 747.

340. — Toxische Aminosäuren. Pharmazie **21**, 444 (1966).

341. TSUJI, H., K. MORITOKI, T. OGAWA, and K. SASAOKA: Fate of 1-aminoproline and urinary excretion of 1-aminoprolyl hydrazone of pyridoxal in rats. Agric. Biol. Chem. **41**, 1413 (1977).

342. VAN EGERAAT, A. S.: Enzymic studies on ninhydrin-positive compounds exuded by root tips of pea seedlings. Plant Soil **47**, 645 (1977).

343. VAN HEJENOORT, J., S. E. BRICA, B. C. DAS et E. LEDERER: Determination de sequence d'acides aminés dans des oligopeptides par la spectrometrie de masse. IX. (A) Acylation avec de nouveaux radicaux mixtes, (B) Peptides contenant des acides amines trifonctionnels. Tetrahedron **23**, 3403 (1967).

344. VAN SLYKE, D. D., R. T. DILLON, D. A. MACFADYEN, and P. HAMILTON: Gasometric determination of carboxyl groups in free amino acids. J. Biol. Chem. **141**, 627 (1941).

345. VARNER, J. E., and G. L. WEBSTER: Studies in the enzymatic synthesis of glutamine. Plant Physiol. **30**, 393 (1955).
346. VICKERY, H. B., G. W. PUCHER, H. E. CLARK, A. C. CHIBNALL, and R. G. WESTALE: The determination of glutamine in the presence of asparagine. Biochem. J. **29**, 2710 (1935).
347. VIRTANEN, A. I.: A review — Studies on organic sulphur compounds and other labile substances in plants. Phytochemistry **4**, 207 (1965).
348. — Antimikrobielle und antithyroide Stoffe in einigen Nahrungspflanzen. Qualitas Plant. Mater. Vegetabiles **18**, 8 (1969).
349. VIRTANEN, A. I., and A. BERG: γ-Glutamyl-alanine in pea seedlings. Acta Chem. Scand. **8**, 1089 (1954).
350. VIRTANEN, A. I., and T. ETTALA: A new γ-glutamyltripeptide in *Juncus* species. Acta Chem. Scand. **12**, 787 (1958).
351. VIRTANEN, A. I., M. HATANAKA und M. BERLIN: γ-L-Glutamyl-S-n-propylcystein in Knoblauch *(Allium sativum)*. Suom. Kemistilehti **B 35**, 52 (1962).
352. VIRTANEN, A. I., and P. K. HIETALA: γ-Hydroxyglutamic acid in green plants. Acta Chem. Scand. **9**, 175 (1955).
353. VIRTANEN, A. I., and E. J. MATIKKALA: The isolation of S-methyl-cysteinesulphoxide and S-n-propylcysteinesulphoxide from onion *(Allium cepa)* and the antibiotic activity of crushed onion. Acta Chem. Scand. **13**, 1898 (1959).
354. — — New γ-glutamylpeptides in onion *(Allium cepa)*. I. γ-Glutamylphenylalanine and γ-glutamyl-S-(β-carboxy-β-methylethyl)-cysteinylglycine. Suom. Kemistilehti **B 33**, 83 (1960).
355. — — Neue γ-Glutamylpeptide in der Zwiebel *(Allium cepa)*. Hoppe-Seyler's Z. physiol. Chem. **322**, 8 (1960).
356. — — New γ-L-glutamyl peptides in onion *(Allium cepa)*. III. Suom. Kemistilehti **B 34**, 53 (1961).
357. — — Structure of γ-glutamyl peptide 4 isolated from onion *(Allium cepa)*. γ-L-glutamyl-S-1-propenyl cysteine sulfoxide. Suom. Kemistilehti **B 34**, 84 (1961).
358. — — Proofs of the presence of γ-L-glutamyl-S-(1-propenyl)cysteine sulphoxide and cycloalliin as original compounds in onion *(Allium cepa)*. Suom. Kemistilehti **B 34**, 114 (1961).
359. — — γ-L-Glutamyl-S-(prop-1-enyl)-L-cysteine in the seeds of chives. Suom. Kemistilehti **B 35**, 245 (1962).
360. VIRTANEN, A. I., and I. MATTILA: γ-L-glutamyl-S-allyl-L-cysteine in garlic *(Allium sativum)*. Suom. Kemistilehti **B 34**, 44 (1961).
361. VOGEL, F. S., L. A. K. KEMPER, S. J. McGARRY, and D. G. GRAHAM: Cytostatic, cytocidal, and potential antitumor properties of a class of quinoid compounds, initiators of the dormant state in the spores of *Agaricus bisporus*. Amer. J. Pathol. **78**, 33 (1975).
362. VOGEL, F. S., S. J. McGARRY, L. A. K. KEMPER, and D. G. GRAHAM: Bacteriocidal properties of a class of quinoid compounds related to sporulation in the mushroom, *Agaricus bisporus*. Amer. J. Pathol. **76**, 165 (1974).
363. WALEY, S. G.: Acidic peptides of the lens. 3. The structure of ophthalmic acid. Biochem. J. **68**, 189 (1958).
364. — Naturally occurring peptides. Advan. Protein Chem. **21**, 1 (1966).
365. WATANABE, T., Y. SHIMA, and T. ICHIHARA: Changes in free amino acids during ripening period of soybean seed. Hokkaido Kyoiku Daigaku Kiyo. Dai-2-Bu, **A 20**, 76 (1970) (in Japanese with English summary).
366. WATANABE, T., M. TSUGAWA, N. TAKAYAMA, and Y. FURUKAWA: Changes of free amino acids of each organ in the development and growth of the kidney bean plant. Hokkaido Kyoiku Daigaku Kiyo, Dai-2-Bu, **B 22**, 45 (1971) (in Japanese with English summary).

367. WATSON, R., and L. FOWDEN: The uptake of phenylalanine and tyrosine by seedling root tips. Phytochemistry 14, 1181 (1975).
368. WEAVER, R. F., K. V. RAJAGOPALAN, and P. HANDLER: Mechanism of action of a respiratory inhibition from the gill tissue of the sprouting mushroom, *Agaricus bisporus*. Arch. Biochem. Biophys. 149, 541 (1972).
369. WEAVER, R. F., K. V. RAJAGOPALAN, P. HANDLER, and W. L. BYRNE: γ-L-Glutaminyl-3,4-benzoquinone. Structural studies and enzymic synthesis. J. Biol. Chem. 246, 2015 (1971).
370. WEAVER, R. F., K. V. RAJAGOPALAN, P. HANDLER, P. JEFFS, W. L. BYRNE, and D. ROSENTHAL: Isolation of γ-L-glutaminyl-4-hydroxybenzene and γ-L-glutaminyl-3,4-benzoquinone: a natural sulfhydryl reagent, from sporulating gill tissue of the mushroom *Agaricus bisporus*. Proc. Nat. Acad. Sci. U.S. 67, 1050 (1970).
371. WEAVER, R. F., K. V. RAJAGOPALAN, P. HANDLER, D. ROSENTHAL, and P. W. JEFFS: Isolation from the mushroom *Agaricus bisporus* and chemical synthesis of γ-L-glutaminyl-4-hydroxybenzene. J. Biol. Chem. 246, 2010 (1971).
372. WEBSTER, G. C.: Peptide-bond synthesis in higher plants. I. The synthesis of glutathione. Arch. Biochem. Biophys. 47, 241 (1953).
373. — Enzymatic synthesis of γ-glutamylcysteine in higher plants. Plant Physiol. 28, 728 (1953).
374. WEBSTER, G. C., and J. E. VARNER: Peptide-synthesis in higher plants. II. Studies on the mechanism of synthesis of γ-glutamylcysteine. Arch. Biochem. Biophys. 52, 22 (1954).
375. — — Peptide bond synthesis in higher plants. III. The formation of glutathione from γ-glutamylcysteine. Arch. Biochem. Biophys. 55, 95 (1955).
376. WELTER, A., J. JADOT, G. DARDENNE, M. MARLIER et J. CASIMIR: L-γ-Glutamyl-2-amino-3-hexanone dans *Russula ochroleuca*. Phytochemistry 15, 1984 (1976).
377. WICKREMASINGHE, R. L., and K. P. W. C. PERERA: Site of biosynthesis and translocation of theanine in the tea plant. Tea Quart. 43, 175 (1972).
378. WIEWIOROWSKI, M., and H. AUGUSTYNIAKOWA: Occurrence of γ-L-glutamyl-L-tyrosine and γ-L-glutamyl-L-phenylalanine in seeds of *Lupinus angustifolius*, and *Lupinus albus*. Acta Biochim. Pol. 9, 399 (1962).
379. WILK, S., and M. ORLOWSKI: Determination of pyrrolidone carboxylate and γ-glutamyl amino acids by gas chromatography. Anal. Biochem. 69, 100 (1975).
380. WILSON, H., and R. K. CANNAN: The Glutamic acid-pyrrolidonecarboxylic acid system. J. Biol. Chem. 119, 309 (1937).
381. WOLOSIUK, R. A., and B. B. BUCHANAN: Thioredoxin and glutathione regulate photosynthesis in chloroplasts. Nature 266, 565 (1977).
382. WU, P. L., and D. A. CALDAS: Conversion of glutamic acid to 2-pyrrolidone-5-carboxylic acid in plant extracts during elution from ion-exchange resins. An. Acad. Brasil. Cienc. 44, 273 (1972).
383. WÜNSCH, E.: Mehrfunktionelle Aminosäuren und ihre Einbeziehung in die Synthese. In: Methoden der Organischen Chemie (Houben-Weyl), 4. Auflage (Ed. E. MÜLLER). Vol. 15,1. Synthese von Peptide, Teil 1, p. 468. Stuttgart: G. Thieme. 1974.
384. YASUMOTO, K., K. IWAMI, and H. MITSUDA: A new sulphur-containing peptide from *Lentinus edodes* acting as a precursor for lenthionine. Agric. Biol. Chem. 35, 2059 (1971).
385. — — — Enzyme-catalysed evolution of lenthionine from lentinic acid. Agric. Biol. Chem. 35, 2070 (1971).
386. — — — Enzymic formation of Shii-ta-ke aroma from non-volatile precursors. (S)-lenthionine from lentinic acid. Mushroom Sci. 9, 371 (1976).
387. YASUMOTO, K., K. IWAMI, H. MIZUSAWA, and H. MITSUDA: Preparation of des-glutamyllentinic acid, a new sulfur-containing amino acid, from lentinic acid and its

capacity to form a complex with pyridoxal phosphate. Nippon Nogei Kagaku Kaishi **50**, 563 (1976) (in Japanese with English summary).

388. YASUMOTO, K., K. IWAMI, T. YONEZAWA, and H. MITSUDA: Anion activation of γ-glutamyltransferase from fruiting bodies of *Lentinus edodes*. Phytochemistry **16**, 1351 (1977).

389. YEMM, E. W.: Cellular oxidations and the synthesis of amino-acids and amides in plants. In: Recent Developments in Cell Physiology (Ed. J. A. KITCHING). London: Butterworth. 1954.

390. ZACHARIUS, R. M.: Composition of the nitrogenous components of the bush bean seed *(Phaseolus vulgaris)* including isolation of δ-acetylornithine. Phytochemistry **9**, 2047 (1970).

391. ZACHARIUS, R. M., C. J. MORRIS, and J. F. THOMPSON: Isolation and characterization of γ-L-glutamyl-S-methyl-L-cysteine from kidney beans *(Phaseolus vulgaris)*. Arch. Biochem. Biophys. **80**, 199 (1959).

392. ZACHARIUS, R. M., and E. A. TALLEY: Elution behavior of naturally occurring ninhydrin-positive compounds during ion exchange chromatography. Anal. Chem. **34**, 1551 (1962).

393. IUPAC-IUB: Nomenclature of α-amino acids. Biochemistry **14**, 449 (1975).

394. HATANAKA, S., and S. KANEKO: γ-L-Glutamyl-L-lathyrine from *Lathyrus japonicus*. Phytochemistry **17**, 2027 (1978).

395. KINOSHITA, T., and S. MINATO: Structural studies on anthglutin, an inhibitor of γ-glutamyl transpeptidase, from *Penicillium oxalicum*. Bull. Chem. Soc. Japan **51**, 3282 (1978).

396. MINATO, S.: Isolation of anthglutin, an inhibitor of γ-glutamyl transpeptidase from *Penicillium oxalicum*. Arch. Biochem. Biophys. **192**, 235 (1979).

397. GMELIN, R.: Personal communication.

398. MOURI, T., T. MURAHARA, H. KAYAMA, S. TSUTSUI, T. KUROKAWA, Y. SHIBATA, N. ISHIDA, S. KAKIMOTO, F. ASAKURA, H. SHIRAHAMA, and T. MATSUMOTO: New inhibitors for the blastgenation of human lymphocytes. Isolation from edible mushrooms. Agric. Biol. Chem. **42**, 2179 (1978).

399. IWAMI, K., K. YASUMOTO, and T. FUSHIKI: A unique γ-glutamyltransferase from fruiting bodies of *Tricholoma shijemi*. Agric. Biol. Chem. **42**, 2175 (1978).

400. MAZELIS, M., and R. K. CREVELING: 5-Oxoprolinase (L-pyroglutamate hydrolase) in higher plants. Plant Physiol. **62**, 798 (1978).

*(Received November 6, 1978)*

# Author Index

By

A. Siegel, Wien

Page numbers printed in *italics* refer to References

Aasen, A. J. *164*
Aberhart, D. J. *111*
Abillon, E. *111*
Acher, A. *114*
Acree, T. E. *267*
Aguilar-Martinez, M. *164*
Ahmad, R. *111, 113*
Aitzetmüller, K. *171*
Akaiwa, A. *120*
Alburn, H. E. *274*
Allerhand, A. *61*
Alper, J. B. *120*
Amar, L. *267*
Anagnostopoulos, C. E. *111*
Anderson, D. G. *267*
Anderson, L. *58*
Anderson, R. B. *116*
Anderson, R. C. 1, 29, 30, 31, *56*
André, D. *168*
Andresen, S. A. *169*
Andressov, K. *275*
Andrewes, A. G. *164, 165*
Anet, F. A. L. *58*
Antia, N. J. *167*
Arai, S. *267*
Archibald, R. M. *267*
Arison, B. H. *282·*
Arnold, W. N. *282*
Arpin, N. *165, 166, 170*
Asakura, F. *285*
Asato, A. E. *170*
Asen, S. *268, 277*
Askew, S. A. 64, *111*
Aslin, S. *280*
Aspinall, G. O. *59*

Atkins, D. 71, *111*
Augusti, K. T. *268*
Augustyniakowa, H. *284*
Austin, D. J. *166*
Austin, S. J. *268, 281*

Baer, H. H. *59*
Baggiolini, E. G. *118*
Bailey, W. F. *58*
Bakker, S. A. *111*
Bal, K. 77, *120*
Bannai, K. *117*
Barber, M. S. *165*
Barbier, M. *168*
Barlow, L. *165*
Barrett, A. G. M. *111, 112*
Barrow, M. V. *268*
Barsley, E. A. *268*
Barta, M. A. *119*
Bartlett, L. 134, *165*
Barton, D. H. R. 2, 16, *56*, 90, 94, 107, *111, 112*
Bassham, J. A. *280*
Basson, R. *119*
Baths, J. D. *115*
Baum, J. *112*
Bauman, P. *113*
Baumert, A. *279*
Bayne, S. *56*
Bazely, N. *118, 119*
Becker, J. E. *116*
Becker, R. R. *276*
Beecham, A. F. *268*
Behre, H. *59*
Bell, E. A. *268*
Bell, P. A. *118*

# Subject Index

By

A. SIEGEL, Wien

Fortschritte der Chemie organischer Naturstoffe

# Progress in the Chemistry of Organic Natural Products

All Volumes and Cumulative Index 1—20 available / Alle Bände und Generalregister 1—20 lieferbar.

*Price reduction for subscribers / Preisermäßigung für Subskribenten: 10%.*

**Special price reduction (20% of the list price) for the Vols. 1—20 plus Cumulative Index. / Vorzugspreis (20% Nachlaß) bei Bezug der Bände 1—20 inklusive Generalregister.**

**Springer-Verlag Wien · New York**